AF746835

WITHDRAWN
University of
Reading Library

90 0526343 1

ALLE ZEIT WACH
1842

David E. Boyce · Peter Nijkamp
Daniel Shefer (Eds.)

Regional Science

Retrospect and Prospect

With 61 Figures

Springer-Verlag
Berlin Heidelberg New York
London Paris Tokyo
Hong Kong Barcelona

Prof. Dr. David E. Boyce, Urban Transportation Center, University of Illinois at Chicago, 1033 W. Van Buren Street, Suite 700-South, Chicago, Illinois 60607, USA

Prof. Dr. Peter Nijkamp, Department of Economics, Free University, De Boelelaan 1105, 1081 HV Amsterdam, The Netherlands

Prof. Dr. Daniel Shefer, Israel Institute of Technology, Transportation Research Institute, Technion City, Haifa 32000, Israel

fe
1–6
43–51
70
191–208
317–321
333–337

ISBN 3-540-53493-8 Springer-Verlag Berlin Heidelberg New York Tokyo
ISBN 0-387-53493-8 Springer-Verlag New York Heidelberg Berlin Tokyo

This work is subject to copyright. All rights are reserved, whether the whole or part of the material is concerned, specifically the rights of translation, reprinting, reuse of illustration, recitation, broadcasting, reproduction on microfilms or in other ways, and storage in data banks. Duplication of this publication or parts thereof is only permitted under the provisions of the German Copyright Law of September 9, 1965, in its version of June 24, 1985, and a copyright fee must always be paid. Violations fall under the prosecution act of the German Copyright Law.

© Springer-Verlag Berlin · Heidelberg 1991
Printed in Germany

The use of registered names, trademarks, etc. in this publication does not imply, even in the absence of a specific statement, that such names are exempt from the relevant protective laws and regulation and therefore free for general use.

Printing: Zechnersche Buchdruckerei GmbH u. Co. KG., D-6720 Speyer
Bookbinding: J. Schäffer GmbH & CO. KG., Grünstadt

2142/7130-543210

SB tech

Preface

Regional science, as an integrating discipline for many scientific analyses concerning the element of space, has a remarkable record of achievements in both breadth and depth. The great many scientific journals in this field and the large number of regional science books mirror the appeal of this discipline, in both the developed and the developing world.

The scientific organization in this area, the Regional Science Association International, is organizing a world conference every four years. The 1989 Regional Science World Conference was held in an interesting setting, viz. kibbutz Ramat Rachel on the outskirts of Jerusalem. The organization of this meeting was undertaken by the Israeli Section of the Regional Science Association, under the supervision of Professor Rafi Bar-El (Development Study Center, Rehovoth). The meeting turned out to be a great success and provided an extremely interesting cross-section of new developments in the area of regional science. It was decided afterward to publish a set of representative papers from this meeting in one volume.

The present book brings together these papers, which were duly selected after a careful screening and refereeing process. The editors wish to thank all referees for their willingness to assist them in the task of producing this volume. Many thanks also go to Rita Hittema for her skillful copy-editing and word-processing of the manuscript.

Chicago		David E. Boyce
Amsterdam	September 1990	Peter Nijkamp
Haifa		Daniel Shefer

Table of Contents

Part C. **ECONOMIC RESTRUCTURING AND SPATIAL DYNAMICS**

Part D. **ADVANCES IN REGIONAL SCIENCE METHODS**

Chapter 1

A PREFACE TO REGIONAL SCIENCE

D.E. Boyce
P. Nijkamp
D. Shefer

1. Prologue

Despite respectable achievements in the past decade, regional science as an established scientific discipline does not yet have a long history. The explicit analysis of the spatial dispersion and cohesion of human activities in our complex society is only very recent, although notable exceptions can be found in previous related scholarly contributions, inter alia in the comparative cost theory of Ricardo or the international trade theory of Heckscher and Ohlin. Also in related disciplines such as economic geography a wide variety of interesting approaches can be found in the pre-war period.

The post-war development of regional science rests to some extent on such path-finding analyses, which were married to individual past contributions by scholars such as Von Thünen, Marx, Hotelling, Weber, Palander, Chamberlain, Christaller and Lösch. It has been the contribution of Isard to connect such separate analyses and to integrate them into a unified framework of regional science.

The evolution of regional science toward a discipline in itself was thus based on a merger of concepts from economics (e.g., general equilibrium theory, input-output analysis, programming theory, production theory), geography (e.g., central place theory, diffusion theory), mathematics and econometrics (e.g., spatial autocorrelation analysis, systems dynamics), and related disciplines such as political science, sociology and decision theory. In more recent years a devolution of regional science toward various sub-disciplines has taken place, such as urban economics, infrastructure economics, evaluation analysis, environmental and resource analysis, and spatial econometrics. However, in all such sub-disciplines three major features appear to be always present, viz. location theory, methods of spatial analysis, and regional development and policy. Therefore these three issues will be discussed in a concise manner here.

2. Location Theory

Location theory makes up the heart of regional science. Both theoretically and empirically numerous contributions have been made in this framework in order to study the 'where' and the 'why there' of human activities, in regard to both firms and households.

From a **methodological** viewpoint much attention has been devoted to the concept of a **region** (e.g., homogeneous region, planning region etc), while more recently many efforts have been made at developing a formal taxonomy and axiomatic typology of regions (e.g, by using fuzzy set analysis and abstract topology). Unfortunately such contributions have remained fairly abstract thus far.

Theoretically, traditional location theory à la Weber and Lösch has been developed much further, by investigating spatial synergies (e.g., industrial complexes), spatial dynamics (e.g., innovation behaviour of firms, migration behaviour of households), and complex spatial systems (e.g., entropy and gravity models, (micro-) simulation analysis). Conventional economic cost concepts are increasingly broadened so as to include also image factors and information access in locational decisions. Following the well-known Tiebout hypothesis, also the locational analysis of public services has gained more importance (e.g., by referring explicitly to equity issues and multi-level policy structures).

The **instruments** for analyzing locational patterns have also shown a profound extension. Notably in the field of discrete choice theory much progress has been made (e.g., by linking it to panel and longitudinal data), inter alia in the area of residential location analysis, migration analysis, transport behaviour, labour market research or location behaviour of firms. Especially the strong formal links between individual utility theory and macro-systems equilibrium have to be mentioned here, so that there is a formal correspondence between micro behaviour and macro functioning of a spatial system. More recently, also the dynamics of such behaviourial-based spatial developments has been analyzed by new methodological advances (e.g., bifurcation theory, chaos and fractal theory).

Finally, from a **practical** viewpoint an enormous variety of contributions to locational analysis can be found, focusing on the spatial choice behaviour of households and their commuting patterns or on the spatial re-orientation of firms as a result of high-tech developments.

3. Methods of Regional Analysis

Regional science has always been marked by a strong quantitative and model orientation. Examples are amongst others: general spectral analysis, spatial autocorrelation analysis, spatial econometrics, input-output models, entropy models, catastrophe analysis, disaggregate choice analysis, spatial interaction models, and non-linear spatial dynamics. Especially (multi-)regional input-output models have proven to be important operational methods in regional analysis, although unfortunately many regional authorities are somewhat hesitant in investing in the construction of regional input-output models. But the current problems of regional resource scarcity and environmental decay will hopefully generate a new interest in regional input-output studies complemented with pollution and resource sectors.

The operational development and use of many such analytical tools is contingent upon the availability of suitable data or, in general, appropriate information systems. Recent computer advances (e.g., in relation to Geographic Information Systems) may provide an important support for improving the current data situation.

In recent years the area of spatial interaction modelling has exhibited a rapid development pace, inter alia by providing a consistent link with multinomial logit analysis, by incorporating non-linear evolutionary or chaos type of models, and linking such models to other advances (e.g., Alonso's general theory of movement, Beckmann's network equilibrium theory).

Finally, some remarks on the treatment of data are in order here. In various cases, traditional 'hard' data measured in a cardinal metric are not available, but 'soft' data measured in a categorical or ordinal way may exist (e.g., a result of measurement imprecision, survey questionnaires, etc.). For such cases a new toolbox has been developed, containing inter alia multidimensional scaling, ordinal regression methods, categorical data analysis, LISREL and PLS models, 'soft' generalized linear models (GLIM), qualitative statistics etc. Each of these techniques aims at treating the available information in a methodologically sound way and to extract the maximum amount of quantitative information from the underlying qualitative data set.

4. Regional Development and Policy

Regional development issues have - with various waves of interest - been of paramount interest for the development of regional science in both the developed and the developing world. The spatial conflict between efficiency and equity lies at the heart of this interest.

Conventional efficiency-oriented approaches such as optimisation analysis, cost-benefit or cost-effectiveness analysis appeared to fail often in evaluating the social significance of regional development programmes. The use of game-theoretic and multi-level policy models led already to an improvement, but could not fulfil all tasks of a balanced evaluation methodology.

The situation has become even more complicated, as in recent years external effects - notably social costs of the environment - are also increasingly incorporated in regional development evaluation. The triangular relationship between efficiency, equity and externality is difficult to cover by means of such conventional instruments. The use of multiple-objective and multi-level optimisation models and multiple criteria evaluation methods has been important to progress in this field, especially because in this framework also qualitative aspects (in terms of both impacts and policy preferences) and interactive decision support elements can be included in the analysis.

Such analyses do not only pertain to problem regions, but to the same extent also to inter- and intra-urban policy questions. Essentially the city is becoming more and more a focal point of (regional) policy analysis, as the city (or the urban agglomeration) may be regarded as the nodal focus in an evolving economic network and economy. Here also an important link with innovation and diffusion theory and with spatial network analysis can be found, as is also witnessed in the revival of (dynamic) models for spatial incubator phenomena. Clearly, urban development policy and regional development policy will increasingly be oriented to one another. The current deregulation tendencies tend to increase even the competitive forces between cities and regions.

5. Emerging Trends

Retrospectively, major novel developments in regional science appear to be multi-faceted, as is witnessed by new directions in macro spatial systems analysis (e.g., spatial interaction models, entropy theory), micro behavioural analysis (e.g., spatial models of discrete choice), evolutionary models of spatial systems (e.g., non-linear dynamic models), policy and programme evaluation analysis (e.g., multi-criteria and conflict analysis), and statistical data analysis (e.g., GIS).

Various new megatrends and driving forces in our society can at present be identified which will no doubt exert a further influence on the direction of regional science thinking. Some examples are:

- the aging process, which - in addition to financial-economic aspects and planning problems concerning the services for the elderly - will have complex implications for spatial mobility patterns.
- the environmental question, which - besides global issues such as acidification and ozonization - will have serious implications for the possibilities to reach a regional (or urban) sustainable development.
- the emerging new technology, which - apart from a re-orientation of production systems - will generate significant shifts in spatial patterns as a consequence of informatics and telecommunications adoption
- the new emerging policy map of Europe, which - even apart from the drastic changes in Eastern Europe - will create a new spatial configuration where in the long run regions may become the focal points (rather than the current nations).
- the new infrastructure policy, which - besides enormous financial

implications - will create a new supply-driven network in which access will turn out to be of paramount importance for cities and regions.

All such evolutionary trends may create a great diversity of new potentials, but also embody a great number of bottlenecks, as is also witnessed by recent experiences around the Channel Tunnel, the French TGV or the German Maglev. Thus the identification of critical success factors in such a dynamic world will be a major task of an alert regional science, by using the theoretical and methodological tool box developed in the past.

This brings us back to the present volume. This volume contains a set of publications prepared by various authors on the occasion of the Third World Congress of the Regional Science Association (Jerusalem, 1989). They contain a cross-section of various recent achievements in the field. They have been systematically grouped around four major themes. This first section did not intend to provide a synthesis of all these contributions, but rather a general background against which these individual studies can be interpreted. In the next section these contributions will be concisely summarized.

6. Summary of the Book

The present volume describes four major areas of research in current regional science. The first field deals with spatial patterns of households and firms. It deals with all three aspects discussed above, viz. locational decisions, analytical tools and policy dimensions.

The first contribution in part A is written by Plane and Rogerson. The authors argue that - despite many advances made in our understanding of migration processes - various new ways have to be explored in order to ensure that future migration research is able to capitalize most effectively on various recent advances. They formulate 'ten commandments' related to the treatment of systems issues in migration modelling and to the overall conduct of inquiry on migration processes.

Following the first contribution, Fischer and Nijkamp provide an overview of major issues in regional labour market analysis. They treat both the demand and supply side of labour markets and point at weak and strong elements in current labour market research, both in terms of methods and policy analysis.

Next, Werczberger discusses the role of endogenous spatial externalities in residential location theory. He proposes and illustrates a typology based on three central facets of the subject, viz. the nature of the externality, the kind of relationship involved and their analytical treatment.

A further contribution to modern housing market analysis is given by Fisch who presents a theoretical analysis of the development of a non-instantaneous and stochastic equilibrium model of the rental housing market. He pays particular attention to transaction costs associated with the extreme variation of housing bundles and the search costs associated with the location of a particular bundle of housing attributes.

The presence of housing and labour markets implies spatial mobility. The chapter written by Kim and Choi attempts to assess the past and current developments in spatial transport patterns and to identify a future research agenda. The authors focus the discussion on two categories: existing transportation issues which will remain challenging research opportunities, and new issues related to transportation systems planning and spatial changes brought about by the new telecommunication technology and computer information systems.

Finally, Spiegel and Spiegel analyze urban land use and the spatial implications of environmental externalities caused by polluting plants. This problem is theoretically studied, starting from a circular monocentric city (including costs of transporting output to the markets) and looking for the best spatial pattern of polluting activities. Moving plants further away from the centre involves additional costs such as commuting and transportation costs. The authors then derive the conditions for optimal plant location, based on an analysis of the bid rent function, which - apart from the location of the polluting plants - appears to be strongly affected by the consumers and workers commuting habits.

Part B of the present volume is addressing the issue of spatial impacts of new technologies as one of the driving forces for spatial dynamics. Also here locational aspects, analytical tools and policy questions appear to be of paramount importance.

The first contribution in this part is written by Shefer. He analyzes the implications of high tech industries for peripheral regions. He argues - based on a case study for Israel - that it is by no means automatically guaranteed that high technology policy is a panacea for lagging areas.

In the same spirit, Giaoutzi and Stratigea argue that new information technologies do not necessarily reduce spatial inequalities. It may even be plausible to expect that interregional discrepancies may increase because of the urban orientation of new information technologies.

Next, Camagni presents a synthesis of arguments on the spatial economic dynamics of technological change by investigating in particular the impacts of innovation networks. The author analyzes the features of the new scientific paradigm by which the field is approached in general economic theory, viz. the new evolutionary paradigm. He then links this approach to the new network behaviour of firms in their struggle for dynamic excellence and focuses particular attention on the role of spatial variables.

The final contribution in part B is written by Romanoff and Levine, who deal with regional development aspects of technological change on the basis of the sequential interindustry model. Using cross-industry information flows for technical diffusion with a time-varying Sequential Interindustry Model (SIM), a better understanding of implications of technical change for regional development processes is offered.

The third part of this book, part C, deals with economic restructuring and spatial dynamics; it is essentially a follow-up of part B and deals again with locational issues, methods and models, and policy questions.

The first contribution of part C is written by Suarez-Villa and it presents a broad overview of the historical process from which entrepreneurship and long-term sectoral and spatial changes emerge. An effort is made to relate micro- and macro-level changes, arising in the long-term process of development, by providing a typology of major innovative entrepreneurial roles. These are then related to four major historical periods, where the shifting sectoral dynamics of each epoch determine the most significant changes in spatial structure and development. The arguments are illustrated on the basis of U.S. data.

Next, Cappellin focuses attention on the European situation and analyzes the consequences of the emerging European internal market and the internationalisation processes for small and medium size enterprises (SMEs). It appears that such an analysis requires different analytical tools compared to those used in standard trade theory. Particular attention is given to the diffusion of innovation for SMEs, where also an attempt is made at identifying various factors that are critical for changes in regional disparities.

A further analysis of Schumpeterian-based evolutionary developments of spatial dynamics and urban hierarchies is provided by Camagni and Diappi. In their contribution on urban dynamics with innovation and synergy effects, the changes in urban ranks and centre functions are analyzed. A new dynamic simulation model is built, allowing a differentiated locational logic for tertiary and manufacturing functions, taking inter alia into account local intersectoral synergies.

The last contribution in part C is written by Weidlich and Sonis, on a dynamic theory and comparative analysis of interregional migration flows. They develop a fully dynamic theory of migratory processes in terms of equations of motion, linking the micro level of individual decisions to the macro level of changing regional populations. Next, they correlate the migratory trends to regional time-dependent key-factors representing the spatio-temporal socio-economic situation.

The final part of this book is devoted to advances in regional science tools. In a broad overview paper, Anselin describes recent developments in quantitative methods in regional science. He identifies major shortcomings and outlines various potentially fruitful directions for future work. Elements discussed are inter alia social accounting matrices, general equilibrium models, spatial econometrics and spatial autocorrelation, scale and aggregation problems, and model validation. The need for appropriate information systems is emphasized.

The next contribution, by Freeman, Talpaz, Fleischer and Laufman, offers a new multiregional input-output model for Israel and the 'Rest of the World'. New features are the Markov chain adjustment of trade flow matrices (in order to represent the impact of middlemen); the adjustment of trade flow matrices (in order to represent the impacts of locating new industries in development regions); the introduction of new technologies; the integration of a quadratic programming algorithm with the multiregional input-output model.

Another contribution to impact analysis is provided by Cesar and Folmer. These authors describe the ex post measurement of environmental policy effects by means of LISREL (Linear Structural Equation) Models. They argue that environmental policy is based on a combination of different instruments for environmental objectives. Such features make the application of LISREL Models with latent variables particularly appropriate to measure effects of environmental policy. The applicability of other measurement methods is also briefly described.

Next, Bailly and Coffey deal with a new issue in regional science, viz. regional medicometry, combining viewpoints and methods from economics, epidemiology, sociology and geography in order to develop a comprehensive analysis of modern health care systems in their regional environments. After identifying the theoretical underpinnings of this approach the authors study the complex interrelations between the medical system and its environment. The importance of health care activities for regional development is emphasized.

Finally, Salzman and Viashima discuss the significance of economic and non-economic components of regional development. Their study presents the results of two exploratory modelling studies which provide some preliminary information on the relative significance of economic vis-à-vis non-economic components of regional progress. It turns out that non-economic elements play a significant role.

The conclusion from the above concise overview is that regional science has been able to base its research on solid methodological grounds and that at the same time it has managed to develop refreshing new analytical contributions. Therefore, we may conclude from the above sample of recent advances in regional science that this discipline has a valuable heritage with a promising future.

Part A. SPATIAL PATTERNS OF HOUSEHOLDS AND FIRMS

Chapter 2

THE TEN COMMANDMENTS OF MIGRATION RESEARCH

D.A. Plane
P.A. Rogerson

1. Introduction

In 1885 Sir Ernest George Ravenstein published his seminal paper titled 'The laws of migration' in the Journal of the Royal Statistical Association. Among other observations, he asserted the preeminence of the economic motivation for understanding geographic patterns of population movement. Since that time, a voluminous literature has developed on the analysis of human migration as influenced by, and influencing, geographic distributions of economic activity. Because a significant portion of all research appearing on migration takes off from Ravenstein's century-old assertion about the primacy of the economic motivation, and because the multidisciplinary field of regional science provides a common meeting ground for scholars from economics, geography, sociology, and the various other disciplines in which the bulk of migration analysis appears, we believe that it is appropriate that migration is still one of the major themes in regional science research all over the world.

In this paper we make ten further assertions. These are not so much about migration processes, per se, but rather about the state-of-the-art of migration research and about desirable future directions it might take. We attempt to focus on certain issues of particular pertinence for those regional scientists whose work either focuses, or touches, on migration; we do not pretend to represent concerns of relevance primarily to migration researchers who eschew the analytical economic/geographic purview of the regional scientist.

Although we aspire to the full breadth of perspective characteristic of regional science, we are cognizant that one of the defining attributes of an academic is a congenital inability to distinguish 'quality' from 'what I do'. The thoughts contained herein are unavoidably products of our own proclivities and the current state of our own research and readings. We recognize, too, that our discussion may be rather more North Americentric than would be optimal.

We pose our ten assertions in the form of 'commandments'. These are, in general, prescriptive rather than proscriptive. Although they represent fundamental principles to which we think migration researchers should adhere, our intent is to provoke discussion rather than to suppress it. We concede that these are contestable assertions. In short form, the commandments may be stated thusly:

1. We should link our research laagers;
2. We should look at the other end of the migration/employment-change chicken;
3. We should turn ever more attention to turnarounds;
4. We should be more careful about the distinctions between stocks, flows, and flows in flows;
5. We should revive our faith in the importance of the integrated analysis of regional systems;
6. We should not forget that we get older one year at a time, and should design our models accordingly;
7. We should have compassion for net migrants - even though there aren't any such animals;
8. We should tailor our modeling approaches to the complexity of the purposes to which they are to be put;
9. We should lobby incessantly for data;
10. We should mobilize on melting pots.

2. The Ten Commandments

(1) Linking the research laagers

One of the strong empirical regularities often noted in migration studies is the existence of 'beaten-path effects' (see, e.g., the general discussion in Hoover and Giarratani, 1984, pp. 278-79). Migration streams, once established, tend to persist over time due to channels of information flow, family ties, social webs, and so forth. So, too, much recent migration research has become channelized into a number of readily discernible tracks of thought.

Our first and paramount commandment is this: we should link together our migration research laagers in the spirit of regional science itself. We assert that greater progress could be made by promoting joint venturing between the inhabitants of the various established camps. We should break down whatever barriers exist to the fertile cross-pollinization of ideas - and even to the few truly new thoughts that bloom on the veldt outside any laager. Our common research enterprise is enriched each time a scholar such as a Weidlich from physics (see, e.g., Weidlich and Haag, 1986) or a Sonis from mathematics (e.g., Sonis, 1980) focuses a vision inherited from a nontraditional migration research field on the analysis of human population movement.

Although in many ways demographic research is noteworthy for the ready availability of seemingly neutral venues for the interchange of ideas, one's academic disciplinary affiliation still asserts a strong influence. Thus, while interdisciplinary population studies centers are something of a prerequisite of status in the demographer's pecking order, and in our country and others the annual meetings of the major population association provide a common meeting ground for those sociologists, economists, planners, and geographers who engage in population studies, the readiest (albeit not totally accurate) typology of the corpus of migration research pertinent to this paper remains one constructed along disciplinary lines.

We may contend, for example, that the economist's approach to migration studies that has been most notable in regional science journals has been typified by structural equations econometric models integrating quite aggregate regional labor market variables - average wages, overall unemployment rates - with equally aggregate migration measures (e.g., total population change, total net migration, total gross in- or gross outmigration). Writing for the Silver Anniversary issue of the Journal of Regional Science, Michael Greenwood (1985) provides a review of recent literature largely by the membership of this research camp. His 1985 paper updates an earlier, widely cited survey (Greenwood, 1975).

Geographers, on the other hand, have typically engaged in greater geographic disaggregation, have advanced alternative conceptualizations of interaction-based systems, and have carried out analyses at a variety of spatial scales of analysis. W.A.V. Clark (1986) nicely summarizes much of the recent geographic literature, classifying several of the predominant underlying themes (see also Rogerson, 1984; White et al., 1989; Plane, 1990).

Continuing with yet one more sweeping generalization, sociologists' mainstream migration papers have been devoid of much attempt at causal explanation, reflecting, rather, a longstanding tradition of concern for accurate measurement and modeling of structural dynamics, as well as a focus on migrants' socioeconomic attributes. To single out a very few representative works from the sociology/demography mold which would be of most direct interest to regional scientists, we should mention the spatial demographic modeling frameworks pioneered by Rogers (e.g., 1975) and the spatial demographic accounts of Rees and Wilson (1977), extended Markovian-style modeling work (such as Ginsberg, 1979 and the duration of residence/cumulative inertia approach tested in, e.g., Myers, McGinnis, and Masnick, 1967), as well as work on model migration schedules (e.g., Rogers, Raquillet, and Castro, 1978; Holmberg, 1984), which represents an extension of methods originally developed for the age-based analysis of mortality and fertility. Picking up a recent issue of Demography (the journal of the Population Association of America) should suffice to illustrate one of the stronger, standard sociological modes of population inquiry. For example, from Volume 25, Number 1 (February 1988), there are a number of good illustrations of a characteristic, multiway crosstabulation style of research (see, e.g., the papers by Fuguitt, Heaton, and Lichter; Dean; Kephart; and Wilson). Although we are not aware of a review piece assessing the 'spatial demography' school of thought on migration research, recent books edited by Woods and Rees (1985) and Congdon and Batey (1989) serve as manifestos.

This first commandment, then, reflects our belief that genetically stronger migration-researcher stock will result from the creative coupling of hitherto distinct strands of thought. Among several encouraging recent signs in this regards are (1) the inauguration through fourteen invited paper sessions at the Toronto 1988 North American RSA Meeting of an international 'network' of population/ migration/regional-labor-market researchers who will participate in organized sessions at future conferences, and (2) the interdisciplinary venture to assess new research directions in migration being spearheaded by Michael Greenwood as his Presidential Panel project for the 1990 Meeting of the Western Regional Science Association (Greenwood et al., 1990). Regional science conferences provide the kind of disciplinary neutered campsites required for linking together the separate migration-research laagers.

Through several of the other commandments we shall suggest some more specific ways in which analytically oriented researchers could contribute to link the migration-research laagers through pioneering new approaches or extending some of the most promising, though still tentative, existing findings. But before delving into nine more commandments, we wish to make one final plea concerning the interface between different research traditions as they intersect in the study of migration.

In a sense, the issue is not just one of different disciplinary bases, but of the entire creative tension between theoretical and empirical study. We contend that danger exists when dredging up a mere empirical regularity can somehow put one beyond the pale of our own regional science variation of 'the glass bead game' (cf. Hesse, 1949). At the same time, we think one must not set forth to joust on the pitted, empirical battleground of migration without riding upon a trusty steed of theory.

We assert that there is a need now for better interchange of ideas between those engaged in empirical investigations of migration and economic change and those persons who possess the expertise to extend our theoretical understanding of these processes. One of the two of us has a senior colleague who sometimes characterizes us with a term that was once in vogue in American geography: 'quantifiers'. While this seems quaint to us flower children of the quantitative revolution (it seemingly presupposes that the intrinsic ideas were already out there merely awaiting transcription into the appropriate matrix notation), the new international division of labor in our field should be of concern. We don't perceive problems with specialization - with some researchers advancing theory and others carrying out empirical exploration - rather, we are concerned that theory should better inform empirical work, and that the results of empirical inquiry should be taken to heart by theorists. Our second commandment also impinges upon the division of labor within the world regional science community.

(2) Looking at the other end of the chicken

Employment change and migration are 'Birds in Egg / Eggs in Bird' phenomena of an Olsson (1985) type. A full-feathered literature has, in fact, been hatched on the 'chicken and egg' (e.g., Muth, 1971; Steinnes, 1978) nature of their relationship. We contend, however, that we need to look more at the other end of the chicken in hopes of locating the controlling mechanisms for migration.

In recent times there has been considerable stress placed on exploring the behavioral aspects of migration decision making. While we do not decry progress in this branch of the field, from the perspective of the regional scientist what has largely been learned is that, in the aggregate, labor force migrants can't fully tell us how they come to make decisions. What is known with high certainty is that, in the aggregate, they respond fairly predictably to changing distributions of job opportunity, both in developed and in lesser developed nation settings.

Although the job-inducing multiplier effect of net in-migration to a region (and the reverse effect of net out-migration) is an interesting area for further study, this is primarily a plea for more regional science research focused directly on the changing distribution of job opportunity. More specifically, there is a need for macro-scale empirical research carried out in such a fashion that the behavioral interface to migration research is a broad and logical one.

Understanding where eggs come from is useful, but to know why the chicken crosses the road we think we need to zoom our micro/macro binoculars in on processes of job location.

Just as there is a need for more cross fertilization of ideas between migration researchers with different disciplinary affiliations, there is a need to link better the ongoing branches of inquiry within regional science. An area in which regional scientists have excelled is their work on understanding the economic bases and locational principles of settlements. Concepts of hierarchical systems have been developed with considerable appeal. Yet there are too few attempts seeking to understand either theoretically or empirically the nature of migration flows within the context of hierarchical settlement systems. A very few noteworthy works come to mind (Hudson, 1975; Huff, 1976; Mulligan's, e.g., 1983 work on population in central place systems) but research on even as overly studied a phenomenon as the 1970s' metro/nonmetro 'turnarounds' of population (see, for example, Brown and Wardwell, 1980) still gets carried out in a largely aspatial context.

One of Ravenstein's assertions about 19th century internal migration patterns in Britain was that migration proceeded in a stepwise fashion up the urban hierarchy. Too little is known about how it proceeds today. There is a wealth of thought on settlement system structure that awaits creative tapping by migration researchers - or, perceived differently, much work still to be done by settlement-system researchers to integrate their theories with those of migration specialists.

Our third commandment takes us back to the issue of the creative tension between theory, as expressed in migration models, and the nature of real-world migration processes.

(3) Turning to turnarounds

In the past couple of decades regional scientists have come to appreciate that the structure of many models precludes the accurate forecasting of 'turnarounds', or changes in the pattern or direction of some variable of interest. This is particularly true in demography, where extrapolative techniques are commonplace. Linear Markov models of migration, for example, lead to the conclusion that regional population shares will change monotonically before asymptotically approaching some equilibrium value. Our third commandment is thus: while advancing knowledge about nonlinear modeling structures, we should also stress empirical monitoring of turnarounds and trends.

It is common in the development of models to construct them so that the future evolution of the state variables reflect past trends. Because only a limited number of parameters are typically made available to model the system at hand, the parameters essentially represent some sort of average past behavior of the system. It should not be surprising that we so often predict that growing regions will continue to grow and that declining regions will continue to decline. Forecasts are often conservative for rapidly changing regions, failing to correctly foresee either nonmonotonic change or the acceleration of recent change.

These observations suggest the importance of monitoring activities. Close monitoring of trends will facilitate the discovery of significant changes within a system. Problems of data availability and measurement method, however, often hinder the implementation of such monitoring. The debate surrounding the so-called metro-nonmetro and core-periphery population 'turnarounds' of the 1970s which occurred in much of the developed world (e.g., Vining and Kontuly, 1978; Hall and Hay, 1980; Vining, Pallone, and Plane, 1981; Vining and Pallone, 1982) was as much about the proper measurement of the phenomena as it was about their underlying causes. The importance for urban policy purposes of careful migration monitoring is now, however, coming to be realized, as attested to by the chapters by Worrall (1989), Forbes (1989), and Nugent and Rampa (1989) in the previously cited book edited by Congdon and Batey.

There are several alternatives available to the researcher or practitioner interested in monitoring. Monitoring change might take place, for example, through (1) updating cross-sectional system parameters, (2) fitting 'causative' matrices (Rogerson and Plane, 1984; Plane and Rogerson, 1986) to track changes in a multistate system, or (3) using time-series analysis to monitor temporal changes in either system variables or parameters.

In addition to the common direction and magnitude of model errors noted above, alternative models often will produce forecasts closer to each other than to the actual variable. This again results from the linear sameness that is either implicitly or explicitly built into the majority of models. The real world features takeoffs, accelerated downspins, and turnarounds; models feature continuations of past trends. These characteristics apply to both simple, extrapolative methods and sophisticated multiequation models. The multiequation models, in fact, require conservative, quasilinear responses to avoid explosive feedbacks. Models with more flexibility and more attention devoted to nonlinearities are required to represent the vagaries of the real world.

Shrunken estimators have not yet been widely employed in social science. The idea is quite simple: future estimates of a variable that is measured across individuals, groups, or regions are moved ('shrunken' or regressed) toward the overall mean in accord with the uncertainty surrounding the forecast. This type of estimator may be thought of as a compromise between the no-change and the constant-rate scenarios. Presumably a small amount of shrinkage would be used in the beginning of a forecast period and a larger amount as time went on; eventually all units would be assigned values equal to the overall mean. One might expect this approach to work well for producing long-run forecasts in those instances where much current change is taking place; large magnitude changes generally dissipate over time.

In migration research, an approach somewhat like that of shrunken estimators is provided by information-theoretic lag models (e.g., Plane, 1982). In this form of model originally suggested by Tony Smith, a priori estimates of place-to-place migration flows are based upon two, or potentially more, base periods of data and the overall information-statistic minimization procedure is used to specify the optimal 'fading away' of the trends embedded within the matrices representative of past system behavior.

Shrunken estimators and information-theoretic lag models, however, do not directly deal with the issues of nonlinear change. The former would also have to be modified into 'enlarged' estimators in order to handle situations where greater change was expected in the near future. It would probably be desirable in such cases to have enlarged estimators in the short-term which would give way to shrunken estimators as the future further unfolded. Observations on the system over several periods of time would allow one to estimate whether particular regions were either accelerating away from the mean or regressing toward it.

A large statistical literature also exists on such topics as: (1) the detection of parameter shifts in statistical distributions such as the normal, the gamma (Kander and Zacks, 1966; Hsu, 1979), and the multinomial (Srivastava and Worsley, 1986); (2) switching regressions, where coefficients change at discrete points in time (Quandt, 1958); and (3) varying parameter regression models (Hildreth and Houck, 1968). This literature has in common a concern with models whose parameters are not constant over time, although the functional specification of the model itself does not change over time. These approaches are concerned with the retrospective estimation of both parameter changes and the time points at which they occurred. Detection of parameter changes may also take place by monitoring observations as they are received (e.g., Bagshaw and Johnson, 1977; Chen, 1978).

Our fourth commandment sounds a call for more general attention to be given to the nature of systemic change in the use of correctly conceptualized economic and demographic variables.

(4) Stocks, flows, and flows in flows

To be more specific about commandments (2) and (3), we further assert that empirical advances on the interface of demographic and economic change must begin by more detailed and careful consideration of the nature of systemic change. The demographer's careful, accounts-based perspective includes a useful concept in its detailed consideration of the nature of stock versus flow variables. In economic modeling this distinction is sometimes lost - to the detriment of the overall modeling enterprise.

A stock variable is a single-point-in-time measure, such as a population count, whereas a flow variable is simply one for which an observational period must be specified. Pertinent flow variables include such measures as net migration to region k, gross movement from region i to region j, births in k, job layoffs in sector s in sector k, and so forth.

Too often models have failed to consider adequately the inter-relationships between labor market and demographic stocks and flows. A prominent case in point is the over-reliance on unemployment variables in regional econometric modeling. Unemployment is a 'stock' measure of the results of the various complex flows actually taking place in a regional labor market; it is these flows - job accessions, quits, new hires, layoffs, and so forth - which are themselves what are truly pertinent to migration. When a 'flow' variable is used, it is often not the best one conceptually. A prominent case in point is the overuse of aggregate employment change rather than less accessible labor market turnover variables (see, e.g., the discussion in Plane and Rogerson, 1985).

We have a related concern about the tenuousness of the foothold that time-series analyses have gained, to date, in the migration literature. This is, in part, attributable to a data-collection tradition centered around infrequent, large-scale censuses, and, in part, to the greater inherent geographic dimensionality of migration than of other regional economic variables. Whereas the preferred, logical view of an internal migration system is as an (n x n x m) matrix of flows from regions i = 1,2,...,n to regions j = 1,2,...,n disaggregated by m migrant attributes, regional employment has been far more frequently modeled based upon data banks of dimension (t x n x s), where t is the number of time periods over which the employment 'stock' variable is measured within s different economic sectors. It is difficult to build a meaningful interface between migration and regional economic studies when the conceptual (as well as empirical) data bases balance only along a knife edge (see also the ninth commandment, below). The regional labor submarkets in which migrants make decisions are probably defined in part by economic sector, but in part also by occupational characteristics.

This, then, is a plea for more time-series consciousness - as well as simply time-series conscience - in migration research. Such attention will aid in developing the interface with regional econometrics, in which time-series work has been the norm. (See, also, the much fuller discussion of state-of-the-art work on this interface given in Isserman, 1986).

A logical part of a broadened time-series perspective in migration studies is a focus not simply on migration 'flows', but also on changes in the level of those flows over time. In demographic parlance, we are asking for attention to be given to 'flows in flows'. Regional growth and decline is intrinsically bound up with the acceleration of interregional streams of population movement, but only a few studies approach the matter from a 'flows in flows' perspective. See Plane (1987) for a review of some relevant current literature and for further development of the case for such a perspective; and see Rogerson and Plane (1984) and Plane and Rogerson (1986) for developments pertaining to a 'causative matrix' extension of traditional Markov models to study such 'flows in flows'.

Our belief in the importance of devoting further attention to the conceptual interface between migration and economic change is extended in the fifth commandment.

(5) Feedbacks, linkage structures, and interregional interdependencies

We should keep the faith of the early period of regional science regarding the significance of the integrated analysis of regional systems. Many models purport to examine and include links between system components (e.g., links between demographic characteristics and labor force characteristics). Comprehensive models comprised of linked submodels have as their raison d'être an emphasis on the feedbacks and relations between system subcomponents. The integrated analysis of regional systems is the subject of a recent, edited collection of papers (Batey and Madden, 1986). As Batey and Madden point out in their introduction, the concept of such integration might legitimately apply in several contexts, including (1) the combined modeling of related phenomena, such as migration and the restructuring of regional labor markets; and (2) the simultaneous modeling of several regions, thereby focusing on the effects of interregional linkages.

Clearly, it would be desirable to be able to produce and use a model of a regional system that was capable of representing multifarious components as well as their interactions. But there are a number of indicators that this has not yet been achieved. Lee's (1973) requiem served as one of the first critical appraisals of models of integrated systems. The difficulties in assembling sufficient data and the necessity of getting each of the components to perform adequately often serve as major impediments to a thorough evaluation of interactions. In many cases there is a consequent lack of attention given to the specific study of linkages and feedbacks. A good case in point is the ECESIS system that we were involved in developing at the U.S. Bureau of the Census to model interactions between demographic change and the economy (see Beaumont et al., 1986). In this model, major attention is given to both demographic accounts (using the Rees-Wilson 1977 approach) and linked models of state economies. Yet these sophisticated submodels are tied together with a single function describing the migration response to regional economic differences. Of course the economic models are driven, to some degree, by population as well, but the point is that although linkages between states are relatively well developed, the demographic and economic models are held together by a single thread. Probably less than five percent of the project, whose objective was to model demographic-economic interaction, was devoted to the 'interaction'. For an empirical analysis of the benefits obtainable through careful specification of economic-demographic interactions in regional econometric models, see Taylor (1986).

Despite the difficulties inherent in attempting to sort out the direction, magnitude, and importance of feedback effects, there has been, of late, somewhat of a revival of interest in large-scale systems models (cf. Batey and Madden, 1986; Boyce, 1988). A new breed of large-scale models has emerged that has benefited from past experience. Examples include the Integrated Transportation and Land Use Package, which has demonstrated its usefulness in planning and policy evaluation, and CATLAS, which rests solidly upon a foundation of behavioral economic theory.

In addition to the examination of feedback effects within the context of large-scale models, it is possible to model them in less resource-intensive ways. Wachter and Lee (1989), for example, reevaluate the previous work of Lee (1974) and Frauenthal and Swick (1983), concluding that while demographic feedback effects occur in the temporal evolution of childbearing, they are insufficient to produce the sustained limit cycles of the Easterlin (e.g., 1973, 1980) type. Keyfitz and Philipov (1981) construct a simple model to evaluate the effects of migration and natural increase on population growth. They find migration to be much more significant than is often realized due to the fact that in-migrants are often persons in cohorts primed for childbearing whereas new borns must wait at least 15 years before additional contributions to population growth are made. In our sixth commandment we shall emphasize the importance of age composition, and the interaction between different age groups, within the context of migration analysis.

To wrap up our discussion of feedbacks and linkages, we note that regional scientists in certain subfields have consistently excelled in treating the interdependencies intrinsic to flow-based systems. Despite the prominent place generally accorded to Ravenstein in tracing the historical development of the gravity model, migration research has not always been conducted with the careful attention devoted to, for example, shopping behavior. The flow of persons from origin region i to destination j is dependent not only on conditions in i and j, but also on those in all other regions of the system. In Plane (1987) the state-of-the-art with regards to treating interregional interdependencies in migration systems is reviewed. Among the noteworthy attempts described therein is work by Willekens and Baydar (1986), Milne (1981), Ledent (1986), and Plane (1984b), who gives a doubly-constrained inverse gravity model treatment of the role played by distance in determining the volume of migration streams. The 'competing destinations' concept of spatial interaction models developed by Stewart Fotheringham also has been applied in recent studies to migration (Fotheringham, 1986, 1987).

Regional scientists ought to continue to extend methods for treating interregional interdependencies and to expose the failings of research which does not faithfully represent the outcomes of interregional competitive processes.

(6) Getting older one year at a time

The lack of time-series analysis tradition in migration research, and in demography in general, is quite perverse given that one of the most predictable empirical regularities we know about is that most of us will get older one year at a time. When aggregated over entire populations, changing age distribution is likely to be one of the most forceful 'drivers' of the dynamic evolution of socioeconomic systems. An age dimension probably contributes significantly through lifecycle mechanisms to shifts in economic geographic patterns. So: we assert that we need to begin paying more attention to this peculiar empirical regularity!

Some recent thoughts by Andrei Rogers on how progress may be made in combined geographic/demographic/economic migration modeling scourges the two of us, personally, for failing to do so; we offer this quote, then, in a self-flagellatory spirit to emphasize our current zeal for this sixth commandment:

> It seems evident that an approach that combines the demographer's strengths in identifying regularities that may be obscured by inappropriate measurement and representation of the migration variable with the economist's strengths in behavioral modeling offers promise as a means for linking demographic and economic change in migration and population redistribution forecasting studies. Some efforts along these lines already have been made, for example, by David Plane and Peter Rogerson, who observe that it 'is only when the temporal change in relative economic attractiveness variables is combined with some measure of the relative strengths of past migration streams that significant gains in forecast accuracy are obtained' (Plane and Rogerson, 1985). Their effort, however, suffers from an inadequate introduction of the demographer's potential contribution. For example, their focus is on the explanatory modeling of Markov transition probabilities and when they do finally disaggregate to consider two age-specific probabilities their approach is to model these independently, ignoring the dependencies and regularities that are exhibited by such probabilities (Rogers, 1985, p. 4).

(7) Compassion for net migrants - even though there aren't any such animals

It has become fashionable to make the observation that 'there is no such thing as a net migrant.' Andrei Rogers (1989) has even written a requiem for the net migrant, clearly illustrating the fallacious interpretations that can occur from taking the uni- rather than multiregional perspective implicit in, e.g., net migration 'rates'. Yet net migration is truly the 'bottom line' for some, but not all, migration research; often the interest in migration is motivated by the need to know how population will change in migrant origin and migrant destination regions. We should not sully attempts to better understand the macro variable, net migration, as the outcome of all the individual level place-to-place flows that make up a migration system.

We contend that we should give equal time to both micro and macro views of migration. In this spirit, we believe too little is known about the aggregate cross-regional variation of gross in-migration and gross out-migration and how this relationship varies over time (see some intriguing, but not fully developed, results in Plane, Rogerson, and Rosen, 1984 and in Mueser and White, 1989).

Let us cite two examples of facts about U.S. interstate net migration that pose challenges for further study. The first is a somewhat troubling empirical observation for many mainline economic papers on migration. In recent years something like two-thirds of all the net interstate streams of population movement in the United States point in the direction of the lower average wage state. While it does not follow necessarily that migrants take pay cuts to move, and, as Rogers (1989) points out, counter-intuitive results with respect to net migration can occur simply through aggregation, in our minds the second commandment's call for further emphasis on job location processes is given greater credence. There are, of course, opposing behavioral motivations for employers and prospective job seekers with respect to wages; migration is the result of a complex interplay of decision processes which, on the migrant side, also involve various commuting options within the local labor market without changing place of residence.

The second empirical reality is that there is extraordinarily significant 'cross-hauling' in developed nation migration patterns when one looks from the viewpoint of total population movement. A long decline in the overall demographic effectiveness of interstate migration in the U.S. (measured as the ratio of net to total migration) was reversed during the post oil-embargo period of the 1970s, but for 1975-80 still stood at a mere 16.8 percent, indicating that for about every six interstate migrants only one of them served to effect state-level populations changes. (See Plane 1984a for a fuller treatment of the demographic effectiveness view of structural change in migration systems.) It is a real challenge to develop behavioral theories robust enough to proxy the richness of the patterns of the real world; migration streams in many cases are in fact the aggregates of the choices of many different decision-makers acting for a wide variety of reasons.

There would also seem to be potential in supplementing neoclassical equilibrium models with disequilibrium models that focus on the behavior of systems away from steady states. There is a developing literature on estimation methods for markets in disequilibrium (e.g., Fair and Jaffee, 1972; Fair and Kelejian, 1974; Maddala and Nelson, 1974; Gourieroux, Laffont, and Montfort, 1980; Anas and Eum, 1984). It would seem desirable to continue this work, to bring its concepts into migration analysis, and to integrate it with concepts of chaotic dynamics and self-organizing systems (Nicolis and Prigogine, 1977; Feigenbaum, 1980) now being explored for applications in economics and geography (e.g., Benhabib and Day, 1981; Rogerson, 1985).

It is important for regional science to continue to strive to understand what motivates the underlying behavior of the individual migrants in a system with widespread crosshauling and a preponderance of net migration streams to lower wage areas, but we have sympathy also for the (albeit sexist) view of P. Vidal de la Blache that 'geography is the science of place, not of men'. We think that real progress can be made in understanding economic/demographic linkages from aggregate, regional structure viewpoints, although spinning behavioral yarns can be very helpful in motivating macroscale analysis. The need for a variety of research approaches from fundamentally differing perspectives is emphasized somewhat differently in our eighth commandment.

(8) Congruous complexity

The ultimate degree of complexity of a migration modeling approach should depend, in part, on the purpose to which the model is to be put. Intuition may tell us that the more complex a model, the better it will be able to mimic the complex processes it seeks to explain. Yet Alonso (1986) has made the point that errors may compound rapidly in complex models. The large errors associated with highly disaggregated data, plus the nature of error propagation in models consisting of more complex mathematical operations, may outweigh potential gains from making the model more realistic. Thus it does not necessarily follow that sophisticated models will produce more accurate forecasts than elementary models. Variable disaggregation and increased complexity of functional form are justified when data quality is adequate, heterogeneity exists in the variables of interest, and the process is understood sufficiently to justify the more complex specification (see, e.g., Grunfeld and Griliches, 1960; Aigner and Goldfeld, 1974; Rogerson, 1983).

In general, more sophisticated models are best suited for understanding relationships between variables and for performing policy-simulation and sensitivity studies; simpler models are often adequate for forecasting.

A related issue is the tradeoff between the costs and returns of data collection (see, for instance, Spencer, 1980). Too often we are completely in the dark regarding the relative merits associated with particular sample sizes and types of data. The influential role of data availability on our ability to carry out research on migration forms the motivation for our ninth commmandment.

(9) Complaining for data

Recently a very real threat existed that the 1990 U.S. census would be bereft of a question on migration. Only the combined efforts of data-user groups convinced the Reaganite zealots at the Office of Management and Budget of the need for such information on 'place of previous residence.' Even with decennial five-year migration data thus preserved, our nation lags considerably behind other developed nations in monitoring internal migration trends (Isserman, Plane, and McMillen, 1982). Threats to the continued availability of migration data, and census and population registry data more generally, have occurred in other nations also. Thus: we should be willing to devote the energy required to lead the efforts for more and better data.

Vigilance is ever necessary; better interaction between the academic research world and the bureaucratic statistical community would help. Especially noteworthy is the need to instill into bureaucrats the need for consistently collected time-series data on place-to-place migration streams disaggregated by migrant characteristics. In devoting one of our commandments to 'data', we assert an intrinsic importance of thinking about data. In a previous paper we stated:

> The nature of the data which are collected exerts a powerful influence on the development of a scientific discipline. In turn, the traditions and academic biases of the scholars in the discipline influence the types of data designated for collection (Plane and Rogerson, 1985, p. 185).

(10) Mobilization on melting pots

At this worldwide gathering of regional scientists we think it is appropriate to end our prescriptive assertions about future migration research with a call for greater emphasis on international migration. Additional regional science research should focus on immigration and the impacts of immigration on internal migration systems. In a recent paper presentation to the Western Regional Science Association, Peter Schaffer (1988) noted that the U.S. is by no means unusual in experiencing largescale foreign immigration. Significant international migration is the rule rather than the exception for many countries. Too little research, to date, has been devoted to international migration from the regional scientist's characteristic perspectives. We would like to see more literature developed along the lines of the analytical, time-series work of Constantinou and Diamantides (1985), as well as from the economic-behavior viewpoint (see, e.g., Schaeffer, 1987).

In addition, too little attention has been paid to the effects on internal population redistribution of international migration. We need more studies (such as Desbarats, 1985) focused on the processes by which foreign migrants - who often enter a country from a single or a very few points of entry - swirl over time across the breadth of national melting pots. These processes have interesting implications for interregional economic restructuring.

3. Conclusions

Despite the impressive quantity of research produced on migration from analytical, regional science perspectives during the century that has passed since Ravenstein's 1885 assertions, we could conclude this paper with Peter Morrison's assertion of 1973 that 'those who know enough to forecast migration know better than to try'. There is still much to be learned about migration processes and their relationship to geographic patterns of economic activity.

The 'ten commandments' that we set forth in this paper give some directions in which we believe accelerated progress lies. Maybe given the rapidity of progress in recent migration research, however, we should be happy if the words of a baseball relief pitcher named Dan Quisenberry prove prophetic: 'I have seen the future, and it is very much like the present, only longer'.

Acknowledgements

An earlier version of this paper was presented at a plenary session of the third World Congress of the Regional Science Association, Kibbutz Ramat Rachel, Israel, April 2-7, 1989. Partial financial assistance for this work was provided through U.S. National Science Foundation awards, SES8810126 and SES8553055. The authors would like to thank the following persons for their helpful comments: Geoffrey Hewings, Michael Sonis, Andrei Rogers, Michael Greenwood; the paper's discussant at the Congress, Walter Weidlich; and coeditors Peter Nijkamp, David Boyce, and Daniel Shefer.

References

Aigner, D. and S. Goldfeld (1974), Estimation and Prediction from Aggregate Data When Aggregates Are Measured More Accurately Than Their Components, Econometrica 42:113-134.

Alonso, W. (1968), Predicting Best with Imperfect Data, Journal of the American Institute of Planners 34:248-254.

Anas, A. and S. Eum (1984), Hedonic Analysis of the Housing Market in Disequilibrium, Journal of Urban Economics 15:87-106.

Bagshaw, M. and R. Johnson (1977), Sequential Procedures for Detecting Parameter Changes in a Time-Series Model, Journal of the American Statistical Association 72:593-600.

Batey, P. and M. Madden (1986), Integrated Analysis of Regional Systems, London, Pion.

Beaumont, P., A. Isserman, D. McMillen, D. Plane and P. Rogerson (1986), The ECESIS Economic-Demographic Model of the United States, in: A.M. Isserman (ed.), Population Change and the Economy: Social Science Theories and Models, Boston, Kluwer-Nijhoff.

Boyce, D. (1988), Renaissance of Large-Scale Models, Papers of the Regional Science Association 65:1-10.

Brown, D.L. and J.M. Wardwell (eds.) (1980), New Directions in Urban-Rural Migration: the Population Turnaround in Rural America, New York, Academic Press.

Chen, R. (1978), A Surveillance System for Congenital Malformations, Journal of the American Statistical Association 73:323-327.

Clark, W.A.V. (1986), Human Migration, Beverly Hills, Sage Publications, Scientific Geography Series, Volume 7.

Congdon, P. and P. Batey (eds.) (1989), Advances in Regional Demography: Forecasts, Information, Models, London, Belhaven Press.

Constantinou, S.T. and N.D. Diamantides (1985), Modeling International Migration: Determinants of Emigration from Greece to the United States, 1820-1980, Annals of the Association of American Geographers 75:352-369.

Dean, K.G. (1988), Interregional Flows of Economically Active Persons in France, 1975-1982, Demography 25:81-98.

Desbarats, J. (1985), Indochinese Resettlements in the United States, Annals of the Association of American Geographers 75:522-538.

Easterlin, R.A. (1973), Relative Economic Status and the American Fertility Swing, in: E.B. Sheldon (ed.), Family Economic Behavior: Problems and Prospects, pp. 170-223, New York, North Holland.

Easterlin, R.A. (1980), Birth and Fortune: The Impact of Numbers on Personal Welfare, New York, Basic Books.

Fair, R. and D. Jaffee (1972), Methods of Estimation for Markets in Disequilibrium, Econometrica 40:497-514.

Fair, R. and H. Kelejian (1974), Methods of Estimation for Markets in Disequilibrium: a Further Study, Econometrica 42:177-190.

Feigenbaum, M. (1980), Universal Behavior in Nonlinear Systems, Los Alamos Science, Summer:4-37.

Forbes, J. (1989), Migration Monitoring and Strategic Planning Policy, in: P. Congdon and P. Batey (eds.), Advances in Regional Demography: Information, Forecasts, Models, pp. 41-57, London, Belhaven Press.

Fotheringham, A.S. (1986), Modelling Hierarchical Destination Choice in Migration, Environment and Planning A 18:401-418.

Fotheringham, A.S. (1987), Hierarchical Destination Choice: Discussion with Evidence from Migration in The Netherlands, Working Paper no. 69, Netherlands Interuniversity Demographic Institute.

Frauenthal, J. and K. Swick (1983), Limit Cycle Oscillations of the Human Population, Demography 20:258-298.

Fuguitt, G.V., T.B. Heaton and D.T. Lichter (1988), Monitoring the Metropolitanization Process, Demography 25:115-128.

Ginsberg, R.B. (1979), Timing and Duration Effects in Residence Histories and Other Longitudinal Data, Regional Science and Urban Economics 9:311-332, 369-392.

Gourieroux, C., J. Laffont and A. Monfort (1980), Disequilibrium Econometrics in Simultaneous Equations Systems, Econometrica 48:75-95.

Greenwood, M.J. (1975), Research on Internal Migration in the United States: A Survey, Journal of Economic Literature 13:397-433.

Greenwood, M.J. (1985), Human Migration: Theory, Models, and Empirical Studies, Journal of Regional Science 25:521-544.

Greenwood, M.J., P. Mueser, D.A. Plane and A. Schlottman (1990), New Directions in Migration Research: Perspectives from Economics, Geography, and Sociology. Presidential Panel Paper presented at the Annual Meeting of the Western Regional Science Association, February 1990.

Grunfeld, Y. and Z. Griliches (1960), Is Aggregation Necessarily Bad?, Review of Economics and Statistics 42:1-13.

Hall, P. and Hay, D. (1980), Growth Centres in the European Urban Systems, London, Heinemann Educational Books.

Hesse, H. (1949), Magister Ludi (translated from the German Das Glasperlenspiel by M. Savill), New York, F. Ungar.

Hildreth, C. and Houck, J. (1968), Some Estimators for a Linear Model with Random Coefficients, Journal of the American Statistical Association 63:584-595.

Holmberg, I. (1984), Model Migration Schedules - The Case of Sweden, Scandinavian Population Studies 6:181-195.

Hoover, E.M. and F. Giarratani (1984), An Introduction to Regional Economics (third edition), New York, Alfred A. Knopf.

Hsu, D. (1979), Detecting Shifts of Parameters in Gamma Sequences with Applications to Stock Prices and Air Traffic Flow Analysis, Journal of the American Statistical Assocation 74:31-40.

Hudson, J.C. (1975), Some Observations on Migration Theory for an Urban System, in: L.A. Kosinski and R.M. Prothero (eds.), People on the Move, pp. 67-74, London, Methuen.

Huff, J.O. (1976), A Hierarchical Model of Population Redistribution Within a Central Place Hierarchy, Geographical Analysis 8:231-254.

Isserman, A.M. (ed.) (1986), Population Change and the Economy: Social Science Theories and Models, Boston, Kluwer-Nijhoff.

Isserman, A.M., D.A. Plane and D.B. McMillen (1982), Internal Migration in the United States: An Evaluation of Federal Data, Review of Public Data Use 10:285-311.

Kander, Z. and S. Zacks (1966), Test Procedures for Possible Changes in Parameters of Statistical Distributions Occurring at Unknown Time Points, Annals of Mathematical Statistics 37:1196-1210.

Kephart, G. (1988), Heterogeneity and the Implied Dynamics of Regional Growth Rates: Was the Nonmetropolitan Turnaround an Artifact of Aggregation?, Demography 25:99-114.

Ledent, J. (1986), Forecasting Interregional Migration: An Economic-

Demographic Approach, in: A.M. Isserman (ed.), Population Change and the Economy: Social Science Theories and Models, pp. 53-77, Boston, Kluwer-Nijhoff.

Lee, D. (1973), Requiem for Large-Scale Models, Journal of the American Institute of Planners 39:163-178.

Lee, R. (1974), The Formal Dynamics of Controlled Populations and the Echo, the Boom, and the Bust, Demography 11:563-585.

Maddala, G. and F. Nelson (1974), Maximum Likelihood Methods for Models of Markets in Disequilibrium, Econometrica 42:1013-1030.

Milne, W.J. (1981), Migration in an Interregional Macroeconomic Model of the United States: Will Net Outmigration from the Northeast Continue?, International Regional Science Review 6:71-83.

Morrison, P.A. (1973), Theoretical Issues in the Design of Population Mobility Models, Environment and Planning 5:125-134.

Mueser, P.R. and M.J. White (1989), Explaining the Association Between Rates of In-migration and Out-migration, Papers of the Regional Science Association 67:121-134.

Mulligan, G.F. (1983), Central Place Populations: A Microeconomic Consideration, Journal of Regional Science 23:83-92.

Muth, R.F. (1971), Migration: Chicken or Egg?, Southern Economic Journal 37:295-306.

Myers, G.C., R. McGinnis and G. Masnick (1967), The Duration of Residence Approach to a Dynamic Model of Internal Migration: A Test of the Axiom of Cumulative Inertia. Eugenics Quarterly 14:121-126.

Nicolis, G. and I. Prigogine (1977), Self-organization in Nonlinear Systems, New York, John Wiley.

Nugent, S. and H. Rampa (1989), Demographic Change at the Small-area Level: Implications for the Planning of Canberra, in: P. Congdon and P. Batey (eds.), Advances in Regional Demography: Information, Forecasts, Models, pp. 58-72, London, Belhaven Press.

Olsson, G. (1980), Birds in Egg / Eggs in Bird, London, Pion.

Plane, D.A. (1982), An Information Theoretic Approach to the Estimation of Migration Flows, Journal of Regional Science 22:441-456.

Plane, D.A. (1984a), A Systemic Demographic Efficiency Analysis of U.S. Interstate Population Exchange, 1935-80, Economic Geography 60:294-312.

Plane, D.A. (1984b), Migration Space: Doubly Constrained Gravity Model Mapping of Relative Interstate Separation, Annals of the Association of American Geographers 74:244-256.

Plane, D.A. (1987), The Geographic Components of Change in a Migration System, Geographical Analysis 19:283-299.

Plane, D.A. (1990), A Perspective from Geography: The Spatial Choice Set, in: New Directions in Migration Research: Perspectives from Economics, Geography, and Sociology, M.J. Greenwood, P. Mueser, D.A. Plane, and A. Schlottman. Paper presented at the Annual Meeting of the Western Regional Science Association, Molokai, Hawaii, February.

Plane, D.A. and P.A. Rogerson (1985), Economic-demographic Models for Forecasting Interregional Migration, Environment and Planning A 17:185-198.

Plane, D.A. and P.A. Rogerson (1986), Dynamic Flow Modeling with Interregional Dependency Effects: An Application to Structural Change in the U.S. Migration System, Demography 23:91-104.

Plane, D.A., P.A. Rogerson and A. Rosen (1984), The Cross-regional Variation of In-migration and Out-migration, Geographical Analysis 16:162-175.

Quandt, R. (1958), The Estimation of the Parameters of a Linear Regression System Obeying Two Separate Regimes, Journal of the American Statistical Association 53:873-880.

Ravenstein, E.G. (1885), The Laws of Migration, Journal of the Royal Statistical Society 48:167-235.

Rees, P.H. and A.G. Wilson (1977), Spatial Population Analysis, London, Edward Arnold.

Rogers, A. (1975), Introduction to Multiregional Mathematical Demography, New York, John Wiley.

Rogers, A. (1985), The Migration Component in Multiregional Modeling: Measurement, Spatial Dynamics, Causal Knowledge, and Forecasting. Boulder, University of Colorado, Institute of Behavioral Science, Population Program, Working Paper WP-85-1.

Rogers, A. (1989), Requiem for the Net Migrant. Paper presented at the North American Meetings of the Regional Science Association, Santa Barbara, California, November.

Rogers, A., R. Raquillet and L.J. Castro (1978), Model Migration Schedules and Their Applications, Environment and Planning A 10:475-502.

Rogerson, P.A. (1983), Comparisons of Aggregate Variable Forecasts Using Aggregate and Disaggregate Models, Socio-economic Planning Sciences 17:373-380.

Rogerson, P.A. (1984), New Directions in the Modelling of Migration, Economic Geography 60:111-121.

Rogerson, P.A. (1985), Disequilibrium Adjustment Processes and Chaotic Dynamics, Geographical Analysis 17:185-198.

Rogerson, P.A. and D.A. Plane (1984), Modeling Temporal Change in Flow Matrices, Papers of the Regional Science Association 54:147-164.

Schaeffer, P.V. (1987), A Dynamical Model of Labor-market Change in International Migrations When Demand for Labor is Exogenous, Environment and Planning A 19:1051-1057.

Schaeffer, P.V. (1988), Guests Who Stay: Foreign Workers and Economic Development. Paper presented at the Annual Meeting of the Western Regional Science Association, Napa, CA.

Sonis, M., Locational Push-Pull Analysis of Migration Streams, Geographical Analysis 12:80-97.

Spencer, B. (1980), Benefit-cost Analysis of Data Used to Allocate Funds, New York, Springer-Verlag.

Srivastava, M. and K. Worsley (1986), Likelihood Ratio Tests for a Change in the Multivariate Normal Mean, Journal of the American Statistical Association 81:199-204.

Steinnes, D.N. (1978), Causality and Migration: a Statistical Resolution of the 'Chicken or Egg' Fowl-up, Southern Economic Journal 45:218-226.

Taylor, C. (1986), The Effects of Refining Demographic-economic Interactions in Regional Econometric Models, in: A.M. Isserman (ed.), Population Change and the Economy: Social Science Theories and Models, Boston, Kluwer-Nijhoff.

Vining, D.R.,Jr. and T. Kontuly (1978), Population Dispersal from Major Metropolitan Regions: an International Comparison, International Regional Science Review 3:49-73.

Vining, D.R.,Jr. and R. Pallone (1982), Migration between Core and Peripheral Regions: a Description and Tentative Explanation of the Patterns in 22 Countries, Geoforum 13:339-410.

Vining, D.R.,Jr., R. Pallone and D.A. Plane (1981), Recent Migration Patterns in the Developed World: a Clarification of Some Differences Between Our and IIASA's Findings, Environment and Planning A 13:243-250.

Wachter, K. and R. Lee (1989), U.S. Births and Limit Cycle Models, Demography 26:99-115.

Weidlich, W. and G. Haag (1986), Stochastic Migration Theory and Migration Phase Transitions, in: D.A. Griffith and R.P. Haining (eds.), Transformations Through Space and Time: an Analysis of Nonlinear Structures, Bifurcation Points and Autoregressive Dependencies, pp. 104-117, Dordrecht, Martinus Nijhoff.

White, S.E., L.A. Brown, W.A.V. Clark, P. Gober, R. Jones, K. McHugh and R.L. Morrill (1989), Population Geography, in: G.L. Gaile and C.J. Willmott (eds.), Geography in America, Merrill, Columbus, Ohio.

Willekens, F. and N. Baydar (1986), Forecasting Place-to-Place Migration with Generalized Linear Models, in: R. Woods and P. Rees (eds.), Population Structures and Models: Developments in Spatial Demography, pp. 203-244, London, Allen and Unwin.

Wilson, F.D. (1988), Components of Change in Migration and Destination-Propensity Rates for Metropolitan and Nonmetropolitan Areas: 1935-1980, Demography 25:129-140.

Worrall, L. (1988), Urban Demographic Information Systems, in: P. Congdon and P. Batey (eds.), Advances in Regional Demography: Information, Forecasts, Models, pp. 25-40, London, Belhaven Press.

Chapter 3

MAJOR ISSUES IN REGIONAL LABOUR MARKET ANALYSIS

M.M. Fischer
P. Nijkamp

1. Introduction

Although there is a substantial body of literature on labour market analysis, most of it ignores the spatial dimension of the labour market. A spatial perspective in analyzing labour market processes is important for several reasons (see Fischer, 1986). First, labour markets are by no means as homogeneous as conventional labour market theories assume. In reality, labour markets are segmented by firm, industry, job type, worker type, occupation and - last but not least - spatial location. In other words, the macro pattern of labour market segmentation is reflected in a spatial segmentation. For instance, weak spatial labour markets are characterized by a high concentration of problem groups on the labour market as well as by a tendency to structural unemployment. The fact that labour markets are spatially segmented raises also important policy issues. Second, most countries are displaying strong spatial variations in the dynamics of unemployment. Certain regions (e.g., peripheral ones, old industrial areas and increasingly also inner cities) are more seriously suffering from unemployment problems than others. There are spatial inequalities, not only in terms of unemployment rates, but also in the range of quality of jobs available. The poor spatial coordination of labour supply and labour demand has recently become an important policy issue (see Clark, 1983b). Third, geographical space exerts a frictional effect on labour market adjustment processes. Frictions of distance and resulting problems of market coordination and information may lead to a less efficient allocation of labour and to structural imbalances on some labour markets (see Ballard and Clark, 1981). Fourth, the spatial distribution of job loss induced by economic restructuring and adjustment processes contrasts with that of new job creation, because of geographic differences in seedbed conditions for new firms. A convincing illustration of the relevance of a spatial perspective to labour markets, based on empirical observations, can be found in Fischer and Nijkamp (1987a).

Especially in an era of national economic stagnation, spatial discontinuities on the labour market are significant sources of inefficiency because spatial mismatches between the demand for and the supply of labour contribute to higher overall national unemployment. Up to now, relatively little is known about the operation and interaction of spatial labour markets. But a deeper theoretical and empirical understanding of spatial labour markets is a prerequisite for public policy aiming at promoting more efficient spatial labour market coordination and allocation. In general, however, regional labour market research is to a large extent using concepts and methods from macro-based labour market analyses, which at best are extended with some regional aspects. This observation also clarifies why in a survey various fairly general regional labour market analyses are included.

The general idea underlying regional labour market research is that spatial disaggregation of the labour market provides additional insight into the observed labour market behaviour, by both identifying the various internal forces operating within individual regional labour markets and revealing the external forces which are transmitted between different spatial labour markets via various economic, social and institutional-political linkages (see also Martin 1981b). Thus, as a first analytical issue the definition and delineation of a spatial or regional labour market is of central importance.

Definitions in the literature on demarcation principles for a spatial or regional labour market vary considerably. Usually the regional labour market is defined as a region within which there is a clear labour market pattern defined by the spatial range of employment opportunities open to a worker without changing his place of residence. Evidently, this spatial range is affected by income, by transport access, by permeability of boundaries, commuter time and the extent of knowledge on alternative employment opportunities. From the viewpoint of the employer the spatial labour market may be defined as the spatial area which contains those (potential) members of the labour force a firm can - theoretically considered - attract (see Lever, 1980). Clearly, these two perspectives do not necessarily lead to an unambiguous result.

Even though there is no single definition which is universally appropriate for all types of regional labour market analysis, three criteria seem to be important as elements of a more general definition of a regional labour market: journey-to-work, the market's employment opportunities and the accessibility to market information about job openings, actual and potential future wages, etc. Based upon these criteria, a regional labour market may then be defined to be a spatially delineated area which fulfils the following boundary requirements (cf. also Hart, 1981):

- Daily journey-to-work across the boundary is insignificant.
- The pecuniary and non-pecuniary commuting and migration costs within the boundary of the region are significantly less than those between this region and any other region of the economy.
- Firms are located in those places where they can obtain the major proportion of their potential labour supply within the boundary.
- Although information about job openings and wages is imperfect, for the intra-regional labour force the search costs within the region are significantly less than those for searching in an alternative region.

Although in practice researchers are tempted to use regional delineations on the basis of available statistical information, it is noteworthy that the decision and demarcation on the design of a spatial framework is a crucial element in multiregional labour market modelling. Surprisingly, relatively little research has been done on this topic up to now. One of the rare exceptions are multiregional labour supply modelling efforts for Austria undertaken by Baumann et al. (1983, 1988). In principle, there are two fundamental approaches to the regionalization problem: the disjoint functional regionalization, and the non-disjoint (overlapping) functional regionalization approach. In the first case the basic spatial units of the study area are aggregated to form disjoint regional labour markets. There are several powerful numerical procedures available which use commuting flows between the basic spatial units as starting points and identify regional labour markets in such a way that cross-commuting is minimized (see Baumann et al., 1983). The second approach leads to a non-disjoint regionalization framework where transition zones can be taken into account (see Baumann et al., 1988). These two studies illustrate that the performance of multiregional labour market models may critically be dependent on the choice of an appropriate spatial framework and of the relative size of the labour market region. This means that it is by no means admissible to ignore possible effects of the choice of a spatial demarcation in multiregional labour market modelling. In the case of forecasting, additionally the question of the stability of a system of spatial labour markets defined on the basis of commuting flows may arise because commuting flows tend to change over time due to changes in the transport systems and other factors.

In the past few years regional labour market analysis has received increasing attention by economic geographers and regional economists. A central question dealt with is the distribution of employment (unemployment, vacancies) and wages over a system of regional labour markets. In particular, the existence and determinants of employment and wage income disparities in space have received a great deal of attention. A substantial part of labour market research is policy-oriented focusing on attempts to equalize regional unemployment and to receive intraregional equilibrium in the sense of low and stable unemployment rates (see Oosterhaven and Folmer, 1985).

Broadly considered, three types of research streams may be distinguished in the field of spatial labour market research. The first concerns labour supply issues, migration and job search behaviour (see Section 2). The second refers to the determinants of labour demand, where the major focus is on actual and likely future employment effects of spatial structural economic and broader technological changes (see Section 3). Finally, the third type deals with the spatial dimensions of the unemployment problem (see Section

4). These three types of spatial labour market research will be briefly discussed in the sequel from a non-technical point of view.

2. Labour Supply Issues

Regional labour supply issues deal with the structure and change of labour supply, spatial dimensions of job search, commuting within and migration between regional labour markets. In the short run these issues deal amongst others with the static allocation of labour supplied by current residents, whereas in the long run such issues also refer to increments or decrements to the supply attributable to adjustments, e.g. through migration.

2.1 General aspects

Labour supply analysis can in principle refer to a variety of different phenomena, for instance, having a job or not (i.e., labour force participation - a binary decision), hours worked per week or year (a continuous decision), etc. (see Heckman et al., 1981 and Schubert et al., 1987 for more details on this issue). In a regional labour market context labour supply is usually defined as the sum of the regional labour force and net commuting (see Schubert, 1982).

Various new evolutions are taking place in regional labour markets, e.g., demographic trends, labour force participation changes, a higher rate of self-employed, a higher skill level, and an increase in part-time jobs. Changes in participation rates and changes in net migration cause a dynamic evolution in the regional labour force. The most significant impact on regional labour supply in advanced economies in recent years has been the result of both changes in the size of the population of working age due to demographic factors and changes in participation behaviour. High birth rates in the 1950s and early 1960s resulted in large numbers of young people entering the labour force in the late 1970s and early 1980s, whereas the number of people approaching retirement age was relatively low. Much more important, however, was the rapidly increasing labour force participation of women, especially of married women. The rise in participation rates has been accompanied by two other important changes in female activity over the life cycle: a decline in the proportion of women responsible for the care of young children and an increase in the number of women carrying the double burden of professional (often part-time work in the service sector) and child-rearing responsibilities. Consequently, a higher proportion of the female workforce is building up long periods of continuous labour market experience, and thus is able to establish and maintain higher skill levels. Over the past two decades, women have made considerable progress in entering the labour market in most industrialized countries. However, there are still significant differences between female participation in various countries (e.g., Sweden vis-à-vis Poland), whilst even within a country there are significant differences (see Fischer and Nijkamp, 1987).

2.2 Labour force participation

Characteristically, spatial analysis has used either cross-section data (often disaggregated by age and sex) in a multiregional setting in order to reveal spatial variations in labour market participation behaviour or time series data in a single region context in order to evaluate the response of regional labour force to short run and long run changes in labour market conditions. Income variables (market wages and non-labour income), monetary costs (such as, e.g., commuting costs), attributes reflecting the uncertainty about the chances to obtain a job and/or preferences between income and leisure are usually included as major determinants or labour force participation.

There are fundamental differences in participation rates between sexes, especially between married men and women. Spatial variations in male participation rates tend to be rather small and may be explainable in terms of the age structure, concentrations of retirement migrants and social class differences in level of education. In analyzing spatial disparities, studies of female participation often describe patterns of association between participation rates and indicators of local opportunities for women's employment. The existence of stronger constraints on spatial mobility for women workers evidently implies that their employment status is more likely to depend on circumstances in specific spatial labour markets (see Gordon and Molho, 1985).

2.3 Migration

In the field of migration analysis there exists a large body of literature. For a long time, migration research was rather descriptive in nature, focusing on the following three questions: who moves, why do they move, and where do they move. In the recent past there has been a change in research focus to a less descriptive and a more analytical approach and to a greater concern with the linkage of migration to housing markets, labour markets, educational facilities and access (see Clark, 1982).

In a labour market context the concern is only with migration of the economically active population, so that wage differentials and job opportunities play an important role. Most of the theoretical and empirical work on mobility and migration of labour has been concerned with interregional migration. It is clear that research on interregional migration is rather diversified; it may pertain to different socio-economic groups, to different types of regions, etc. One important analytical distinction which will be made here concerns the level of aggregation. Broadly speaking, two approaches may be distinguished: macro-approaches and micro-approaches (see Figure 1). Macro-approaches

Three major lines of inquiry may be distinguished at the macro-level. First, there is a continuing concern with the analysis of migration flows based on gravity-type or entropy maximizing models, Markov chain models (MacKinnon, 1975; Rogerson, 1979) and simple statistical analysis of the structure of migration flows including inter alia the work by Slater, 1976 and Tobler, 1981. Such studies are usually descriptive rather than explanatory in nature.

A second stream of macro-research has utilized a demographic accounting framework. This type of research is concerned with population change. But of course migration is a central component in the framework. Prominent examples are provided by Rogerson (1980) and Wilson and Rees

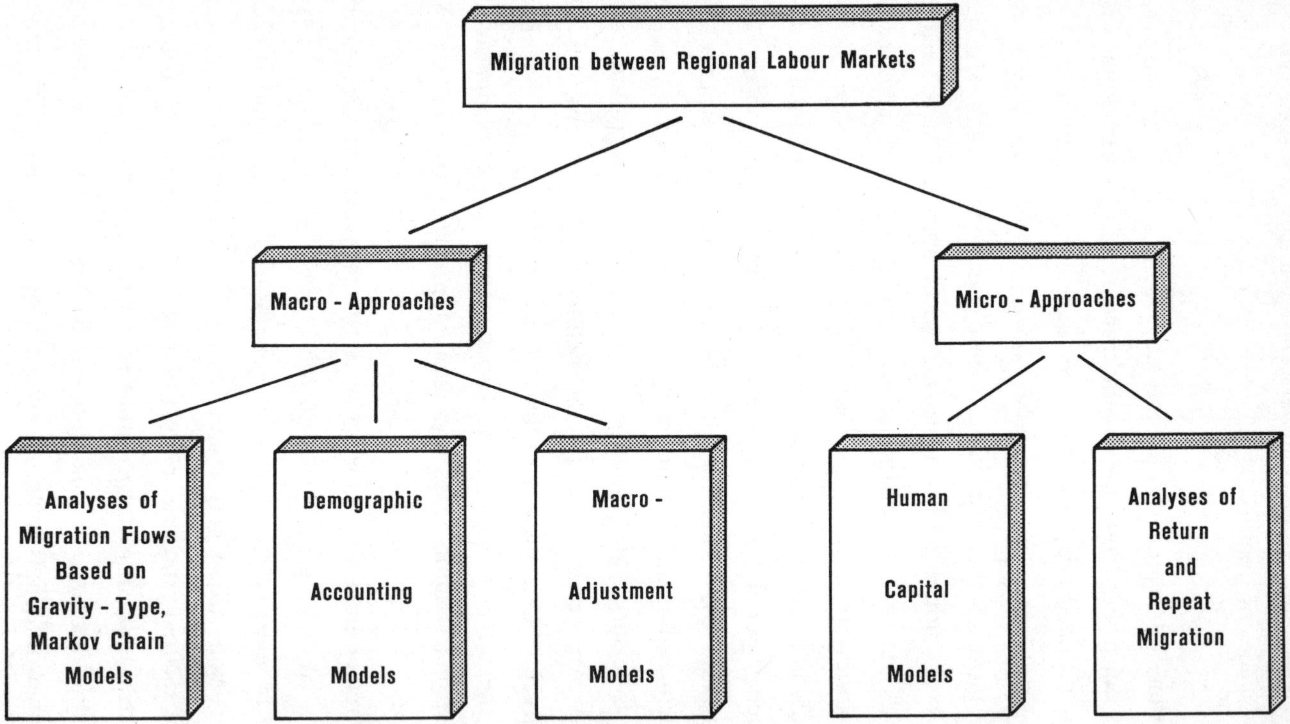

Figure 1. The Structure of Interregional Migration Research
(Source: Clark 1982, p. 11)

(1977). In the context of demographic models there is also increasing attention for life-cycle migration (see, e.g., Van Wissen and Rima, 1988).

A third line of inquiry at the macro-level deals with explanations, either in terms of economic and amenity explanations or in terms of macro-adjustment models. Macro-adjustment models view migration essentially as an equilibrating mechanism which optimally allocates demand and supply of labour over the regional labour markets in the long run, and thus equalizes the regional wage rates. This neoclassical view implies that workers will migrate out of low wage, high unemployment regional labour markets and move to high wage, low unemployment regional labour markets. The majority of empirical analyses dealing with labour migration adopts this perspective and characterizes migration as an economic push-pull process, where regional wages and employment characteristics are considered to be key determinants. But the underlying assumptions of perfect information, labour homogeneity and zero transaction costs are heroic and conflict with the reality of decision making (see Fischer and Nijkamp, 1987c). Moreover, empirical evidence does not always support this hypothesis. Evidence from American and European studies suggests that regional immigration rates often behave as predicted by theory, but for out-migration rates there tends to be either no significant spatial variation or the variation is in contrast with prior expectation. Very little research has been done up to now in analyzing the temporal dynamics of gross migration. The question whether out-migration lags in-migration or whether out- and in-migration are closely synchronized in time is still an unresolved question (see Clark, 1983c).

In the recent past, it has been increasingly realized that models of labour migration have to take into account behavioural factors and constraints which influence the migration decision. Distance, information about job offers, skill constraints and home ownership (especially where the house is located in a depressed regional labour market) tend to be the major constraints whereas economic conditions, demographic characteristics and quality of life aspects usually play an important role in creating migration flows.

Two competing approaches available to analyze the labour migrant departure process enable an explanation of why out-migration may be high in the case of fast growing spatial labour markets. These approaches refer to the processes of heterogeneity and non-stationarity which both result in the same macro result in terms of gross migration flows. The heterogeneity approach suggests that different internally homogeneous categories show variations in the propensity to move. The non-stationarity approach is often integrated into mobility models in the form of the concept of cumulative inertia which suggests that the propensity to move declines with increasing duration of residence at a specific location (Gleave and Sellens, 1984). An example may be found in Hyman (1974) who presents a migration model in which migration decisions are characterized by both types of processes and moreover by past memories of individual migrations and personal histories. These latter aspects are especially important in the case of migration related to internal labour market adjustment processes.

Micro-approaches

Within the class of micro - or individual - oriented approaches two broad topics may be distinguished, viz. human capital models, and repeat and return migration.

The human capital approach postulates a relationship in which human capital investment (especially in education and on-the-job-training) results in higher wage rates, because in a competitive labour market workers are paid according to their marginal productivity and they increase their productivity by undertaking human capital investments. The theory predicts that in the long run, wage differentials between occupations in a competitive labour market will depend on the training required and will be just sufficient to compensate for the costs of this investment.

Human capital models of migration consider migration as an investment to increase the productivity of human resources. The model of migration originally suggested by Sjaastad (1962) and modified later on in several ways assumes that an individual chooses migration as a strategy if the expected benefits exceed the costs of migration. Because the benefits of migration can only accrue over a period of time which renders migration an investment it can be considered as an investment in human capital. It is worthwhile to mention that benefits and costs of migration are not limited to economic variables (Clark, 1982). This human capital approach suggests that an individual responds to the present value of net gain which is given as the difference between the expected utility of destinations k and l minus the costs incurred of moving from k to l, summed over the expected length of the remaining lifetime adjusted by the discount rate where the discount rate reflects uncertainty about the future. In other words, an individual will choose destination l when the present value of the net gain from migrating from k to l is greater than any other destination (Clark, 1982).

This approach has received a great deal of attention in the past but has been criticized too because of the difficulty involved in operating the concept of life time earnings. The human capital model approach has also been extended to consider migrants who repeat their primary migration by moving again and others who repeat their migration and return to their initial location. DaVanzo and Morrison (1981) argue that the prospective migrant can be viewed from the concept of location-specific capital (such as, e.g., home ownership) in the case of repeat migration and from the concept of imperfect information in the case of return migration (for a more detailed discussion see Clark, 1982).

2.4 Commuting

Commuting may be the result of two different forces: a choice of a dwelling given a job, or a choice of a job given a residential location. So far much less attention has been paid to modelling commuting change as important spatial labour market adjustment process. The models usually rely on gravity concepts. Moreover, commuting and labour migration have usually been analyzed separately. One of the exceptions dealing with these two types of labour supply mobility in an integrated framework is given in Evers and van der Veen (1985). This study is also worth mentioning because it uses an aggregated nested multinomial logit approach to model several choices involving workplace and residential location (see also Fischer and Maier, 1986, in this context).

Another interesting example is contained in Gordon and Molho's (1985) model of the operation of female labour markets in the London metropolitan region which links migrational flows, changes in commuting, changes in unemployment and participation rates, new house construction and house price changes. The results obtained confirm that for all economically active females (both married and unmarried women) there tends to be significantly lower levels of inter-area mobility than for their male counterparts. In the study of commuting flows restricted female commuting fields were identified. In particular, the growth in part-time employment by married women tends to shorten the commuting radius of such workers.

Finally, it has to be added that there is also a link between the increase in labour force participation and the commuting radius, especially in case of multi-workers households. A new job of one of the members of the family in a new place does often not lead to a migration, but rather to an increase in commuting distance. This interwoven connection would certainly deserve further investigation.

2.5 Occupational mobility

Also the relationship between occupational mobility and migration needs further investigation. Up to now, there are only few studies focusing on this link. Gleave's rather detailed analysis of labour mobility in the mid 1970s is one of the exceptions. This study has shown that the labour force tends to be polarized. A higher than expected proportion of workers are either non-mobile or both geographically and occupationally mobile. Semi-skilled and unskilled workers display low rates of migration, but high rates of occupational mobility (see Gleave, 1983). Especially in case of short-cycle employment (e.g., in the tourist sector) the throughflow may be very high.

In analyzing occupational mobility over the period 1966-1975 in the US, Rosenberg found - according to Gleave and Sellens (1984) - that occupational mobility is associated with upward social mobility at a time of unemployment expansion and thus enables many workers to move from the secondary to the primary labour market. In times of recession, however, market conditions impose downward mobility. It is interesting to mention that younger workers who achieve upward mobility are more likely to sustain this attainment than older workers. This points to the fact that young workers are often underprivileged in the labour market primarily as a consequence of their age and not because they are disproportionately characterized by less satisfactory attributes.

2.6 Job search

Spatial mismatches (e.g., interregional unemployment rate differentials) call for a behavioural explanation at the disaggregate spatial level. In recent years there has been an explosion in the amount of research undertaken in the general area of job search. Economic search models attempt to characterize worker's behaviour by assuming that the worker collects job information in the best available manner; taking into account learning effects. This evidently depend on the objectives of the worker and the spatial environment in which he operates. The latter includes aspects such as the distribution of wage offers, the costs of search, knowledge of market conditions, the contradicting terms, etc. On the basis of objectives and constraints, job search behaviour can be modelled. Most formal search models limit the job search to a single characteristic, namely the wage rate which is usually considered to be constant over time and restrict the degree of uncertainty involved in the search process. The actual job offer is, for example, considered to contain complete information about layoff risks, career prospects, working conditions, etc. (see Burdett and Mortenson, 1980). Moreover, the models generally suppress risk aversion and assume that individuals making a job search attempt to maximize their expected income net of search costs (see e.g., Parsons, 1975; Lippman and McCall, 1976, for an overview). Up to now, present labour market search models have only partially succeeded in capturing distinctive aspects of the labour market, mainly due to the restrictive or unrealistic assumptions made. It is for instance often assumed that the labour market is atomistic in nature, characterized by the optimizing behaviour of firms and workers, and constrained only by imperfect information. In particular, the simplified treatment of information and uncertainty is a crucial failure of many existing job search models. First, individuals are not necessarily equally well informed about job openings, wages, etc. Potential jobs, for example, may be found through informal social networks. Second, information concerning the availability of jobs, their wages, skill requirements and location is spatially biased towards the spatial core of the labour market and tends to filter down to the periphery.

Only quite recently attempts have been made to develop spatial models of search. Most of the spatial models of search introduce space as a component of the search process and link job search theory and migration theory (for an overview see Miron, 1978 and Rogerson, 1982). In these models some fundamental assumptions on the sequence of job search, the job acceptance and the migration decision have to be made. For example, David (1974) assumes that migration is a prerequisite for job search in a spatial labour market different from that in which the home of residence is located. In order to analyze migration behaviour more realistically, it would be necessary to consider both possibilities, the possibility of migration prior to search and also the possibility of a successful long distance search without prior migration.

Rogerson and MacKinnon (1981) examine the effects of search and migration, but at a very high level of aggregation. All jobs are assumed to be identical and all workers to have identical skills. The migration decision is based on perceived job vacancy distributions. Amrhein and MacKinnon (1985) relax these homogeneity assumptions. The refined multiregional job search and migration model assigns workers with heterogeneous skills in a Pareto optimal fashion to various jobs in each of the spatial labour markets. No effort, however, is made to take job tenure and career related aspects into account. The model does not approach the individual level of decision making, but does approach an intermediate scale of aggregation of different categories of jobs and workers.

This model and most other spatial models of job search and migration assume market competition of imperfect information in order to explain why spatial mismatches occur (see, for example, Curry, 1985). In outlining an alternative job search theory where notions of contract theory play a central role, Clark (1987) argues that in the present job search models the notion of information is not taken seriously enough. Information is not only imperfect about job openings and wage offers, but also indeterminate.

Moreover, in most search models, the competitive aspects of job search are neglected. Mattson and Weibull's (1981) attempt to model the intraurban job search process as a function of market competition and job accessibility belongs to the very few exceptions. This work is also worth mentioning because it both combines a search-theoretic perspective as such with the multinomial logit approach and presents an interesting empirical test using a sample of 600 job openings in the Stockholm region (see Rogerson, 1987 for more details). The empirical results obtained underline that competition may show a strong and uneven effect on job searcher's prospects.

Up to now, there has been only little research into actual job search behaviour. Much more work needs to be done to get deeper insights into the ways the volume, pattern and efficiency of actual job search activity are shaped and affected by different spatial labour market conditions (see also Martin, 1986).

3. Labour Demand Issues

The second major type of regional labour market research, viz. the analysis of the determinants of the demand side, is much less developed. Clearly, demand differences may be an important source of intra-regional variation in labour market characteristics. The spatial dimensions of recruitment behaviour of firms, however, are hard to trace from empirical studies, although it is well-known that the search space of firms is limited. In any case, more information on regional labour market effects of economic or technological restructuring is available. Hence the major focus here is on the analysis of actual or likely future employment effects of regional economic and technological changes, changes in market positions of firms, etc.

3.1 General aspects

The dynamics of firms has been studied quite thoroughly in industrial economics. Especially in recent years there has been considerable interest in structural changes, caused amongst others by innovations and reflected amongst others in the rapid rise of the business service sector and of small and medium size enterprises (see also Giaoutzi et al., 1988).

The relationship between technological change and the labour market is a complex one which cannot be adequately analyzed at the macro level. Research in this area is therefore increasingly directed towards both the meso-scale of spatial labour markets and sectors, and the micro-scale of firms. Cities and regions are affected in different ways by new technologies, depending upon their locational profile, their manpower potential and other aspects of the labour market concerned.

In general, a distinction has to be made between product and process innovations. The first class tends to be oriented towards large agglomerations and hence various R&D and skilled labour in central areas. The latter class is often found in peripheral and/or intermediate areas and is usually more related to routine-types of labour force (see Davelaar, 1989).

Technological change in a spatial context has three different major aspects, the generation of new technologies, their diffusion patterns and their socio-economic impacts (see Stoneman, 1984). These aspects and their relevant spatial labour market consequences will be briefly discussed (for more details see Gillespie, 1983 and Nijkamp, 1986a).

3.2 Generation of new technologies

The generation of new technology refers basically to the invention of a new product, production system or management at a specific location. Central in the discussion in this context is the breeding place or incubator hypothesis which states that large (especially metropolitan) agglomerations - through their diversified labour markets, their accessibility as nodes on a spatial network and their flexible industrial structure - may act as the seedbeds for technological innovations (see e.g., Andersson and Johansson, 1984b; Davelaar, 1989). In this framework, innovation may be seen - in accordance with Piatier (1984) - as covering a succession of operations, i.e., the transition from the idea to its materialization, followed by a result (i.e., the product itself or one of its inputs or the method of its production, etc.).

According to this view, a metropolitan region has potentially a comparative advantage as a place of introducing new economic activities. Naturally, this advantage can be lost over time. Malecki (1983), for example, presents empirical evidence that many large cities have lost their innovative potential while medium sized cities which are not too far away from large cities, exhibit a boom process due to the design and use of new technologies (see also Storper, 1986).

Technology diffusion is the subsequent application of an innovation after its initial commercialization. It includes adoption by other users as well as more extensive use by the original innovation. Moreover, it encompasses all the actions at the level of the firm or organization taken to exploit the economic benefits of the innovation. Most of the economic benefits of new technologies come from the diffusion rather than the development of the technology.

3.3 Diffusion of new technologies

The diffusion and adoption of new technologies is an important issue in economic geography and regional economics. Diffusion processes of new phenomena have both a temporal and a geographical dimension. In a temporal context, the cumulative pattern of adoption of a new technology usually conforms quite closely to a cumulative normal distribution function of an S-shaped (logistic) curve, while the spatial distribution at a certain point in time may be considered as a result of an expansion or hierarchical diffusion process or a combination of both types. Various factors such as e.g. the technical applicability and profitability of the invention, the availability of physical and human capital, management attitudes as well as size, structure and organization of the potential adopter, market pressures for diffusion and the regional selection environment (the presence or absence of a stimulating and innovative entrepreneurial climate) influence the speed of diffusion in time and space. Diffusion and adoption of innovation is of crucial importance for regional growth, because regions that lag behind in terms of adoption tend to face increased problems of industrial decline.

Product and process diffusion have often been studied at the establishment level in terms of the characteristics of the innovation and of the adopter of the innovation. Only few studies have taken account of both the temporal and the spatial dimension of the diffusion process. A notable exception is the study of regional innovation levels in Great Britain undertaken by the Centre for Urban and Regional Development Studies, University of Newcastle upon Tyne. This study involved a national questionnaire survey of 807 establishments and looked in more detail at rates of adoption (defined as the introduction of a new or improved product to factories in the period 1973-1977) in three subsectors of the electronics and mechanical engineering industry (scientific and industrial instruments, metalworking machine tools, radio and electronic components) in the UK (see Thwaites et al., 1981, Oakey et al., 1982).

This micro-oriented study indicates spatial variations in the establishment acceptance of product innovations, to the advantage of the South East of the UK and to the detriment of the peripheral areas. Moreover, the study shows a strong association between location, size and ownership of establishments and innovation and reveals the importance of R&D efforts on the factory site for innovations. The group of firms which have severely low rates of product innovations are single plant independent firms in development regions which more likely reflect the less favourable economic environmental influences than all other industrial plants in the same area. Thus, the results of this study imply that regional industrial environments influence the level of product innovation in manufacturing firms.

Process innovations (e.g., computerized controlled machine tools, the incorporation of microprocessors in products and production processes) do not reveal the same degree of spatial variations. Multiside organizations do not disproportionately influence the pattern of regional parity in process innovation levels (see Thwaites et al., 1982, Oakey et al., 1982). The studies undertaken by the Centre for Urban and Regional Development Studies illustrate that interesting insights into the relationship between technological change, industrial dynamics and regional development can be achieved by this type of research.

3.4 Effects of new technologies

The effects of technological change are multidimensional in nature. They range from locational and environmental impacts to labour market effects. There is a considerable debate about the effects of technological change on employment. In particular, the debate was focused on the consequence of the new information technologies, i.e. all new technologies related to the storage, processing, communication and dissemination of information. Also the issue whether these new technologies strengthen geographical concentration of economic activities or on the contrary, encourage decentralization has attracted many studies (see for a survey, Giaoutzi and Nijkamp, 1987 and, for a characterization of the micro-electronic revolution and its impact on labour and employment, Fischer, 1989b).

A key feature of these micro-electronics based technologies is their pervasiveness in generating a whole range of new applications over a wide range of sectors. This pervasiveness is induced by other features such as a dramatic decline in costs of transmitting, storing and processing information; a new capability to integrate and control industrial and service activities; a reduction in the time-lag between inventions and (commercial) innovations compared to earlier technologies; and a large potential for productivity increase in relation to all factor inputs. All these characteristics shape the impact on employment and its structure. The introduction of these new technologies creates imbalances of various kinds, and sets in motion an adaptation process which affects not only the number of jobs, but also their distribution by occupation, industry and region.

The factual impact of new technologies on overall employment in different regional sectors is hard to assess with great precision. There are serious methodological flaws involved in identifying and measuring employment implications. First, there is an analytical problem of isolating employment effects induced by the introduction of a new technology from effects resulting from all other changes taking place at the same time. A second major difficulty refers to the choice of an appropriate methodological approach for measuring the effects. The case study approach neglects any compensating effect and thus overstates the negative effects of technological change while the introduction of new products and services is by and large ignored in partial or general equilibrium approaches which moreover lack the detailed information obtained by micro-approaches. Finally, there are significant technical problems of quantifying employment effects of new technologies because of inadequacies in industrial and occupational statistics. In spite of these problems there is a wide range of studies in several countries which have attempted to assess and to forecast the impacts on employment and labour (see Fischer, 1988b for a more detailed survey).

All these studies show more or less severe methodological deficiencies. Most ignore the productivity effects on inputs other than labour. Only few take explicitly into account the international dimensions of the technological change-employment relationship. Relatively little attention is given to the factors influencing the speed and patterns of application of new technologies. The overall impact on employment in any particular country, however, depends crucially on the domestic rate of diffusion. The slower the new technologies diffuse relative to other countries, the more likely it is that other countries will gain a permanent advantage via the indirect employment effects.

Most studies devote a great deal of effort to identifying direct effects, especially the negative implications induced by process innovations. There have been only few attempts to assess the importance of compensating effects. Especially price and income effects are rarely taken into account. These shortcomings certainly do not arise from a lack of understanding the factors and the dynamic nature of the technological change-employment relationship, but are associated with the methodological and statistical problems mentioned above (see Brainard and Fullgrabe, 1986). The few studies addressing the issue of current employment effects at the macro-level tend to agree in that employment and unemployment levels in recent years were not significantly influenced by technological change (see, e.g., the OECD (1982) report on the effects of micro-electronics, robotics and jobs based on studies from thirteen countries). Shifts in demand patterns, international competition and economic growth patterns appeared to be much more important (Fischer, 1988b).

In the long run the impact of the new technologies on overall employment crucially depends on the question whether these technologies raise the productive potential of national (and regional) economies and provide a basis for sustained growth in output, employment and income. But the extent to which potential growth becomes translated into actual growth of output, employment and income is not solely determined by technological change, but is depending also upon factors such as the overall rate of economic growth, the degree of adaptability at the enterprise level, the flexibility in product and factor markets, industrial relations systems, education and training, trade and investment across countries.

Quite recently, increasing attention is being paid no only to quantitative, but also to qualitative effects of new technologies on employment, such as changes in occupational structure, changes in work tasks, job skills and the work environment. The impact on the overall occupational distribution of jobs is also difficult to assess due to several reasons. Changes in job content that do occur do not get quickly reflected in job titles. On the other hand job titles may change while the content of work may remain the same. Moreover, there are other changes in the sectoral mix which may be quantitatively more important.

Particular emphasis is often placed on the upskilling-deskilling issue (see also the discussion on the Braverman deskilling hypothesis). Some studies have noted a deskilling effect of micro-electronics based technologies or both a deskilling effect in some and an upskilling effect in other industries. Especially occupations which are largely involved in simpler information processing tasks (e.g. stenography, typing, filing) are predominantly vulnerable to displacement. On the other hand, there is no doubt that new information technologies will generally lead to an increasing demand for higher-level skills of programmers, engineers and maintenance personnel. Up to now, however, not very much is known about the restructuring of occupational hierarchies and skill requirements. The same is true with respect to effects on the informal sector and on the household economies. Especially regionally discriminating effects of such developments have unsatisfactorily been studied.

4. Regional Unemployment Disparities: Causes and Dynamics

The third major area in current regional labour market research refers to the causes and dynamics of regional unemployment disparities. This type of labour market research has received increasing attention in geography and regional economics in the recent past for at least two reasons. First, most advanced economies have been experiencing very high rates of unemployment because of a combination of global recession in demand for goods and services, a lack of international competitiveness, problems of adjustment of structural change, and shifts in labour supply. Second, these countries have displayed sharp spatial variations in unemployment. On the one hand, those regional economies specializing in depressed sectors such as shipbuilding, coal, steel, textiles, and with older manufacturing cities have faced severe adjustment difficulties in the light of deindustrialization, with consequent high levels of unemployment. On the other hand, some of the more diversified cities and their metropolitan regions have enjoyed a substantial expansion of new jobs mainly in service industries, but also in manufacturing. Regional fluctuations, however, reflect not only spatial variations in rates of job creation and job redundancy, but also changes in the occupational structure of labour demand and supply, as well as in commuting and migration (see Gleave, 1987).

Shifts on regional labour markets are thus often non-uniform, whilst they are also influenced by the reward system of labour. The reward system is not the result of anonymous market forces, but of institutional actions. For instance, the efficiency wage hypothesis provides an explanation for the existence of an equilibrium wage above the market clearing wage level (see Yellen, 1984), a situation which may even differ for different regions.

Furthermore, it is noteworthy that not only the level of unemployment may differ for various regions in an economy, but also the duration of unemployment (see Gorter et al., 1989). Moreover, the timing of unemployment response in regional labour markets is an important component to analyze links between public policies, changing regional economic structure, and resulting variations in unemployment characteristics. Typically the studies analyze the relationship between national and regional unemployment in general and the lead/lag structure of unemployment response in particular by means of exploratory statistical analysis. If a region consistently reacts earlier than others it is usually termed a leading region, whereas conversely, if a region regularly reacts later than other regions or the national average, then it is described as a lagging region. The study of the spatial distribution of such leads and lags is will developed, e.g. in various equilibrium models of interregional unemployment rate differentials.

The conventional approach to analyze the relationship between national and regional unemployment is a basic descriptive model put forward by Brechling (1967) and subsequently used in a wide variety of settings and at different spatial scales (see, for example, King et al., 1972; Jeffrey, 1974; King and Clark, 1978; Frost and Spence, 1981). This approach measures the behaviour of individual regions relative to the national series and is based upon a regressional framework involving a set of time regressions in which the regional unemployment rate is related to structural, cyclical and seasonal components of unemployment.

This methodological approach has been used extensively in the past (see, for example, King and Clark, 1978; Frost and Spence, 1981) and with some confidence in its ability to identify relationships between national and regional unemployment. But its adequacy and efficacy have been questioned more recently. In particular, criticism has focused on two major shortcomings. First, the implicit assumption of a symmetry in the lead-lag relationship has been criticized (see Johnston, 1983). This assumption implies that if a region leads the nation into a depression then it will also lead it out and if a region lags into the depression then it will also lag in the recovery stage. This assumption built in the classical mode of analysis is unwarranted both from a theoretical and an empirical point of view. The second objection refers to the fact that in nearly all applications the assumption of non-autoregression is violated. The presence of serial autocorrelation among the error terms has several serious implications. Even if the least squares estimates are unbiased in this case, they are likely to be inefficient. Moreover, the estimated variances seriously underestimate the true variances. The estimated variances, however, play an important role in constructing confidence intervals, testing hypotheses and computing t-ratios. Thus, even though the estimated parameters appear to be quite reliable (small variances) they are in fact extremely unreliable in the case of serially correlated errors. All these reasons suggest that generalization concerning the patterns of national and regional unemployment relationships based upon the classical mode of analysis may be spurious (for more details see Fischer and Petz, 1988).

In response to these problems and deficiencies of the classical mode of analysis an alternative mode of analysis has been put forward by Clark (1979) and more recently by Fischer and Petz (1988): the transfer function-noise model approach. This alternative mode of analysis (see Figure 2) shows some very attractive features. It can be used to describe a wide range of types of non-stationarity, seasonality and dynamic lag structure and to analyze the dynamic (or lagged) relationship by means of more general lag and error structures. The approach is not only more flexible in terms of identifying the relationships between national and regional unemployment in comparison with the classical mode. It also provides a great deal of flexibility in terms of characterizing the process which might be responsible for the auto-correlation among the error terms (for more details see Fischer and Petz, 1988).

Several studies have revealed the existence of response-time differences in several countries. Most of them are exploratory in nature, but very few attempts have been made to relate them to the economic mecha-

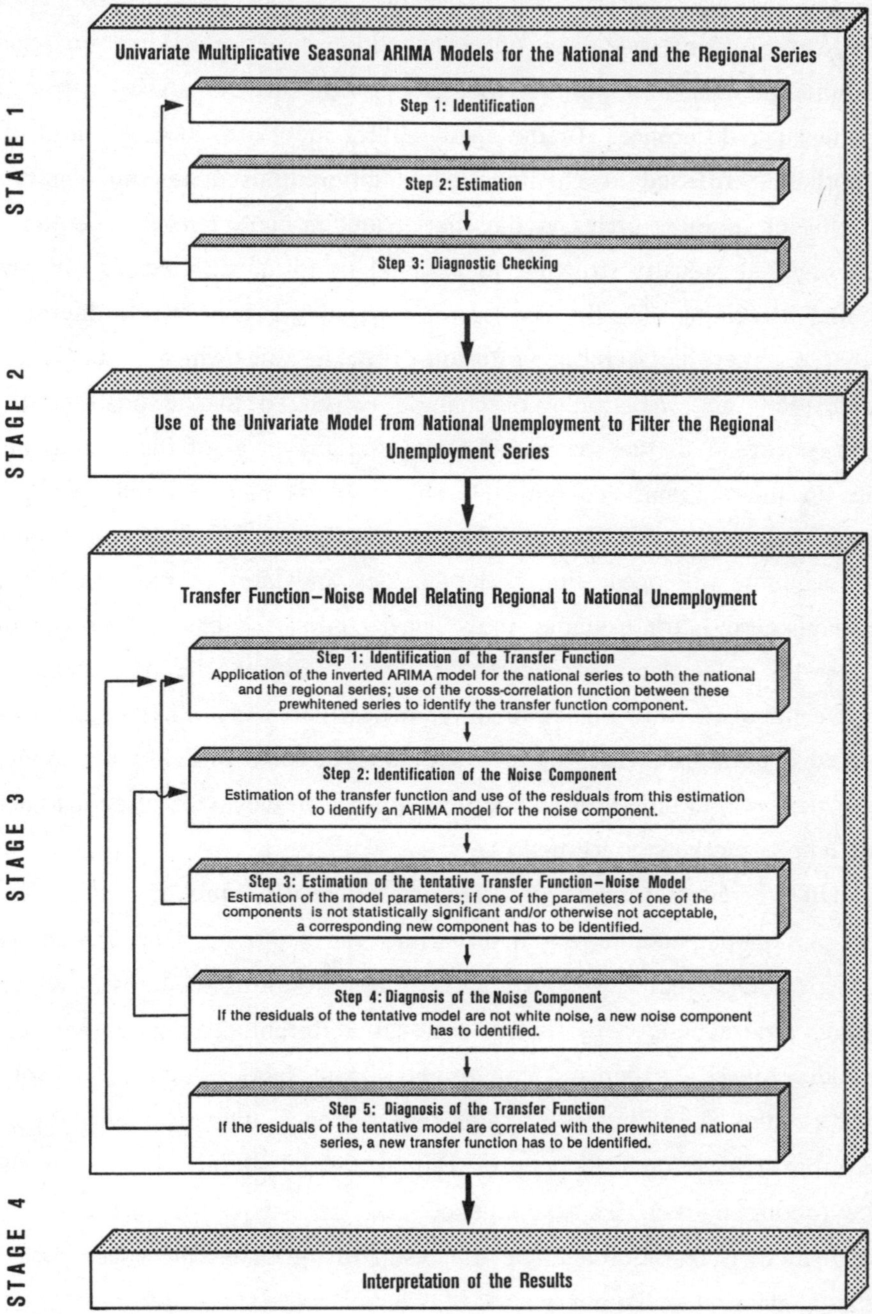

Figure 2. The Transfer Function-Noise Model Approach to Analysing the Relationship between National and Regional Unemployment (Fischer and Petz 1988)

nisms which cause them. An explanation of the more or less persistent differences among regional unemployment rates is one of the greatest challenges to regional scientists, since conventional theory of long run regional labour market behaviour predicts that labour flows would adjust to and, thus, eliminate such differences. In the (geographic) literature, at least three types of hypotheses are used to explain regional differentials in unemployment and the timing of business cycles at the spatial labour market level. The first one relates regional industry structure to regional performance (see e.g., Thirlwall, 1966 and Brechling, 1967). The second hypothesis concerns a hierarchical effect (i.e. the hierarchical diffusion process) in which the level of unemployment and the timing of changes is related to the position of the regions concerned in the spatial hierarchy and the level of integration of the regions in the national economy. Hereby it is assumed that changes in the demand for labour are transmitted through the system with lags where upturns in the economy will occur first in large cities and later in the smaller cities. King et al. (1969), for example, draw heavily upon the basis of central place and growth pole theory to establish and interpret the transmission of economic impulses from higher to lower order places in both time and space. The third hypothesis focuses upon a structural effect in which some industries lead or lag the national trend, and upon the employment structure of the spatial labour markets concerned.

In the regional economic literature different schools of thought on the subject of regional unemployment disparities can be found. One school argues that aggregate demand pressures are unevenly diffused across the space economy. Consequently, persistent regional differentials are considered to reflect differences in demand for products and labour. Another school of thought on this issue basically argues that regional unemployment disparities derive from a basic dualism between capital and labour and the segmentation of the labour market. The emergence and persistence of regions of high unemployment is considered to be the result of an historical process in which the spatial division of labour plays an essential role. These different points of view indicate the range of factors around which conceptual and empirical research has been developing. Up to now, however, there is no such thing as a fully formulated theory of spatial labour market adjustment.

Another important area in which significant contributions have been made to the debate on regional unemployment refers to the wage-unemployment relationships, especially in the context of the validity of the so-called Phillips-curve. Geographical work in this area is largely concentrated on the empirical analysis of the relationship in specific regional labour markets. Much empirical evidence has been provided to suggest that the regional variation in wages within advanced economies has much more to do with the industry mix of regions, the pressure of demand in the labour market, the spread of multiplant companies paying the same wages across the country, the degree of unionization and the national wage bargaining process than with regional variations of unemployment (see Martin, 1981c).

Very recently, increasing emphasis has been laid on the wage-change interdependence among regional labour markets. The most widely used approach in such studies has been to postulate the regional wage leadership hypothesis. This hypothesis states that wage increases achieved by workers in one or more leading regions effectively determine the rate of wage change in the rest of the economy. Leading regions have usually been defined as those which experience the greatest labour demand pressures (lowest unemployment) and/or the highest wage (or wage changes) (see Martin, 1981c). Martin (1981c) reviewing several conceptual and empirical formulations of this hypothesis points to the necessity to define the concept of a leading region on a much wider range of labour market characteristics including institutional and structural aspects.

Finally, it is worth mentioning that the spatial wage interdependence can influence the degree and direction of the impact of economic policy and labour market policy measures.

5. Outlook

This survey has been designed to cover briefly some of the current major interests and issues of research, from a non-technical point of view in contrast to other recent essays in the field (see Isserman et al., 1986, and Schubert et al., 1987). This field appears to be a rapidly expanding area and to offer a rich potential for new research.

Nevertheless, it is at the same time necessary to emphasize that the nature and operation of regional labour markets can not be satisfactorily analyzed in isolation from the broader working and performance of the economy and its macro-social context. The political, social and economic problems associated with the dysfunction of the labour market owe several of their causes to spatial aspects as well as to factors outside the national control. In this respect, spatial research and cross-national comparisons like those reported in Fischer and Nijkamp (1987a) are extremely important. As institutional aspects are more or less invariant within countries, the impact of institutional factors on regional labour markets can only be effectively analyzed by cross-national comparative studies of the functioning of such markets.

In this context, much more emphasis would have to be placed on two weaker points in current regional labour market analysis:

- the systematic collection and monitoring of relevant disaggregate data by means of panel and longitudinal methods and
- a more careful analysis of the behavioural aspects of the demand side; this is necessary for a more balanced view on the causes of spatial mismatches on the labour market.

References

Amrhein, C.G. (1985), Interregional Labour Migration and Information Flows, Environment and Planning A 17:1111-1126.

Amrhein, C.G. and R.D. MacKinnon (1985), An Elementary Simulation Model of the Job Matching Process within an Interregional Setting, Regional Studies 19:193-202.

Andersson, A.E. and B. Johannson (1984a), Industrial Dynamics, Product Cycles and Employment Structure, Working Paper WP-84-9, Laxenburg, IIASA.

Andersson, A.E. and B. Johannson (1984b), Knowledge Intensity and Product Cycles in Metropolitan Regions, Working Paper WP-84-13, Laxenburg, IIASA.

Ballard, K.P. and G.L. Clark (1981), The Short-Run Dynamics of Interstate Migration: A Space-Time Economic Adjustment Model of In-migration to Fast Growing States, Regional Studies 15:213-228.

Bartels, C.P.A. (1977), The Structure of Regional Unemployment in The Netherlands: An Exploratory Statistical Analysis, Regional Science and Urban Economics 7:88-102.

Baumann, J.H., M.M. Fischer and U. Schubert (1983), A Multiregional Labour Supply Model for Austria: The Effects of Different Regionalisations in Multiregional Labour Market Modelling, Papers of the Regional Science Association 52:53-83.

Baumann, J., M.M. Fischer and U. Schubert (1988), A Choice-Theoretical Labour-Market Model: Empirical Tests at the Mesolevel, Environment and Planning A 20:1085-1102.

Brainard, R. and Fullgrabe, K. (1986), Technology and Jobs, STI Review 1:9-46.

Brechling, F.P.R. (1967), Trends and Cycles in British Regional Unemployment, Oxford Economic Papers (New Series) 19:1-21.

Burdett, K. and D. Mortensen (1980), Search Layoffs and Labour Market Equilibrium, Journal of Political Economy 88:652-672.

Clark, G.L. (1979), Predicting the Regional Impact of a Full Employment Policy in Canada: A Box-Jenkins Approach, Economic Geography 55:213-226.

Clark, G.L. (1983a), Fluctuations and Rigidities in Local Labor Markets, Part 1: Theory and Evidence, Environment and Planning A 15:165-185.

Clark, G.L. (1983b), Spatial Labor Markets and the Distribution of Transaction Costs, Environment and Planning D 1:305-322.

Clark, G.L. (1983c), Interregional Migration, National Policy and Social Justice, Totowa, New Jersey, Rowman and Allenheld.

Clark, G.L. (1987), Job Search Theory and Indeterminate Information, in: M.M. Fischer and P. Nijkamp (eds.), Regional Labour Markets. Analytical Contributions and Cross-National Comparisons, pp. 169-188, Amsterdam, North-Holland.

Clark, W.A.V. (1982), Recent Research on Migration and Mobility: A Review and Interpretation, Progress in Planning 18:1-56.

Curry, L. (1985), Inefficiencies in the Geographical Operation of Labor Markets, Regional Studies 19:203-216.

DaVanzo, J. and P.A. Morrison (1981), Return and Other Sequences of Migration in the United States, Demography 18:85-101.

Davelaar, E.J. (1989), Incubation and Innovation, Ph.D. Thesis, Free University, Amsterdam.

David, P.A. (1974), Fortune, Risk and the Microeconomics of Migration, in: P.A. David and M.W. Reder (eds.), Nations and Households in Economic Growth, pp. 21-88, New York, Academic Press.

Dijk, J. van and H. Folmer (1985), Entry of the Unemployed into Employment. Theory, Methodology and Dutch Experience, Regional Studies 19:243-256.

Evers, G.H. and A. van der Veen (1985), A Simultaneous Non-Linear Model for Labour Migration and Commuting, Regional Studies 19:217-230.

Fischer, M.M. (1986), Why Spatial Labour Market Research? Environment and Planning A 18:1417-1420.

Fischer, M.M. (1988a), Innovation, Diffusion and Regions, in: A.E. Andersson, D. Batten and C. Karlsson (eds.), Knowledge and Industrial Organization, Berlin et al., Springer. (in press).

Fischer, M.M. (1988b), The Micro-Electronic Revolution and its Impact on Labour and Employment, in: R. Cappelin and P. Nijkamp (eds.), Technological Change and Spatial Development, Gower, Aldershot. (in press).

Fischer, M.M. and G. Maier (1986), Regional Labour Supply Mobility and the Logit Model: Testing Some Alternative Probability Choice Structures by Means of Monte Carlo Simulation, in: M.M. Fischer and F.-J. Kemper (eds.),

Modelle für diskrete Daten und raumbezogene Probleme, pp. 38-75, Vienna, Österreichisches Institut für Raumplanung.

Fischer, M.M. and P. Nijkamp (eds.) (1987a), Regional Labour Markets. Analytical Contributions and Cross-National Comparisons, Amsterdam, North-Holland.

Fischer, M.M. and P. Nijkamp (1987b), Structure of Regional Labour Markets: Problems and Perspectives, in: M.M. Fischer and M. Sauberer (eds.), Gesellschaft - Wirtschaft - Raum, pp. 150-161, Vienna, AMR.

Fischer, M.M. and P. Nijkamp (1987c), Labour Market Theories: Perspectives, Problems and Policy Implications, in: M.M. Fischer and P. Nijkamp (eds.), Regional Labour Markets. Analytical Contributions and Cross-National Comparisons, pp. 37-52, Amsterdam, North-Holland.

Fischer, M.M. and P. Nijkamp (1987d), Current Trends as Regional Labour Markets, in: I. Gordon (ed.), Unemployment, Regions and Labour Markets: Reactions to Recession, pp. 15-43, London, Pion.

Fischer, M.M. and P. Nijkamp (1987e), Spatial Labour Market Analysis: Relevance and Scope, in: M.M. Fischer and P. Nijkamp (eds.), Regional Labour Markets. Analytical Contributions and Cross-National Comparisons, pp. 1-33, Amsterdam, North-Holland.

Fischer, M.M. and G. Petz (1988), The Timing of Unemployment Response in Austrian Regional Labour Markets. The Classical and an Alternative Mode of Exploratory Statistical Analysis. Paper presented at the 28th European Congress of the Regional Science Association, Stockholm, August 23-26, 1988 (Sistemi Urbani 10, in press).

Frost, M. and N. Spence (1981), The Timing of Unemployment Response in British Regional Labour Markets, 1963-1976, in: R.L. Martin (ed.), Regional Wage Inflation and Unemployment, pp. 208-231, London, Pion.

Giaoutzi, M. and P. Nijkamp (eds.) (1988), Informatics and Regional Development, Aldershot, Gower.

Giaoutzi, M., P. Nijkamp and D. Storey (eds.) (1988), Small and Medium Size Enterprises and Regional Development, London, Routledge.

Gillespie, A. (ed.) (1983), Technological Change and Regional Dynamics, London, Pion.

Gleave, D. (1983), The Relationship between Labour Mobility and Technical Change, London, The Technical Change Centre.

Gleave, D. (1987), Dynamics in Spatial Variations in Unemployment, in: M.M.

Fischer and P. Nijkamp (eds.), Regional Labour Markets. Analytical Contributions and Cross-National Comparisons, pp. 269-288, Amsterdam et al., North-Holland.

Gleave, D. and R. Sellens (1984), An Investigation into British Labour Market Processes, Environment and Planning Committee Paper 3, Economic and Social Research Council.

Gordon, I.R. (1985), The Cyclical Sensitivity of Regional Employment and Unemployment Differentials, Regional Studies 1:95-110.

Gordon, I. and I. Molho (1985), Women in the Labour Markets of the London Region: A Model of Dependence and Constraint, Urban Studies 22:367-386.

Gorter, C., P. Nijkamp and P. Rietveld (1989), Duration of Unemployment, Applied Economics, forthcoming.

Hamilton, F.E.I. (1987), Industrial Organisation and Regional Labour Markets, in: M.M. Fischer and P. Nijkamp (eds.), Regional Labour Markets. Analytical Contributions and Cross-National Comparisons, pp. 289-312, Amsterdam et al., North-Holland.

Hart, R.A. (1981), Regional Wage-Change Transmission in the Regions of the United Kingdom, in: R.L. Martin (ed.), Regional Wage Inflation and Unemployment, pp. 17-45, London, Pion.

Heckman, J.J., M.R. Killingsworth and T. MaCurdy (1981), Empirical Evidence on Static Labour Supply Models: A Survey of Recent Developments, in: Z. Hornstein, J. Grice and A. Webb (eds.), The Economics of the Labour Market, pp. 75-122, London, Her Majesty's Stationery Office.

Herzog, H. and A. Schlottmann (1984), Labor Force Mobility in the United States: Migration, Unemployment, and Remigration, International Regional Science Review 9:43-58.

Hyman, G. (1974), Cumulative Inertia and the Problem of Heterogeneity in the Analysis of Geographical Mobility, CES Research Paper, RP 11.

Issaev, B., P. Nijkamp, P. Rietveld and F. Snickars (eds.) (1982), Multi-regional Economic Modelling: Practice and Prospect, Amsterdam, North-Holland.

Isserman, A. (1985), Economic-Demographic Modeling with Endogenously Determined Birth and Migration Rates: Theory and Prospects, Environment and Planning A 17:25-45.

Isserman, A., D. Plane, P. Rogerson and P. Beaumont (1985), Forecasting

Interstate Migration with Limited Data: A Demographic-Economic Approach, Journal of the American Statistical Association 80:277-285.

Isserman, A., C. Taylor, S. Gerking and U. Schubert (1986), Regional Labour Market Analysis, in: P. Nijkamp (ed.), Handbook of Regional and Urban Economics. Volume 1: Regional Economics, pp. 543-580, Amsterdam, North-Holland.

Jeffrey, D. (1974), Regional Fluctuations in Unemployment within the U.S. Urban Economic System: A Study of the Spatial Impact of Short Term Economic Change, Economic Geography 50:111-123.

Johnston, R.J. (1983), On the Symmetry of Leads and Lags in Studies of Regional Unemployment, Regional Studies 17:105-112.

King, L.J. and G.L. Clark (1978), Regional Unemployment Patterns and the Spatial Dimensions of Macro-Economic Policy, Regional Studies 12:283-296.

King, L.J., E. Cassetti and D. Jeffrey (1969), Economic Impulses in a Regional System of Cities: A Study of Spatial Interaction, Regional Studies 3:213-218.

King, L.J., E. Cassetti and D. Jeffrey (1972), Cyclical Fluctuations in Unemployment Levels in U.S. Metropolitan Area, Tijdschrift voor Economische en Sociale Geografie 63:345-352.

Lever, W.F. (1980), The Operations of Local Labour Markets in Great Britain, Papers of the Regional Science Association 44:37-55.

Lever, W.F. (1987), New Trends in the Supply and Demand Patterns of Labour in Western Economies, in: M.M. Fischer and P. Nijkamp (eds.), Regional Labour Markets. Analytical Contributions and Cross-National Comparisons, pp. 249-267, Amsterdam et al., North-Holland.

Lippmann, S.A. and J.J. McCall (1976), The Economics of Job Search: A Survey, Economic Inquiry 14:155-189 and 347-368.

MacKinnon, R.D. (1975), Controlling Interregional Migration Processes of a Markovian Type, Environment and Planning A 7:781-792.

Maier, G. (1985), Cumulative Causation and Selectivity in Labour Market Oriented Migration Caused by Imperfect Information, Regional Studies 19:231-242.

Maier, G. and M.M. Fischer (1985), Random Utility Modelling and the Labour Supply Mobility Analysis, Papers of the Regional Science Association 58:21-33.

Maier, G. and U. Schubert (1984), Struktur und Probleme ökonometrischer interregionaler Arbeitsmarktmodelle - mit einigen ausgewählten Ergebnissen für Österreich, in: G. Bahrenberg and M.M. Fischer (eds.), Theorie und Quantitative Methodik in der Geographie, pp. 48-91, Bremen, University of Bremen.

Malecki, E.J. (1983), Technology and Regional Development: A Survey, International Regional Review 8:89-123.

Martin, R.L. (ed.) (1981a), Regional Wage Inflation and Unemployment, London, Pion.

Martin, R.L. (1981b), Introduction, in: R.L. Martin (ed.), Regional Wage Inflation and Unemployment, pp. 1-16, London, Pion.

Martin, R.L. (1981c), Wage-Change Interdependence Amongst Regional Labour Markets: Conceptual Issues and Some Empirical Evidence for the United States, in: R.L. Martin (ed.), Regional Wage Inflation and Unemployment, pp. 96-135, London, Pion.

Martin, R.L. (1986), Getting the Labour Market into Geographical Perspective, Environment and Planning A 18:569-572.

Mattson, L. and J. Weibull (1981), Competition and Accessibility on a Regional Labour Market, Regional Science and Urban Economics 11:471-497.

Miron, J. (1978), Job Search Perspectives on Migration Behaviour, Environment and Planning A 10:519-536.

Nijkamp, P. (ed.) (1986a), Technological Change, Employment and Spatial Dynamics, Berlin and Heidelberg, Springer.

Nijkamp, P. (1986b), The Triangle of Industrial Dynamics, Labour Markets and Spatial Systems, in: P. Nijkamp (ed.), Technological Change, Employment and Spatial Dynamics, pp. 1-17, Berlin and Heidelberg, Springer.

Nijkamp, P. and P. Rietveld (1987), Technological Development and Regional Labour Markets, in: M.M. Fischer and P. Nijkamp (eds.), Regional Labour Markets. Analytical Contributions and Cross-National Comparisons, pp. 117-138, Amsterdam et al., North-Holland.

Oakey, R.P., A.T. Thwaites and P.A. Nash (1982), Technological Change and Regional Development: Some Evidence on Regional Variations in Product and Process Innovation, Environment and Planning A 14:1073-1086.

OECD (1982), Micro-Electronics, Robotics and Jobs. ICCP Series, no. 7, Paris, Organisation for Economic Co-operation and Development.

Oosterhaven, J. and H. Folmer (1985), An Interregional Labour Market

Model Incorporating Vacancy Chains and Social Security, Papers of the Regional Science Association 58:141-155.

Parsons, D.O. (1975), Models of Labour Market Turnover: A Theoretical and Empirical Survey, Unpublished Paper, Ohio State University.

Piatier, A. (1984), Long Waves and Industrial Revolutions, Collaborative Paper CP-84-54, Laxenburg, IIASA.

Rogers, A. (1979), Migration Patterns and Population Redistribution, Regional Science and Urban Economics 9:275-310.

Rogers, A. (1980), Introduction to Multistate Mathematical Demography, Environment and Planning A 12:489-498.

Rogerson, P. (1979), Prediction: A Modified Markov Chain Approach, Journal of Regional Science 19:469-478.

Rogerson, P. (1982), Spatial Models of Search, Geographical Analysis 14:217-228.

Rogerson, P. (1987), Competition Among Applicants for Job Openings, in: M.M. Fischer and P. Nijkamp (eds.), Regional Labour Markets. Analytical Contributions and Cross-National Comparisons, pp. 229-245, Amsterdam et al., North-Holland.

Rogerson, P. and R. MacKinnon (1981), A Geographical Model of Job Search Migration and Unemployment, Papers of the Regional Science Association 48:89-102.

Rogerson, P. and R. MacKinnon (1982), An Interregional Model with Source and Interaction Information, Environment and Planning A 14:445-454.

Rouwendal, J. and P. Nijkamp (1987), Regional Economic Research on Labour Markets, in: M.M. Fischer and P. Nijkamp (eds.), Regional Labour Markets. Analytical Contributions and Cross-National Comparisons, pp. 95-115, Amsterdam et al., North-Holland.

Schubert, U. (1982), Remo - An Interregional Labour Market Model of Austria, Environment and Planning A 14:223-249.

Schubert, U, S. Gerking, A.S. Isserman and C. Taylor (1987), Regional Labour Market Modelling: A State of the Art Review, in: M.M. Fischer and P. Nijkamp (eds.), Regional Labour Markets. Analytical Contributions and Cross-National Comparisons, pp. 37-52, Amsterdam et al., North-Holland.

Sjaastad, L. (1982), The Costs and Returns of Human Migration, Journal of Political Economy, Supplement 70:80-93.

Slater, P.B. (1976), A Hierarchical Regionalization of State Economic Areas Based upon Migration Flows, Regional Studies 10:123-132.

Stoneman, P. (1984), The Economic Analysis of Technological Change, Oxford, Oxford University Press.

Storper, M. (1986), Technology and New Regional Growth Complexes, in: P. Nijkamp (ed.), Technological Change, Employment and Spatial Dynamics, pp. 46-75, Berlin and Heidelberg, Springer.

Thirwall, A.P. (1966), Regional Unemployment as a Cyclical Phenomenon, Scottish Journal of Political Economy 13:205-219.

Thwaites, A.T. (1982), Some Evidence of Regional Variations in the Introduction and Diffusion of Industrial Products and Processes Within British Manufacturing Industry, Regional Studies 16:371-381.

Thwaites, A.T., A. Edwards and D.C. Gibbs (1982), Interregional Diffusion of Production Innovations in Great Britain, Volume I, Centre for Urban and Regional Development Studies, University of Newcastle upon Tyne.

Thwaites, A.T., R.P. Oakey and P.A. Nash (1981), Industrial Innovation and Regional Development. Final Report to the Department of the Environment. Centre for Urban and Regional Development Studies, University of Newcastle upon Tyne.

Tobler, W. (1981), A Model of Geographical Movement, Geographical Analysis 13:1-20.

Wilson, A. and P. Rees (1977), Spatial Population Analysis, London, Arnold.

Wissen, L.J.S. and A. Rima (1988), Modelling Urban Housing Market Dynamics, Amsterdam, North-Holland.

Yellen, J. (1984), Efficiency Wage Models of Unemployment, American Economic Review, May 1984, pp. 100-108.

Chapter 4

ENDOGENOUS SPATIAL EXTERNALITIES IN RESIDENTIAL LOCATION THEORY

E. Werczberger

1. Introduction

This essay contributes an interpretative discussion of residential location theory, focusing on one particular but central aspect of the subject: the role of spatial externalities.[1] The essence of the city has been described as 'a complex system of spatial interdependence between its constituent elements: households, businesses, industries, and public institutions' (Papageorgiou, 1982). Neoclassical residential location theory has, from its beginning, emphasized one form of interdependence - access to the predetermined city center - analyzing it as the principal force affecting residential spatial structure. Yet distance to the CBD, or even to employment, explains only part of the variation in locational preferences and thus of housing prices. Other factors, most of which can be interpreted as external effects of urban activity, have been shown to be at least as important for residential location choice (see, e.g., Ball, 1979; Bartik and Smith, 1987). It suffices to mention location factors such as neighborhood prestige, industrial pollution and school quality. All of these have a significant effect on residential preferences and thus on the spatial differentiation of the urban system. Without their explicit consideration, processes such as the generation of ethnic ghettoes and gentrification are all but impossible to explain.

The discussion is divided into four parts. The next section begins with an examination of the nature of spatial externalities and of their implications for residential location theory. In the remainder, a typology is developed

[1] The paper is not intended as a survey, however, for which the reader is referred to a number of excellent reviews of recent progress in theory and modelling. See, in particular, the contributions by Muth (1985), Harris (1985), Fujita (1986), Kain (1987), Miyao and Kanemoto (1987), Straszheim (1987), Bertuglia et al. (1987) and Berechman and Small (1988).

based on three central aspects of spatial externalities: their nature, the kind of relationship involved, and their analytical treatment. For each aspect, the possible types of interdependencies and approaches are further classified and illustrated.

2. Residential Location Factors and Spatial Externalities

2.1 Spatial externalities

Externalities are generated when an economic agent affects the consumption or production possibilities of other agents or influences their welfare or profits but does not compensate them (or, alternatively, is reimbursed by them) for his actions (Miyao and Kanemoto, 1987).[2] The intensity of the effects of spatial externalities, as opposed to non-spatial effects, declines with distance from their source. Thus, they cause most of the differences in attraction between residential sites and act as principal factors for the spatial differentiation of urban and especially residential areas.

The variables which affect the attractivity of alternative locations, and control residential location choice, can be divided into two major groups: site quality and location factors. The determinants of site quality comprise both the attributes of the specific parcel under consideration and its improvements. They include characteristics such as the shape of the lot, the availability of infrastructure, as well as the size of the dwelling unit and its condition. Location factors, on the other hand, are properties which are due to the location of a site relative to other land uses in the region. They are caused by the presence or by the actions of other activities, i.e., by the spatial externalities to which these give rise. They include effects such as access to schools, proximity to prestigious neighborhoods and nuisances from pollution (Smith, 1977; Diamond and Tolley, 1982; Freeman, 1977).

2.2 The standard neo-classical model

As a point of departure it may be useful to consider the classical monocentric model used in urban economics (Alonso, 1958; Straszheim, 1987).

---

[2] For a comprehensive rigorous discussion, see Cornes and Sandler (1986).

Briefly, its central features are the assumption of an exogenously given, immobile single centre, which serves as the principal destination of commuter and shopping trips. Access to the center generates spatial externalities in the form of savings in commuting or travel cost, which are internalized by the urban actors and result in a gradient of land rents. Thus, the possibility is created of substituting lower land cost and lower density for access to the center, causing the density gradient between the CBD and the periphery. Moreover, depending on the relative magnitude of the income elasticity of demand for housing (or land) versus the income elasticity of transportation costs, the rich will concentrate in the center or in the suburbs. Since the location of the center is assumed to be predetermined and not affected by the spatial distribution of the households, the externality characteristic of access can be ignored in the analysis.

2.3 Analytical problems associated with endogenous externalities

The presence of endogenous externalities complicates the analysis of the residential structure in several ways. As one of the main causes of market failure, it prevents the competitive market from achieving a Pareto optimum allocation. Therefore, the socially optimum allocation of households generally differs from the market equilibrium and must be studied separately. Moreover, it will generally only be achieved if the externalities are internalized by their sources, typically as a result of public intervention such as planning, taxation, or change in property rights (Cornes and Sandler, 1986). Endogenous externalities also involve non-convexities which may result in multiple and possibly disjoint sets of equilibria (Starrett, 1972). These create the possibility of inferior local solutions and thus provide an additional economic justification for planning and for public intervention in the market.

In many instances the location or activity level of the sources of an externality can, at least in the short run, be considered exogenous to the system analyzed. The locational behaviour of households can then be studied in the framework of partial equilibrium. This is, of course, what neo-classical residential location theory does by assuming a predetermined and usually single employment center (Alonso, 1958; etc.). The simplification avoids the need for simultaneously determining the location decision of both households and their employers. Moreover, if the analysis is limited to a monocentric urban pattern, equilibrium conditions are relatively simple. In the long run, however, the sources of the externality are liable themselves to be affected by the spatial distribution of households. The externalities to which they give rise thus become inherently endogenous. In that case, partial equilibrium analysis is prone to yield erroneous results about long-run development.

2.4 Endogenous spatial externalities and urban structure

Because of its emphasis on access to the center, neoclassical residential location theory has been limited in its analysis to certain aspects of the spatial differentiation of the urban population. Its major contributions are to the understanding of patterns related to the rent gradient or to concentric growth. It has been less successful with processes, which are the result of endogenous spatial externalities, such as the spatial concentration of ethnic groups or the sectoral development (Hoyt, 1939). These can only be understood if endogenous spatial externalities are explicitly taken into account.

Consider, e.g., suburbanization and gentrification, probably the two most important current urban developments. The standard explanation for suburbanization is, of course, the decline of commuting costs to the center. However, the process is intensified and accelerated by endogenous externalities. The extension of the labour market through residential suburbanization attracts also industrial and service employment into the suburbs (Niedercorn and Kain, 1962; Steinnes, 1982; Palumbo and Hutton, 1987). This leads to an increase in cross-commuting between outlying communities, and in turn reinforces the suburbanization of the labour force. Gentrification, on the other hand, represents a reversal of long-run trends in neighborhood decline (see, e.g. Smith, 1979; for a different view see Le Roy and Soustelle, 1983). Negative neighborhood effects caused by low income residents and physical deterioration tend to prevent higher status groups from settling in these areas. This may create a rent gap between current property values and the market potential in the absence of these externalities. If the gap becomes large enough, less prejudiced 'pioneers' may move into the neighborhood to reap the potential capital gain. A cumulative process thus begins which may eventually lead to a complete replacement of the original population.

Both processes, suburbanization and gentrification, involve the combined effect of exogenous stimuli and endogenous changes, which are modified or exacerbated by spatial externalities. An analysis which ignores the endogenous character of externalities is thus unlikely to correctly predict the change or its rate of development. The more realistic but far more difficult alternative to partial equilibrium analysis is to deal with externalities, i.e. their sources, their level and their location, as endogenous to the urban system. A conceptual and formal framework is then needed, which focuses on the interdependence between sensitive households and the sources of external effects.

The analytical problems posed by such interactions and the multiplicity of possible equilibria due to the presence of endogenous spatial externalities also necessitates consideration of the dynamics of the urban system. The actual spatial pattern attained is then construed as the result of both exogenous historical accidents and the evolution of the system. Thus Hoyt (1939) explains the evolution of homogeneous sectors as the consequence of exogenous amenities existing at the outset and the cumulative impact of neighborhood effects on the development. Similarly, because of the multiplicity of disjoint solutions, mathematical models which explicitly account for the effect of endogenous externalities require a dynamic or at least pseudo-dynamic model structure (Werczberger and Berechman, 1988).

3. The Nature of Spatial Externalities

It would not be very useful to provide an exhaustive enumeration of all the types of spatial externalities which theoretically can affect residential location choice. Instead this and the following sections propose a typology of spatial externalities. It is based on three aspects, considered to be most relevant for an understanding of their role in residential location: their nature, the kind of relationship they involve, and the types of approaches used for their formal analysis.

For a characterization of the nature of spatial externalities two criteria seem to be most pertinent: the **type of source** or agent which causes the external effect, and the **type of the effect**.

Spatial externalities have been classified according to the type of their source into three categories (Miyao, 1987): (a) **economic activities** managed mostly for profit, such as industry and business; (2) **households**; and (3) **communities**, mainly local government and non-profit organizations. The three categories of sources differ by the decision criteria and factors which determine their location and the level of externality generated. It is because of this difference that a uniform treatment of the different types of spatial externalities has not been possible hitherto.[3]

The second criterion suggested for the classification is the type of effect which is caused by the externality and affects the desirability of a given location. Three classes of location factors have been distinguished, each of which is crucial for residential location choice (Smith, 1988): (1) **access**, which relates to the availability of factors such as employment or services and is operationally defined as distance or travel time to specific destinations such as employment or the CBD; (2) **neighborhood effects**, which are due to the proximity of certain populations or activities and involve subjective and objective consequences such as nuisances, prestige, and desirability; and (3) **local public goods** affecting the attractivity of whole communities as residential locations.

Joint consideration of the two criteria - type of source and type of effect - yields nine combinations of source and externality types, viz. location factors. Figure 1 suggests one or two typical examples for each possible pair. Only some of these have seriously been studied in the context of residential location theory: access to employment, pollution from industry and neighborhood effects from households. The neglect of the other factors is probably due to the difficulty in quantifying their magnitude and effect. In the following, we try briefly to illustrate these combinations, relating them to selected studies in the literature.

---

[3] Transportation is in this context not considered as a separate source of externalities but as an additional external effect due to the interaction between urban activities.

	Types of sources		
Types of Externalities	Economic Activities	Households	Community
Access	Employment	Ethnic groups	Schools, Museums
Neighborhood effects	Pollution Nuisances	Prestige	Open space, Infrastructure
Public good	Improvement of tax base	Social support	Services taxation

Figure 1. Examples of the Different Forms of Externalities Associated with the Various Types of Sources

3.1 Externalities due to economic activity

In most neoclassical and spatial interaction models, access to employment is considered as the principal location factors (see Alonso, 1958, Lowry 1964, etc.). Employment location is then typically assumed to be exogenously given. One of the few studies which treat employment location as an endogenous factor is Fujita's analysis (1988) of the interaction between communication related firms and their employees. Similarly, in some quasi-dynamic versions of Lowry-type models, the interaction between employment and residential location has been made explicit (Prastacos, 1988; Werczberger, 1987). In the more recent nonlinear dynamic models, households and employment are modeled as choosing their respective locations as a function of the spatial distribution of the other activity (see, e.g., Lombardi and Rabino, 1984). Thus, the interdependence between employment and population can be shown to lead to a continuing process of relocation of urban land uses.

Neighborhood effects generated by economic activity mainly take the form of nuisances due to pollution. Models used to analyze these effects usually assume that all pollution sources are exogenously determined and concentrate in the city center (Kanemoto, 1987). We are not aware of studies which deal with the interaction between polluters and housing when the location of both activities is endogenous.

Public goods (or public bads) generated by economic activity affect the community as a whole. Typical examples would be their impact on the economic base of communities, the municipal budget or employment opportunities. Each of these effects is commonly assumed to affect the interurban location choice of households. They play therefore a major role in public policy but are rarely, if ever, studied in the context of residential location theory, except in a policy context.

3.2 Households as the source of neighborhood effects

The importance as a location factor of access to households -relatives, friends, or families from the same origin - has been emphasized by many migration studies. Because of the exorbitant data requirements these factors are difficult to incorporate into operational models of location choice. They may, however, be introduced into micro- simulation studies which concentrate on the locational career of individual households.

Neighborhood effects associated with certain population groups are the most widely studied type of externality caused by households. They may be caused by attitudes such as racist prejudice, attraction to high socio-economic status, or attraction to people of similar ethnic background. Most of the research has been based on a static micro-economic paradigm and simplifies the problem by assuming a linear city and the location of one of the household groups as given (Bailey, 1959; Kanemoto, 1980 and 1987; Papageorgiou, 1982; and others). Efforts have been made to extend the analysis to two-dimensional space (Loury, 1976; Werczberger and Levin, 1986) and to spatial-interaction models (Werczberger and Berechman, 1988). Only in a dynamic framework, however, it is possible to observe how endogenous neighborhood effects affect the rate of development or change its direction as suggested in the sectoral model (Hoyt, 1939; Werczberger, 1984; Clarke and Wilson, 1981; Werczberger, 1987).

Examples of local public goods generated by the households in a neighborhood are, e.g., social support systems involving the feeling of community and mutual trust (Suttles, 1969). In formal analysis, they are probably best handled together with the neighborhood effects discussed above.

3.3 The community as the source of externalities

Among the most important residential location factors are local public goods, including services and amenities such as air pollution control, school quality, and crime rate (Bartik and Smith, 1987; Diamond and Tolley, 1982). Decisions about these policies are the outcome of the political process, which in democracies is affected by the social structure and composition of the residents and hence endogenous to the urban structure. Three different types of local public goods have been distinguished (Fujita, 1988). They roughly correspond to the three types of externalities mentioned above: (1) **public facilities provided at discrete locations**, such as schools and hospitals, whose accessibility and thus contribution decline with distance from their location; (2) **spatially continuous strictly local public goods**, such as parks or street lighting, in which the supply may vary in space and enjoyment is limited to households residing at or near that location; and (3) **pure local public goods**, for which the benefit or externality level is homogeneous within the boundaries of the community and vanishes outside.

Most of the emphasis in research on local public goods has been on theoretical or policy issues, such as the existence of a Pareto optimum equilibrium, the optimum level of supply or taxation, and the optimum location of the facilities (Tiebout, 1956; Wildasin, 1987; Revelle, 1987). Little attention has been paid to their effect on residential location choice (Cremer, 1990). There are, of course, inherent difficulties in handling local public goods as endogenous externalities. The main problem is the lack of well developed behavioral decision making models which would allow expressing the supply of local public goods as a function of location or population mix.

4. The Kind of Relationship Involved

Another way of looking at externalities is by characterizing the kind of relationship in which they are involved; in particular, the number of interacting activity types and the effect of the externality on the utility of sensitive households.

Location processes involving spatial externalities can be characterized according to whether one or several groups are involved. This suggests a distinction between **intragroup** and **intergroup** effects. If there is only one group whose members compete for space, models of congestion or crowding are implied (Fujita, 1987). On the other hand, if the relationship involves two or more population groups, processes of integration and segregation arise, as analyzed in the various models of racial prejudice (Bailey, 1959; Yinger, 1976, etc.). In the context of residential location theory, this point of view is relevant mainly for the analysis of neighborhood effects between households. Externalities affecting location choice which are generated by economic activities or communities involve, by definition, more than one type of activity.

Equally important with regard to the effect of externalities is, of course, the direction or sign of the effect of the externality on the utility of the household. Depending on the sign, the result is **attraction** or **aversion** between the source and the households sensitive to the externality.

Considering the two criteria together, we obtain the two-way classification shown in Figure 2 (Miyao, 1981). **Ceteris paribus**, of the four possible elements in the matrix, two provide incentives for segregation - intergroup aversion and intragroup attraction - and two for integration-intergroup attraction and intragroup aversion (Papageorgiou, 1978). Intragroup attraction implies that the individual actors derive benefits from the proximity of households which are similar to them. As a consequence, they concentrate in certain areas, such as ethnic enclaves, generally at higher densities than otherwise expected (Werczberger and Berechman, 1988). Intragroup aversion, on the other hand, arises in the presence of negative externalities, such as crowding or congestion. It should lead to a lower density than anticipated in the absence of externalities. The result of intergroup effects is the converse. Thus intergroup attraction should result in integration except, of course, if the relationship between the two populations, say black and white households, is not symmetric (Schelling, 1972). Intergroup aversion, on the other hand, is the usual case of negative neighborhood effects, such as between industrial and residential areas of two unfriendly ethnic groups.

	attraction	aversion
intergroup effects	integration	segregation
intragroup effects	segregation	integration

Figure 2. The Four Types of Neighborhood Effects

5. Approaches to Formal Analysis

Any formal analysis of spatial externalities requires a choice with regard to three topics: (1) the identification of the relevant source, (2) a definition of the distance between source and receptor, and (3) the functional form of the spatial attenuation of the effect.

If there is more than a single potential source of externalities, the first problem is the identification of the activity which gives rise to the external effect. Only two of the theoretical possibilities seem to have been used in residential location models: (a) The simplest approach is the **a priori** assignment of a single predetermined source. In the case of commuting, e.g., it is access to the place of employment which is measured. The relevant service level, in the case of local public goods, is the level in the locality in which the site is located. (b) If there are several sources of externalities which jointly affect the location, their effect is usually assumed to be additive. The contribution of each source is then weighted by some function of the distance (Papageorgiou, 1978). This form has been used mainly in the analysis of neighborhood effects, as for the impact of pollution and of neighborhood prestige. Additional functional forms could be conceived, such as multiplicative or optimizing functions of the contribution of individual sources (Fujita and Smith, 1990). Thus the best or nearest school might be considered relevant for location choice. However, there is no empirical evidence at this point which would justify the choice of one approach over the other.

When there is more than one source of externality which affects location choice, a measure must be chosen of the generalized distance between the sources and the site of the sensitive household. The alternative measures commonly used are based on three different concepts of the character of the effect (Kanemoto, 1987; Fujita, 1988): (a) The concept of **local externalities** has been used when households are only affected by sources which are located at the same location or in the same neighborhood (See, Rose-Ackerman, 1975; or Courant and Yinger, 1977). Their effect is taken to depend on the relative proportion of the two activities or groups in the relevant area. This measure has been used for studying the stability of integrated allocations, such as Schelling's analysis of the tipping point (1972) and related problems (Yinger, 1976; Kern, 1981). (b) The concept of **border effects** is useful only if the competing populations are completely segregated. In that case, the distance of the location from the border separating the two groups can be considered as the relevant measure (Rose-Ackerman, 1975; Yinger, 1976; Schnorre, 1976; see also Courant and Yinger, 1977, for a critique). The measure has been applied for the analysis of the rent levels and changes in the ghetto boundary. (c) Particularly in the case of **neighborhood effects**, the sources are spatially dispersed while their effects are additive. The actual distance between the source of the externality and the site of the sensitive household is then the most convenient measure for estimating the contribution to the overall effect. (Yelling, 1974; Kanemoto, 1980; Werczberger and Berechman, 1988; and others).

The functional form of the spatial decline of the effect has been shown to be critical for the resulting location pattern (Fujita, 1988; Loury, 1976; Werczberger and Levin, 1986). Functions used for the analysis of the attenuation of externalities in space include the negative exponential function (Werczberger, 1984), linear functions (Fujita, 1988), a binary step function

(Baily, 1956), and limitation of the effect to the immediately adjacent sites (Loury, 1976) or to the site itself (Yinger, 1976). However, with the exception of access and the diffusion of air pollution, there is very little empirical evidence on the shape of the function or on parameter values. In the absence of adequate data, the choice is inevitably based on analytical or computational convenience.

6. Conclusion

This essay suggested that location factors encountered in residential location theory can be conceptualized as spatial externalities. For a full understanding of urban spatial structure and development the sources of the externalities and their level must be treated as endogenous variables. Significant progress has been made in dealing with external effects resulting from household location and pollution associated with economic activity. Yet there is much left for future efforts dealing with the interdependence between residential location and the endogenous processes associated mainly with employment location and the supply of local public goods.

One of the important developments in location theory during the past two decades has been the convergence of different strands of residential location theories. Very different conceptual approaches, such as the gravity-type spatial interaction model and the discrete choice model, have been shown to be formally equivalent to the multinomial logit model and could be embedded within the neoclassical framework of constrained utility maximization (Anas, 1982). There is place for a similar effort to unite the treatment of the different strands of the analysis of spatial externalities in a unified framework. It would have to involve a combined formal treatment of three different issues: the process by which the location of externality sources is determined, the criteria by which their intensity is decided, and the functional form of their decline with distance. The differences between the various types of externalities could then be expressed in the appropriate choice of parameters. Thus, Fujita and Smith (1990) show the mathematical equivalence between different types of additive-interaction effects. However, before an integrated model can be formulated, more significant progress will have to be made in understanding the interdependence between household location and each of the sources of external effects.

References

Alonso, W. (1964), Location and Land Use, Cambridge, Harvard University Press.

Anas, A. (1982), Residential Location Markets and Urban Transportation, New York, Academic Press.

Bailey, M.J. (1959), Note on the Economics of Residential Zoning and Urban Renewal, Land Economics 35:288-292.

Ball, M. (1973), Recent Empirical Work on the Determinants of Relative House Prices, Urban Studies 10:213-233.

Bartik, T.J. and V.K. Smith (1986), Urban Amenities and Public Policy, in: P. Nijkamp (ed.), Handbook of Regional and Urban Econmics, volume 1, pp. 357-406, Amsterdam, North Holland.

Berechman, J. and K.A. Small (1988), Modeling Land Use and Transportation: An Interpretive Review for Growth Areas, Environment and Planning A 20:1285-1309.

Bertuglia, C.S. et al. (1987), An Historical Review of Approaches to Urban Modelling, in: C.S. Bertuglia et al (eds.), Urban Systems: Contemporary Approaches to Modelling, London, Croom Helm.

Clarke, M. and A.G. Wilson (1983), The Dynamics of Urban Spatial Structure: Progress and Problems, Journal of Regional Science 23:1-18.

Cornes, R. and T. Sandler (1986), The Theory of Externalities, Public Goods and Club Goods, Cambridge, Cambridge University Press.

Courant, P.N. and J. Yinger (1977), On Models of Racial Prejudice and Urban Residential Structure, Journal of Urban Economics 4:272-291.

Cremer, H. (1990), Residential Choice and the Supply of Local Public Goods, Journal of Urban Economics 27:168-187.

Diamond, D.B. and G.S. Tolley (1982), The Economics of Urban Amenities, New York:Academic Press.

Fujita, M. (1986), Urban Land Use Theory, in: J.J. Gabszewicz, J.-F. Thisse, M. Fujita and U. Schweitzer (eds), Location Theory, pp. 73-150, Chur, Harwood.

Fujita, M. (1988), Spatial Interaction and Agglomeration in Urban Economics, Working Paper # 126, Dept. of Regional Science, University of Pennsylvania, Philadelphia.

Fujita, M. and T.E. Smith (1990), Additive-Interaction Models of Spatial Agglomeration, Journal of Regional Science 30:51-74.

Hoyt, H. (1939), The Structure and Growth of Residential Neighborhoods in American Cities, Washington D.C., Government Printing Office.

Kain, J.F. (1987), Computer Simulation Models of Urban Location, in: P. Nijkamp (ed.), Handbook of Regional and Urban Economics, Volume II, pp. 847-876, Amsterdam, North Holland.

Kanemoto, Y. (1980), Theories of Urban Externalities, Amsterdam, North Holland.

Kanemoto, Y. (1987), Externalities in Space, in: Y. Kanemoto and T. Miyao (eds.), Urban Dynamics and Urban Externalities, Chur, Harwood.

Kern, C.R. (1981), Racial Prejudice and Residential Segregation: The Yinger Model Revisited, Journal of Urban Economics 10:164-172.

Le Roy, S.G. and J. Soustelle (1983), Paradise Lost and Regained: Transportation Innovation, Income and Residential Location, Journal of Urban Economics 13:67-89.

Lombardi, S.T. and G.A. Rabino (1984), Nonlinear Dynamic Models for Spatial Interaction: Some Empirical Results, Papers and Proceedings of the Regional Science Association 55:83-101.

Loury, J.C. (1976), The Minimum Border Length Hypothesis Does Not Explain the Shape of Black Ghettoes, Journal of Urban Economics 5:147-153.

Lowry, I.S. (1964), A Model of Metropolis, RM-4035-RC, Rand Corporation.

Miyao, T. (1987), Dynamic Urban Models, in: P. Nijkamp (ed.), Handbook of Regional and Urban Economics, Vol II, pp. 877-925, Amsterdam, North Holland.

Miyao, T. and Y. Kanemoto (1987), Urban Dynamics and Urban Externalities, Chur, Harwood.

Muth, F.R. (1969), Cities and Housing: The Spatial Pattern of Urban Residential Land Use, Chicago, University of Chicago Press.

Muth, F.R. (1985), Models of Land Use, Housing, and Rent: An Evaluation, Journal of Regional Science 25:4:593-606.

Niedercorn, J.H. and J.F. Kain (1962), An Econometric Model of Metropolitan Development, Papers and Proceedings of the Regional Science Association.

Palumbo, G.P. and P. Jutton (1987), On the Causality of Intra-urban Location, Journal of Urban Economics 22:1-13.

Prastacos, P. (1986), An Integrated Land-Use-Transportation Model for the San Francisco Region: 1. Design and Mathematical Structure, Environment and Planning A 18:307-322.

Papageorgiou, Y. (1987), Spatial Externalities, Annals of the Association of American Geographers 68: 465-475 and 476-492.

Papageorgiou, Y. (1982), Externality Diffusion, Sistemi Urbani 1/2:17-39.

Revelle, C. (1987), Urban Public Facility Location, in: P. Nijkamp (ed.), Handbook of Regional and Urban Economics, volume II, pp. 1053-1095, Amsterdam, North Holland.

Rose-Ackerman, S. (1975), Racism and Urban Structure, Journal of Urban Economics 2:85-103.

Schelling, T.C. (1972), A Process of Residential Segregation: Neighborhood Tipping, in: A.H. Pascal (ed.), Racial Discrimination in Economic Life, Lexington MA, Lexington Books.

Schnorre, A.B. (1976), Racial and Ethnic Price Differentials in the Urban Housing Market, Urban Studies 13:107-120.

Smith, L.B., K.T. Rosen and G. Fallis (1988), Recent Developments in Economic Models of the Housing Market, Journal of Economic Literature 26:29-64.

Smith, N. (1979), Toward a Theory of Gentrification, Journal of the American Planning Association 45:538-548.

Smith, V.K. (1977), Residential Location and Environmental Amenities: A Review of the Evidence, Regional Studies 11.

Starrett, D.A. (1972), Fundamental Non-Convexities in the Theory of Externalities, Journal of Economic Theory 4:180-199.

Steinnes, D.N. (1982), Do People Follow Jobs or Jobs Follow People, Urban Studies 19:187-192.

Straszheim, M. (1987), The Theory of Urban Residential Location, in: P. Nijkamp (ed.), Handbook of Regional and Urban Economics, vol. II, pp. 717-758, Amsterdam, North Holland.

Stull, W.J. (1974), Land Use and Zoning in an Urban Economy, American Economic Review 64:337-347.

Suttles, G.D. (1969), The Social Order of the Slum: Ethnicity and Territory in the Inner City, Chicago, The University of Chicago Press.

Tiebout, C.M. (1956), A Pure Theory of Local Expenditures, Journal of Political Economy 64:416-424.

Werczberger, E. (1984), Modeling the Segregation of the Urban Population, Environment and Planning B, Planning and Design 11:15-28.

Werczberger, E. (1987), A Dynamic Model of Urban Land Use with Externalities, Regional Science and Urban Economics 17:391-410.

Werczberger, E. and J. Berechman (1988), Incorporating Neighborhood Effects into Spatial Allocation Models, Environment and Planning A 20:595-607.

Werczberger, E. and E. Levin (1986), Spatial Consequences of Neighborhood Effects in a Linear City, Working Paper no. 89. The Center of Urban and Regional Studies, Tel-Aviv University.

Wildasin, D.E. (1987), Theoretical Analysis of Local Public Economics, in: P. Nijkamp (ed.), Handbook of Regional and Urban Economics, vol. II, pp. 1131-1178, Amsterdam, North-Holland.

Wingo, L. (1961), Transportation and Urban Land Use, Baltimore, John Hopkins University Press.

Yinger, J. (1976), A Note on the Length of the Black-White Border, Journal of Urban Economics 3:370-382.

Yellin, (1974), Urban Population Distribution, Family Income and Social Prejudice, Journal of Urban Economics 1:21-47.

Chapter 5

THE EFFECT OF TRANSACTION COSTS IN THE HOUSING MARKET

O. Fisch

1. Introduction

This paper contends that the locational limitations of tenant's information about housing quality, have profound effects upon the market structure, making the housing rental market a special case to analyze. Symmetrically, we contend that the limitations of landlord's information on the extensive variation on tenant's housing preferences and the short-run limitations of the landlord's options to modify the quality attributes of his standing stock of housing units also have profound effects upon the market structure.

Tenants are periodically making choices among housing units. We recognize that tenants lack full information [Akerlof (1980), Stiglitz (1985)] about the implicit rents of the multitude of quality attributes of the housing units, but we want to press that the tenants' information is worst about the extensive quality variation of the standing stock of housing units, simply because this information is more difficult (and very costly) to obtain. This information includes no only the myriad of quality attributes of the housing unit as physical shelter (architectural attributes), but also information in which it must be included all the neighborhood attributes specifically accruing from the location of the housing unit [Sweeney (1972)].

In particular, and as a significant departure from the extensive literature, we are going to contrast symmetrically monopoly power for a consumer good that increases inversely with the tenant's knowledge about the location of bundles of housing quality attributes, vis-à-vis monopsony power for a consumer good that increases inversely with the landlord's certainty about finding a 'demand' for his particular housing bundle.

It is also significant in an empirical way, not only because housing expenditures are between 25 and 30% of total expenditures in the market of consumer goods, but also because the results obtained here will give an insight into the working of a market of heterogeneous goods with transaction costs [Coase (1937), Coase (1960), Alchian and Demsetz (1972)] accruing from search to locate the 'good' in one side of the market and to locate the 'consumer' on the other side of the market.

In order to obtain a meaningful market equilibrium, we construct a partial microsimulation equilibrium model in which the short-run stochastic search behavior of both tenants and landlords is examined and the dispersion of rents in the housing market is endogeneously determined.

We assume that tenants already know the rent and the location of some bundles of housing attributes with a fuzzy description of each bundle, either given by advertising or by brokers acting in the market. We call this fuzzy description prior information: it is incomplete information, but it is enough information to orient the tenant's sequential search to locate his preferred housing bundle.

In this context we define search specifically to include any way of evaluating these housing bundles subject to the following restrictions: first, the tenant must transport himself to the location of the bundle; second, the tenant must inspect the housing unit in order to complete his prior (incomplete) information about the bundle of quality attributes; third, the tenant evaluates the utility of the inspected bundle of quality attributes; and fourth, the tenant must compute his reservation rent for that particular bundle and negotiates the final contractual rent prior to signing his lease.

Correspondingly, we define the tenant's search costs as the cost in leisure time consumed in: first, in the time consumed in the assessment of the prior (incomplete) information of the different bundles; second, in the time consumed in his physical transfer (i.e., transportation) to the location of a specific bundle; third, the time consumed in inspecting the bundle in order to complete his information about the quality attributes of the housing unit; and finally, the time consumed in the evaluation of the utility of the bundle, in the computation of the reservation rent and in the negotiation of the contractual rent. We consider here leisure time at the beginning of the search as a dedicated asset [Williamson (1985)] being depleted by the search process.

There are many possible models that can reflect price adjustment in a highly organized market like the rental housing market. In selecting the particular model presented here, the central purpose was to reflect, at extremum, how the heterogeneity of the good exchanged in this specific market necessarily creates conditions of transaction costs. Given the physical durability of the housing unit, the heterogeneity of the good involved is the result of the landlord's (risky) attempt to 'accommodate' the observed heterogeneity of consumer's housing preferences at the time of building the unit and at the time of making decisions about different maintenance schedules [Fisch (1977); Fisch (1986)]. The location fixity of the (risky) bundle climaxes the heterogeneous nature of the good exchanged in this market, necessarily creating the conditions of geographical search. It is shown how, at equilibrium, the presence of tenant and landlord searches are the central reason for housing rent dispersion.

Here we do this by modeling certain totally decentralized, nontâtonment process of rent adjustment by means of which an appropriate new equilibrium concept is investigated: in contrast to short-run adjustment of quantity in the market for homogeneous goods, the quality attributes of the heterogeneous (durable) product are not adjusted in the housing market in the short run.

Our approach here will depart from the standing equilibrium theory and stability analysis, introducing first an adjustment process that per se depletes dedicated assets of the agents in both sides of the market (i.e., it depletes the tenant's endowment of leisure time and it depletes the landlord's endowment of management time), and proceeding to explore the necessary conditions for that process to converge the short-run equilibrium (steady state or stochastically steady state); second, the notion that tenants and landlords adjust their expectations and behavior in response to altered opportunities; and third, in the housing market, as in most markets, there is no nonparticipating auctioneer, and rents are set either by demand or by supply.

Our approach will also depart from the recent literature on dis-equilibrium economics, by changing the central assumptions of that literature to the following:

A1) In order to focus more directly on the impact of heterogeneous quality, we assume that the implicit (hedonic) rents of housing attributes are completely known to both tenants and landlords: and are continuously and costlessly updated by both (see Appendix).

A2) Complete information about a specific bundle of housing quality attributes is acquired by the tenant only at the location of the bundle.

A3) The existence of a tenant preferring a specific bundle of housing quality attributes is totally uncertain to the landlord.

We are omitting price search not because it is not relevant to our inquiry or because the cost of acquiring price information is insignificant. We just want to concentrate on the locational dimension of the housing bundle, a dimension that substantively adds heterogeneity to a heterogeneous good. We consider that our approach to disequilibrium economics is more appealing because this adjustment process, at difference with the literature on price search and optimal stopping rules, is designed to reflect some realistic features and then consider the long-run position of the housing rental market.

2. Implicit Rents and Explicit Contracts

The behavioral rules of landlords and tenants in the rental housing market were tentatively set in the previous section, through different assumptions. These assumptions enable us to characterize behavior by mimicking behavior inthe homogeneous good market. Given a known vector of estimated (implicit) hedonic rents $\mathbf{r}_E$ [Rosen (1974)] applying througout the market, the tenant's preferred bundle of housing attributes $\mathbf{a}_T$, together with x*, the composite consumption good (whose price is normalized to unity) is given by

$$(1) \qquad \{x^*, \mathbf{a}_T\} \leftarrow \mathrm{Max}[u_T(x,\mathbf{a}) \,|\, x + \langle \mathbf{a}, \mathbf{r}_E \rangle = y_T]$$

where y_T is the tenant's income and $r_T(\mathbf{a}_T) = \langle \mathbf{a}_T, \mathbf{r}_E \rangle$ is the tenant's encumbered housing rent. For any landlord's $\mathbf{a}_L \neq \mathbf{a}_T$, the tenant's reservation rent $r_T(a_L)$ is computed implicitly by

(2) $$r_T(a_L) \leftarrow u_T[y_T - r_T(a_L), a_L] = u_T[x^*, a_T]$$

If the landlord's asking rent is

(3) $$r_L(a_L) \equiv \langle a_L, r_E \rangle = r_T(a_T)$$

then it follows that

(4) $$a_L \neq a_T \leftrightarrow r_L(a_L) \geq r_T(a_L)$$

(5) $$a_L = a_T \leftrightarrow r_L(a_L) = r_T(a_L)$$

and we can state the following:

Proposition 1: In the presence of perfect information, costless and instantaneous location of housing attributes and one-to-one matching between the standing stock and the tenants' preferred bundles , the housing rental market clears with a unique transacted rent r^* and all the reservation rents are equal to the asking rents.

Under the assumption in Proposition 1, we can explain the residuals in the statistical estimation of the implicit rents of housing as the result of a short-run lack of market coordination of bundles of attributes between supply and demand of rental housing. But, it is in the relaxing of some or all of the assumptions where reality of the market process can be incorporated into the analysis. In order to increase the specificity of the behavioral rules, we have to define the time frame of the decision making process.

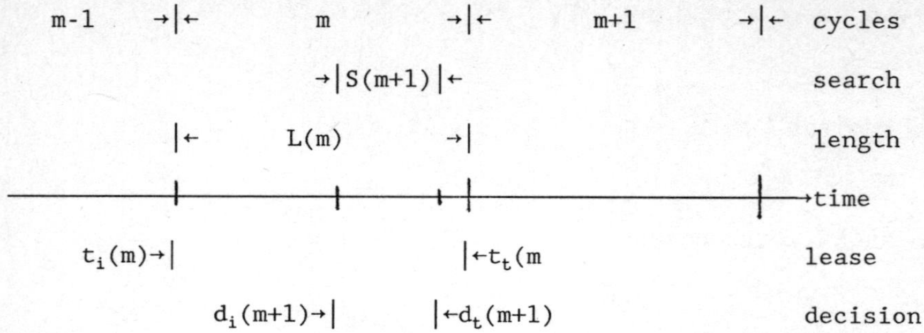

Figure 1. Lease Cycles of the Housing Unit in the Rental Market

Figure 1 gives a description of this time frame, in which we define graphically:

m = the lease cycle of the housing unit

t_i = the initial time of the lease

t_t = the terminal time of the lease

d_i = the initial decision time of the lease

d_t = the terminal decision time of the lease

L = the time length of the lease

S = the time length of the search

In this framework, the following identities hold:

(6) $$t_t^{+}(m) \equiv t_i^{-}(m+1)$$

(7) $$L(m) \equiv t_t(m) - t_i(m)$$

(8) $$S(m) \equiv d_t(m) - d_i(m)$$

The overlap of cycles is framed by

(9) $t_i(m) < d_i(m+1) < d_t(m+1) < t_t(m)$

If current time t is such that $t \in [d_i(m+1),d_t(m+1)]$, then the landlord and the tenant are active in the housing market. At any particular current time t, the standing stock of housing units and the tenants of the entire housing rental market are divided into two disjoint sets: all the landlords and the tenants for which

$$t \in [d_i(m+1),d_t(m+1)]$$

for all m, belong to the active market; all the housing units and the tenants for which

$$t \in (d_t(m),d_i(m+1))$$

for all m belong to the inactive market.

All the transacted rents r^* in the inactive market and their associated housing bundles $\mathbf{a}_L$ are continuously used in updating the current estimated vector of implicit rents $\mathbf{r}_E(t)$ that is used in the active market at current time t. This updating is analyzed in the appendix of this paper.

A. Breach of the rental contract

As explained in the preceeding section, the housing market cycle is institutionally framed in a tight time schedule: the result is a lease arrangement for a time length L(m), with an initial contractual time $t_i(m)$, a terminal contractual time $t_t(m)$ and a decision (transaction)time $d_t(m)$, for which

(10) $d_t(m) < t_i(m) < t_t(m)$

For any current time t within the mth cycle, $d_t(m) < t < t_t(m)$, the lease specifies a contractual rent, $r^*(\mathbf{a}_L) = r[\mathbf{r}_E(d_t),\mathbf{a}_T(d_t)]$, and the services of a bundle housing attributes $\mathbf{a}_L$.

Given $\mathbf{r}_E(d_t)$, there is at $t > d_t$ additional information on implicit rents from a sample set $\{r^*(s),\mathbf{a}_L(s) \mid S\epsilon[d_t,t)\}$ of the contractual rents r^* and their transacted bundles $\mathbf{a}_L$ of housing attributes, for which we assume a linear statistical specification

$$(11) \qquad r^*(s) = <\mathbf{a}_L(s), \mathbf{r}_E(s)> + e(s) \quad \forall s \in [d_t, t)$$

where e is a random error term and the implicit rents update is given by

$$(12) \qquad \mathbf{r}_E(t) = \mathbf{r}_E(d_t) + \Delta\{r^*(s), \mathbf{a}_L(s) \mid s \in [d_t, t)\}$$

where Δ is an update vecotr in increments (positive and/or negative) for the elements of the vector $\mathbf{r}_E(d_t)$. The tenant's review of the current contractual rent $r^*(d_t)$ starts with

$$(13) \qquad \{x^*(t), \mathbf{a}_T(t)\} \leftarrow \mathrm{Max}[u_T(x, \mathbf{a}; t) \mid x + <\mathbf{a}, \mathbf{r}_E(t)> = y_T(t)]$$

If $\mathbf{a}_L(d_t) \neq \mathbf{a}_T(t)$, the tenant's rent uneasiness is given by

$$(14) \qquad r_T(\mathbf{a}_L; t) \leftarrow u_T[y_T(t) - r_T(\mathbf{a}_L; t), \mathbf{a}_L] = u_T[x^*(t), \mathbf{a}_T(t)]$$

If $r^*(d_t) \leq r_T(\mathbf{a}_L;t)$ the tenant's review process stops. If $r^*(d_t) > r_T(\mathbf{a}_L;t)$, then the tenant considers breaching the contract and evaluates

$$(15) \qquad \{x^{*D}(t), \mathbf{a}_{TD}(t)\} \leftarrow \mathrm{Max}[u_T(x, \mathbf{a}; t) \mid x + <\mathbf{a}, \mathbf{r}_E(t)> = y_T - D(t)]$$

where D(t) are the certain damages the tenant faces in breaching the contract. It follows that $\mathbf{a}_{TD}(t) < \mathbf{a}_T(t)$.

Proposition 2a: The tenant decides at current time t to search, in spite of facing damages, if

(17) $u_T[y(t) - r^*(d_t),\ a_L(d_t)] < u_T[x^{*D}(t), a_{TD}(t)]$

Proposition 2b: In finding any $\mathbf{a}_L(t) \neq \mathbf{a}_{TD}(t)$, the tenant decides to breach the current contract iff

(18) $u_T[y(t) - r^*(t), a_L(d_t)] < u_T[y_T - D(t) - r^*(a_L(t), a_L(t)]$

B. Renewal of the rental contract

If at current time $t \epsilon [d_i(m), d_t(m)]$, then the tenant and the landlord face the decision to renew for the mth cycle the contract covering the current (m-1)th cycle of the housing unit: i.e., both tenant and landlord are active agents in the housing makret. The current contract has a contractual rent $r^*(d_t(m-1))$, transacted one period (L(m-1) ago. In the meantime, all the transactions in the housing market fed the updating process of the estimated implicit rents as in equation (1) and a new updated vector $\mathbf{r}_E(t)$ is arrived at (equation (12)). The landlord computes his new asking rent $r_L(\mathbf{a}_L;t) = \langle \mathbf{a}_L, \mathbf{r}_E(t) \rangle$ and the tenant's review starts with equation (13).

Proposition 3a: The lease is renewed for the new mth cycle with a contractual rent $r^*(t) = r_L(\mathbf{a}_L;t)$ iff $\mathbf{a}_T(t) = \mathbf{a}_L$.

If $\mathbf{a}_T(t) \neq \mathbf{a}_L$, then the tenant considers moving and evaluates

(19) $\{x^{*M}(t), \mathbf{a}_{TM}(t)\} \leftarrow \mathrm{Max}[u_T(x, \mathbf{a};t) \,|\, x + \langle \mathbf{a}, \mathbf{r}_E(t) \rangle = y_T(t) - M]$

where M are the moving costs the tenant faces in moving. The tenant's reservation rent of the current bundle is given by

(20) $r_{TM}(\mathbf{a}_L;t) \leftarrow u_T[y_T(t) - r_{TM}(\mathbf{a}_L), \mathbf{a}_L;t] = u_T[x^{*M}(t), \mathbf{a}_{TM}(t)]$

Proposition 3b. If $a_T(t) \neq a_L$, then, when the tenant faces moving costs, the lease is renewed iff $r_{TM}(a_L;t) \geq r_L(a_L;t)$ with the final contractual rent $r^*(t) \in [r_L(a_L;t), r_{TM}(a_L;t]$ for the mth cycle; if the rent is set by the landlord, then $r^*(t) = r_L(a_L;t)$; if rent is set by the tenant, then $r^*(t) = r_{TM}(a_L;t)$.

If the landlord faces rent loss in case of vacancy when the current tenant moves out before a new tenant moves in, then he computes his asking rent $r_L V(a_L;t) = \langle a_L, r_E(t) \rangle - V_E(t)$, and the results of Proposition 3b still applies. If follows that

$$(21) \quad \langle a_L, r_E(t) \rangle \in [r_{LV}(a_L;t), r_{TM}(a_L;t)]$$

and both transaction costs, moving and vacancy, will affect the future updating of the value of the vector of implicit (hedonic) rents.

3. Tenant's Search Time vs. Landlord's Management Time

If follows from the preceding section that at time $t \in [d_i, d_t; m]$, if $r_{TM}(a_L) < r_{LV}(a_L)$, both tenant and landlord are active agents in the rental housing market for the next mth cycle. Both are searching, the tenant searches to 'locate' the desired housing bundle, and the landlord tries to 'locate' the tenant whose desired bundle matches the one he owns.

At decision time t, the tenant in making a decision about the whole length of the mth cycle, computes

$$(22) \quad \{x^*(m), a_T(m), £^*(m)\} \leftarrow \text{Max}[u_T(x, a, £; m) \mid x + \langle a, r_E(t) \rangle = y_T(m)]$$

$$(23) \quad y_T(m) = [£(m) - £(m)] \; w_T(m)$$

where £ is the total time endowment for the cycle m, $£^*$ is the optimal allocation of that endowment for leisure for the same mth cycle, and w_T is his (exogeneous) contracted wage. Thus $£^*(m)$ is his leisure time endowment at $t_i(m)$, the initial time of the lease, ahead of the current decision time t, i.e., $t_i(m) > t = d_i(m)$. At the current decision time, during the current (m-1)th cycle, at the time when the tenant is starting to be active in the housing market for his mth cycle lease, his initial leisure time endowment $£^{*}(m-1)$ was depleted to a level $£(d_i(m) = £_o(m-1) << £^{*}(m-1)$.

After his first search, his initial endowment is decreased by the duration of the search s_1, i.e., $£_1(m-1) = £_0(m-1) - s_1$; and in general, after the kth search, the level of the endowment is given by $£K(m-1) = £_{K-1}(m-1) - s_K$. For $s_K = s_{K+!}$, we assume that the disutility of search time increases with the depletion of the endowment, i.e.,

(24) $-u_T(s_{K+!}) > -u_T(s_K)$.

During the kth search, the tenant locates the bundle of housing attributes and after he completes his information about bundle a_L, he computes his reservation rent as

(25) $r_{TK}(a_L) \leftarrow u_T[y_T(m)-r_{TK}(a_L), a_L, £_K(m-1)] =$

$$u_T[x^{*}(m), a_T(m), £_K(m-1)-s_{K+1}]$$

It follows from (29) that $r_{TK+1}(a_L)>r_{TK}(a_L)$, i.e., the reservation rent for the same bundle of housing attributes increases with the time spent by the tenant in the housing market.

Symmetrically, the landlord starts to act in the housing market for the mth cycle of his housing unit with an endowed management time $\mathfrak{z}_0(m-1)$. After the jth inspection time, his endowment is down by the jth wait, i.e., $\mathfrak{z}_j(m-1) = \mathfrak{z}_{J-1}(m-1) - w_J$. For $w_J = w_{J+1}$, we assume again that

(26) $-u_L(w_{J+1}) > -u_L(w_J)$

In general, after the jth inspection, the landlord computes his asking rent from

(27) $r_{LJ}(a_L) \leftarrow u_L[r_{LJ}(a_L), \Im_J(m-1)] =$

$$u_L[<a_L, r_E(t)>, \Im_J(m-1) - w_{J+1}]$$

It follows from (26) that $r_{LJ+1}(\mathbf{a}_L) < r_{LJ}(\mathbf{a}_L)$; in other words, the asking rent for the same bundle of housing attributes decreases with the time spent by the landlord in the housing market.

Proposition 4a. (Stochastic Equilibrium). If at current time t, the tenant in his kth search inspects the $\mathbf{a}_L$ bundle that was inspected j times by other active tenants, and iff $r_{LJ}(\mathbf{a}_L) < r_{TK}(\mathbf{a}_T)$, then the housing bundle is transacted, with the final contractual rent $r^*(m) = r_{LJ}(\mathbf{a}_L)$ if rent is set by the supply $r^*(m) = r_{LJ}(\mathbf{a}_L)$ if rent is set by the demand.

As we can see from Proposition 4a, the contractual rent, if set by demand, is independent of the marginal utility of the landlord's management time, and depends only upon the reservation rent distribution of the tenants, the depletion of their endowment by their acting in this market and the stochastic nature of their arrival at the location of the landlord's housing unit. If the contractual rent is set by the supply, it will be independent of the marginal utility of the tenant's leisure time, and depends only upon the distribution of asking rents of the active landlords in the rental market, the depletion of their management time by their acting in this market and the stochastic nature of locating their 'demand' for their housing bundles.

At the same transaction time t we define a set of reservation rents $\{\aleph(t)\}$ <- $\{r_{TK}(\mathbf{a}_L) > r_{LJ}(\mathbf{a}_L); T=1,2,.....,N;\ K=1,2,...\}$, from all the active tenants who would have a reservation rent larger than the asking rent, iff all the tenants would have complete information on $\mathbf{a}_L$. This set defines the reservation rents of all latent tenants for which the transaction condition holds for a particular landlord at transaction time.

Symmetrically, at the same time we define a set of asking rents $\{\beth(t)\}$ <- $r_{LJ}(\mathbf{a}_L) < r_{TK}(\mathbf{a}_L); L=1,2,.....,N;\ J=1,2,...\}$, from all the active landlords with the same bundle $\mathbf{a}_L$ who would have an asking rent smaller than the reservation rent, iff the tenant would have complete information on all $\mathbf{a}_L$'s.

This set defines the asking rents of all latent landlords for which the transaction condition holds for a particular tenant.

Proposition 4b. If the contractual rent is set by the demand, then the equilibrium condition $r^*(m) = \text{Sup}\{\aleph(t)\}$ should be changed into $r^*(m) \leq \text{Sup}\{\aleph(t)\}$. If the contractual rent is set by the supply, then the equilibrium condition $r^*(m) = \text{Inf}\{\beth(t)\}$ should be changed into $r^*(m) \geq \text{Inf}\{\beth(t)\}$.

If we assume either an infinitely large market, where exhausting all the options for both landlords and tenants is not feasible; or, if the market is small, there is no memory in the system, then from (26) and (27) we can state

Proposition 4c. In a market where the number of tenants is never larger than the number of housing units in the standing stock, if the adjustment process starts in desequilibrium, i.e., $\mathbf{a}_T \neq \mathbf{a}_L$ for all L and all T, and the contractual rent is set by the supply, then for every cycle m and sample s, $\Delta\{r^*(s).\mathbf{a}_L(s) \mid s \in m\} < 0$, and in the long run, $\lim_{t\to\infty} \mathbf{r}_E = 0$.

4. Monte Carlo Simulation

In this section we show the existence of a non-instantaneous equilibrium for our model, by constructing a microsimulation example with the properties described in the preceeding section. This exercise also helps to further characterize the heterogeneous nature of the housing market, when rents and hedonic rents are unique and costlessly known to both sides of the market, but the search of the housing attributes' bundles and their location is costly to both sides of the market in terms of assets depletion.

To generate a numerical solution, it is necessary to give a specific different form to the utility function of each individual of the n tenants in the active housing market. In order to concentrate on the central core of our model, we assume that every tenant has the same income, and when maximizing their utility function with a common set of prices and hedonic rents, we generate a 'demand' vector of n different bundles of housing attributes $\mathbf{a}_{TJ}$ (j = 1,2,..,n), that are, just by design, having the same encumbered rent, i.e.

(28) $$<\mathbf{a}_{Tj}, \mathbf{r}_E> = <\mathbf{a}_{Tj+1}, \mathbf{r}_E> \qquad j = 1,2,...,n$$

We assume further that matching exactly the 'demand' vector, there are n landlords offering n different bundles of housing attributes, generating a 'supply' vector $\mathbf{a}_{Li}$ (i = 1,2,...,n), indices ordered in such a way, that

(29) $$\mathbf{a}_{Li} = \mathbf{a}_{Tj} \qquad \text{iff } i = j$$

The ordering of indices facilitates the reading of the simulation output. It follows from (28) that all the initial asking rents are

(30) $$<\mathbf{a}_{Li}, \mathbf{r}_E> = <\mathbf{a}_{Li+1}, \mathbf{r}_E> \qquad i = 1,2,..,n$$

and from (4) and (5) of section 2, it follows that without transaction costs, we have as initial condition of the simulation run, k = 0

(31) $$r_{Lo}(\mathbf{a}_{Li}) = r_{Tjo}(\mathbf{a}_{Li}) = C \qquad \text{iff } i = j$$

(32) $$r_{Lo}(\mathbf{a}_{Li}) > r_{Tjo}(\mathbf{a}_{Li}) \qquad \text{iff } i \neq j$$

For the numerical analysis presented here we choose n = 11 and C = \$ 3,000. The initial reservation rent matrix $r_{Tkj}(\mathbf{a}_{Li})$ for i,j = 1,2,..,n and kk = 0 was computed and the results presented in Table 1. For n very large, no matrix is computed: at each Monte Carlo run only the appropriate cells are computed from a library function.

At each simulation run two vectors of n random integers out of n integers (without replacement) is generated: one vector is related to the tenants' indices, the other to the landlords' indices. Both vectors articulates the random search, assigning the jth tenant to the ith landlord in the kth run.

Once assigned, the model computes the tenant's reservation rent $r_{Tjk}(a_{Li})$ and the landlord's asking rent $r_{Lk}(a_{Li})$ in the kth run, $k > 0$.

If the following inequality holds

(33) $r_{Tjk}(a_{Li}) > r_{Lk}(a_{Li})$

then a transaction occurs, with the contractual rent arbitrarily defined here as

(34) $r(i,j) = \inf\{r_{Tjk}(a_{Li}), r_{Lik}(a_{Li})\}$

and the ith landlord and the jth tenant are out of the active market, moving into the inactive market at the end of the kth run.

If the following inequality holds

(35) $r_{Tjk}(a_{Li}) < r_{Lk}(a_{Li})$

then, the ith landlord and the jth tenant stay in the active market during the (k+1)th run. For all the numerical experiments but one, we used the following specific forms for the reservation and asking rents

(36) $r_{Tjk}(a_{Li}) = r_{Tjo}(a_{Li}) + 2(k-1)^{1.3}$

(37) $r_{Lk}(a_{Li}) = \langle a_{Li}, r_E \rangle - 2(k-1)^{1.2}$

with $r_{Tjo}(a_{Li})$ given by Table 1.

In order to observe the effects of increased level and variability of the transaction costs, we used the following specific form for the fourth experiment:

(38) $r_{Tjk}(a_{Li}) = r_{Tjo}(a_{Li}) + (n+1-j)^{k-1}$

Table 1: Reservation Rents at k = 0

i\j	1	2	3	4	5	6	7	8	9	10	11
1	3000	2993	2972	2937	2889	2828	2752	2661	2552	2424	2271
2	2993	3000	2993	2972	2938	2891	2830	2754	2661	2549	2415
3	2971	2993	3000	2993	2973	2939	2892	2830	2752	2657	2541
4	2935	2972	2993	3000	2993	2973	2940	2892	2829	2749	2649
5	2885	2937	2973	2993	3000	2993	2973	2939	2891	2826	2743
6	2821	2888	2938	2973	2993	3000	2993	2973	2938	2888	2821
7	2743	2826	2891	2939	2973	2994	3000	2993	2973	2937	2885
8	2649	2749	2829	2892	2940	2973	2993	3000	2993	2972	2935
9	2541	2657	2752	2830	2892	2939	2973	2993	3000	2993	2971
10	2415	2549	2661	2754	2830	2891	2938	2972	2993	3000	2993
11	2271	2424	2552	2661	2752	2828	2889	2937	2972	2993	3000

(39) $r_{Lk}(a_{Li}) = \langle a_{Li}, r_E \rangle - i^{k-1}$

If we compare the equilibrium results of the six Monte Carlo experiments presented here, we can see the relationship between transaction costs and 'size' of the housing market on the one hand; and the relationship between transaction costs and different degree of heterogeneity of the market on the other hand.

Comparing the first with the second experiment, when the 'size' of the market increases with 83%, the average loss in transacted rents to the landlord increases with 140%. Comparing the second with the third experiment, we can see the effects of bundle variability, 'size' of the market held constant when the 'distance' between neighboring bundles of attributes increases with 100%, the average loss of transacted rents to the landlord increases with 40%. These losses have dramatically increased in the fifth and sixth experiment.

In comparing the fourth experiment with the first, we can observe the effects of different speed of assets depletion.

Equilibrium Results
(see Tables 2 to 7)

First Experiment n = 11
i,j=1,2,3,4,5,6,7,8,9,10,11

i	j	k	Transacted rent	Asking rent next cycle
1	1	3	2995	2991
3	2	4	2993	2990
4	3	3	2995	2990
5	4	3	2995	2989
6	5	4	2993	2989
2	6	15	2953	2988
7	7	6	2986	2988
8	8	3	2995	2987
10	9	6	2986	2987
9	10	3	2995	2986
11	11	5	2989	2986
Average			2988	2988

Second Experiment n = 6
i,j=1,2,3,4,5,6

i	j	k	Transacted rent	Asking rent next cycle
1	1	3	2995	2995
2	2	4	2993	2995
3	3	2	2998	2995
4	4	2	2998	2995
6	5	4	2993	2995
5	6	3	2995	2995
Average			2995	2995

Fifth Experiment n = 6
j=1,3,5,7,9,11
i=2,4,6,8,10,11

i	j	k	Transacted rent	Asking rent next cycle
2	1	3	2995	2986
4	5	4	2993	2987
6	3	10	2972	2988
8	7	4	2993	2989
10	9	4	2993	2990
11	11	5	2989	2991
Average			2989	2989

Fourth Experiment n = 11
i,j=1,2,3,4,5,6,7,8,9,10,11

i	j	k	Transacted rent	Asking rent next cycle
3	1	4	2999	3181
1	2	2	2998	3115
4	3	3	2991	3049
5	4	3	2984	2984
2	5	3	2975	2918
8	6	2	2964	2852
6	7	2	2993	2787
9	8	2	2992	2721
10	9	3	2919	2655
7	10	3	2900	2590
11	11	4	1669	2524
Average			2853	2853

Third Experiment n = 6
i,j=1,3,5,7,9,11

i	j	k	Transacted rent	Asking rent next cycle
1	1	3	2995	2995
3	3	5	2989	2994
5	5	2	2998	2993
7	7	2	2998	2992
9	9	5	2989	2991
11	11	5	2989	2990
Average			2993	2993

Sixth Experiment n = 6
j=1,2,3,4,5,6
i=6,7,8,9,10,11

i	j	k	Transacted rent	Asking rent next cycle
6	6	2	2998	3026
7	5	5	2979	2969
8	4	15	2953	2912
9	2	41	2833	2799
10	3	41	2833	2856
11	1	64	2712	2742
			2884	2884

Table 2: First Experiment n = 11

k=1

j	7	2	11	4	10	3	5	9	8	1	6
i	6	1	10	5	7	9	8	2	11	3	4
L	3000	3000	3000	3000	3000	3000	3000	3000	3000	3000	3000
T	2993	2993	2993	2993	2937	2752	2940	2660	2935	2971	2973
r(i,j)											

k=2

j	9	11	3	4	6	8	1	7	5	2	10
i	5	4	8	7	10	9	11	6	3	1	2
L	2998	2998	2998	2998	2998	2998	2998	2998	2998	2998	2998
T	2893	2651	2831	2941	2893	2995	2273	2995	2975	2995	2551
r(i,j)											

k=3

j	3	9	6	8	10	5	1	11	2	7	4
i	4	6	10	8	9	7	1	2	11	3	5
L	2995	2995	2995	2995	2995	2995	2995	2995	2995	2995	2995
T	2998	2943	2896	3005	2998	2978	3005	2420	2429	2897	2998
r(i,j)	2995			2995	2995		2995				2995

k=4

j	5	6	11	9	7	2
i	6	11	7	2	10	3
L	2993	2993	2993	2993	2993	2993
T	3001	2835	2893	2668	2946	3001
r(i,j)	2993					2993

k=5

j	7	9	11	6
i	2	7	11	10
L	2989	2989	2989	2989
T	2842	2985	3012	2903
r(i,j)			2989	

k=6

j	6	9	7
i	2	10	7
L	2986	2986	2986
T	2907	3009	3016
r(i,j)		2986	2986

k=15

j	6
i	2
L	2953
T	2953
r(i,j)	2953

Table 3: Second Experiment n = 6

k=1

j	1	5	3	4	2	6
i	3	4	5	1	6	2
L	3000	3000	3000	3000	3000	3000
T	2971	2993	2973	2937	2888	2891
r(i,j)						

k=2

j	1	5	2	4	6	3
i	2	6	5	4	1	3
L	2998	2998	2998	2998	2998	2998
T	2995	2995	2939	3002	2830	3002
r(i,j)				2998		2998

k=3

j	5	2	1	6
i	2	6	1	5
L	2995	2995	2995	2995
T	2943	2893	3005	2998
r(i,j)			2995	2995

k=4

j	5	2
i	6	2
L	2993	2993
T	3001	3008
r(i,j)	2993	2993

Table 4: Third Experiment n = 6

k=1

j	1	9	5	7	3	11
i	5	7	9	1	11	3
L	3000	3000	3000	3000	3000	3000
T	2885	2973	2892	2752	2552	2541
r(i,j)						

k=2

j	1	9	3	7	11	5
i	3	11	9	7	1	5
L	2998	2998	2998	2998	2998	2998
T	2972	2974	2754	3002	2273	3002
r(i,j)				2998		2998

k=3

j	9	3	1	11
i	3	11	1	9
L	2995	2995	2995	2995
T	2757	2557	3005	2976
r(i,j)			2995	

k=4

j	9	11	3
i	11	3	9
L	2993	2993	2993
T	2980	2549	2760
r(i,j)			

k=5

j	11	9	3
i	11	9	3
L	2989	2989	2989
T	3012	3012	3012
r(i,j)	2989	2989	2989

Table 5: Fourth Experiment n = 11

k=1

j	7	2	11	4	10	3	5	9	8	1	6
i	6	1	10	5	7	9	8	2	11	3	4
L	2999	2999	2999	2999	2999	2999	2999	2999	2999	2999	2999
T	2994	2994	2994	2994	2938	2753	2941	2661	2936	2972	2974
r(i,j)											

k=2

j	9	11	3	4	6	8	1	7	5	2	10
i	5	4	8	7	10	9	11	6	3	1	2
L	2991	2989	2997	2996	2994	2992	2999	2993	2995	2998	2990
T	2897	2657	2833	2946	2893	2996	2272	2999	2982	3004	2559
r(i,j)						2992		2993		2998	

k=3

j	3	9	6	10	5	1	11	4
i	4	10	8	7	2	11	3	5
L	2991	2919	2964	2900	2975	2999	2879	2984
T	3057	2996	2989	2962	3038	2272	2621	3042
r(i,j)	2991	2919	2964	2900	2975			2984

k=4

j	11	1
i	11	3
L	1669	2999
T	2999	3792
r(i,j)	1669	2999

Table 6: Fifth Experiment n = 6

k=1

j	1	5	7	9	11	3
i	6	10	8	2	4	11
L	3000	3000	3000	3000	3000	3000
T	2821	2830	2993	2661	2649	2552
r(i,j)						

k=3

j	9	11	3	1	7	5
i	4	6	8	2	11	10
L	2995	2995	2995	2995	2995	2995
T	2833	2826	2834	2998	2894	2831
r(i,j)				2995		

k=4

j	5	7	9	11	3
i	4	8	10	6	11
L	2993	2993	2993	2993	2993
T	3001	3001	3001	2829	2560
r(i,j)	2993	2993	2993		

k=5

j	3	11
i	6	11
L	2989	2989
T	2950	3012
r(i,j)		2989

k=10

j	3
i	6
L	2972
T	2973
r(i,j)	2972

Table 7: Sixth Experiment n = 6

k=1

j	1	5	3	4	2	6
i	8	9	10	6	11	7
L	3000	3000	3000	3000	3000	3000
T	2649	2892	2661	2973	2424	2994
r(i,j)						

k=2

j	1	5	2	4	6	3
i	7	11	10	9	6	8
L	2998	2998	2998	2998	2998	2998
T	2745	2754	2551	2832	3002	2831
r(i,j					2998	

k=8

j	2	5	4	1	3
i	10	7	9	11	8
L	2979	2979	2979	2979	2979
T	2574	2998	2855	2296	2851
r(i,j)		2979			

k=15

j	4	2	3	1
i	8	10	11	3
L	2953	2953	2953	2953
T	2954			
r(i,j)	2953			

k=41

j	3	2	1
i	10	9	11
L	2833	2833	2833
T	2898	2898	
r(i,j)	2833	2833	

k=64

j	1
i	11
L	2712
T	2712
r(i,j)	2712

5. Concluding Remarks

The theoretical contribution presented here support the development of a non-instantaneous and stochastic equilibrium model of the rental housing market, in the presence of perfect and costless information about implicit rents but with transaction costs associated with the extreme variation of housing bundles and with the location of the bundle of housing attributes.

Housing attributes and location search time belong taxonomically to the four different types of asset specificity: site specificity; physical asset specificity; human asset specificity and dedicated asset specificity. Asset specificity belongs to the predictive core of transaction costs economics.

The model is sufficiently rich to produce endogeneous distribution and dispersion of transaction rents, and, being based on individual utility-maximizing behavior, yields decision rules for both tenants and landlords in the presence of search costs. Also, this model was purposely designed to allow a simple symmetric examination of the depletion of one very important asset: the leisure/management time.

The new asking rents for the next leasing cycle gives structure to the issue of moving/vacancy costs in renewing contracts.

Appendix

In the update of estimated housing implicit rents $\mathbf{r}_E$ we use a weighted least-squares estimator to specify the general function

(A1) $$\mathbf{r}_E(t) = \mathbf{r}_E(t_0) + \Delta\{r^*(t), \mathbf{a}_L(t) \,|\, \forall t \geq t_o\}$$

We specify for every t a stochastic linear relationship, with e as an error term as in equation (11). We define the vectors

(A2) $$\mathbf{r}^*(t) = [r^*(t_0), \ldots\ldots\ldots, r^*(t)]$$

(A3) $$\mathbf{e}(t) = [e(t_0), \ldots\ldots\ldots\ldots, e(t)]$$

and we define the transpose matrix

(A4) $$\mathbf{A}^T(t) = [\mathbf{a}_L(t_0), \ldots\ldots\ldots, \mathbf{a}_L(t)]$$

then it follows that the updating sample can be defined as

(A5) $$\mathbf{r}^*(t) = \mathbf{A}(t)\ \mathbf{r}_E(t) + \mathbf{e}(t)$$

We assume that $E(\mathbf{e}) = 0$, $E<\mathbf{e}, \mathbf{e}^T> = \mathbf{N}$ and $E<(\mathbf{r}_E(t)-\mathbf{r}_E(t_0)), (\mathbf{r}_E(t)-\mathbf{r}_E(t_0))^T > = \mathbf{M}$, where **N** and **M** are known square positive matrices. The weighted least square estimate of $\mathbf{r}_E(t)$ is given by

(A6) $$\text{Min } \mathbf{z} = \tfrac{1}{2}[(\mathbf{r}_E(t)-\mathbf{r}_E(t_0))\ \mathbf{M}^{-1}\ (\mathbf{r}^*(t)-\mathbf{r}_E(t_0))^T + (\mathbf{r}^*(t)-\mathbf{A}(t)\mathbf{r}_E(t))^T\ \mathbf{N}^{-1}\ (\mathbf{r}^*(t)-\mathbf{A}(t)\mathbf{r}_E(t))]$$

(A7) $$\partial \mathbf{z}/\partial \mathbf{r}_E(t)=0=[\mathbf{M}^{-1}(\mathbf{r}_E(t)-\mathbf{r}_E(t_0))\ -\mathbf{A}^T(t)\mathbf{N}^{-1}(\mathbf{r}^*(t)-\mathbf{A}(t)\mathbf{r}_E(t)]$$

$$\text{(A8)} \quad \mathbf{r}_E(t) = \mathbf{r}_E(t_0) + \Delta(t)\ (\mathbf{r}^*(t) - \mathbf{A}(t)\mathbf{r}_E(t))$$

$$\text{(A9)} \quad \Delta(t) = \mathbf{P}\ \mathbf{A}(t)\ \mathbf{N}^{-1}$$

$$\text{(A10)} \quad \mathbf{P}^{-1} = \mathbf{M}^{-1} + \mathbf{A}^T(t)\ \mathbf{N}^{-1}\ \mathbf{A}(t)$$

where **M** is the error covariance matrix before the recent market sample; and **P** is the error covariance matrix after the rent market sample. As continuous sampling decreases the uncertainty of our knowledge of the housing hedonic rents, it follows from the statistical linear filter theory that $(\mathbf{r}_E)^T\ \mathbf{P}\ \mathbf{r}_E < (\mathbf{r}_E)^T\ \mathbf{M}\ \mathbf{r}_E$.

From (A10) and through appropriate matrix reduction, we have

$$\text{(A11)} \quad \mathbf{M} = \mathbf{B}\ \mathbf{P}\ \mathbf{B}^T + \mathbf{C}\ \mathbf{N}\ \mathbf{C}^T$$

In a statistically stationary process $\lim_{t->\infty} \mathbf{M}(t) = \mathbf{M}$ and $\lim_{t->\infty} \mathbf{P}(t) = \mathbf{P}$, i.e., stochastically steady state, and the 'new' information on implicit rents is balanced by the 'old' information on implicit rents.

If we can express $\mathbf{N} = q[n_{ij}]$, and if $\lim_{t->\infty} q(t) = 0$, then the best estimate of the implicit rents tends to the average of all measurements, and the initial estimate $\mathbf{r}_E(t_0)$ influences the best current estimate; if $\lim_{t->\infty} q(t) = \infty$, then the influence of the initial estimates fades and the best current estimate is from the latest segment of the active market, i.e.,

$$\text{(A12)} \quad \mathbf{r}_E(t) = (\mathbf{A}^T(t)\ \mathbf{A}(t))^{-1}\ \mathbf{A}^T(t)\ \mathbf{r}^*(t).$$

References

Akerlof, G. (1980), A Theory of Social Custom, of Which Unemployment May Be One Consequence, Quarterly Journal of Economics 99:749-775.

Alchian, A. and H. Demsetz (1972), Production, Information Costs and Economic Organization, American Economic Review 62:777-795.

Coase, R.H. (1937), The Nature of the Firm, Economica 4:386-405.

Coase, R.H. (1960), The Problem of Social Cost, Journal of Law and Economics 3:1-44.

Fisch, O. (1977), Dynamics of the Housing Market, Journal of Urban Economics 4:428-447.

Fisch, O. (1986), A Neoclassical Model of Housing Quality Growth: An Interpretation, Journal of Regional Science 26,1:103-119.

Rosen, S. (1974), Hedonic Prices and Implicit Markets, Journal of Political Economy 82,1:34-53.

Stiglitz, J.E. (1985), Information in Economic Analysis: A Perspective, Economic Journal (Supplement) 95:21-40.

Sweeney, J.L. (1974), Quality, Commodity Hierarchies and Housing Markets, Econometrica 42:147-167.

Williamson, O.E. (1985), The Economic Institutions of Capitalims, Free Press.

Chapter 6

TRANSPORTATION, SPATIAL INTERACTION, TELECOMMUNICATION AND INFORMATION SYSTEMS: A RESEARCH AGENDA

T.J. Kim
K. Choi

1. Introduction

The general view in a mixed economy is that some goods and services are produced privately and some, such as transportation, are produced publicly. Private institutions, such as households and entrepreneurs, produce and consume goods and services in pursuing their parochial interest, while the public sector attempts to broaden public interests. Thus, at least two parties are involved in the decision-making processes regarding transportation: the public sector that constructs new transportation systems, improves their capacities, and regulates services and prices; and the private sector that chooses the locations of production, modes of transportation, and routes of shipment. At the same time, all forms of transportation influence our lives and give us cause for concern for our environment, health, and safety. Although transportation is intimately interwoven with the daily lives of all individuals and organizations in our society, it is easy to overlook its significance until it fails us in some way (Schofer and Boyce, 1985).

The amount of investment in transportation infrastructure in any country, particularly in developed countries, is immense. The impact of transportation on efficiency, productivity, and competitiveness of a nation's economy is also tremendous. In the United States, the transportation industry, including service delivery and manufacturing, contributes more than 20% of the Gross National Product (GNP).

In the light of these and other indications that suggest the critical importance of transportation to the quality of life, this paper attempts to assess the past and current developments in transportation, particularly in regard to its spatial interaction, telecommunication and intelligent information systems, in order to better identify future research agenda. Discussions on future research opportunities will be focused on the following two categories:

1. existing transportation issues which continue to pose as challenging research opportunities;
2. new issues brought about particularly by the advancing telecommunication and information technology.

An excellent collection of reports and a summary of the state-of-the-art research opportunities on issues in the first category can be found in a special issue of Transportation Research (Boyce, 1985). In this issue, the results of the March 1985 Evanston Workshop on Transportation Research are reported, including assessments on recent developments in transportation systems and discussions of research opportunities in the future.

Rather than duplicating what has already been analyzed and suggested carefully, this paper will briefly review issues in the first category and will focus on those in the second one. The paper is organized as follows. In the next section changes in technology and life styles in relation to transportation are discussed. In Section 3, past and current research in transportation systems from the supply side perspective are assessed and future research opportunities are discussed. In Section 4, research on transportation and spatial interaction from the demand side is reviewed and future research opportunities are identified. Following brief discussions on transportation and artificial intelligence in Section 5, research and transportation, telecommunication, and artificial intelligence systems is analyzed and future research directions are suggested in Section 6. Section 7 summarizes the paper briefly.

2. Changes in Technology and Lifestyles

Present urban and regional systems are complex. Spatial analysis models cannot produce useful recommendations to decision makers unless they are founded with an understanding of the continuously changing spatial structures and the behaviour of the actors involved in the decision-making process.

Research on models and other analytical tools is important for understanding, predicting, and evaluating the performance of spatial systems. At the same time, research on changes in a spatial system is essential to guide the development of useful tools and to provide a theoretical and logical foundation for the choice and application of tools in a particular spatial setting. A good example is the applicability of monocentric urban models analyzing current socio-economic issues in metropolises where the role of CBD has been diminished by the increasing economic role of suburban centers (Richardson, 1988). Thus, it is critically important to understand changes in lifestyles and their impacts on spatial structures to develop sound prescriptions for urban problems.

Continuous increase in the female labor force participation rate, the declining birth rate, changing household size, shifted economic status among social and ethnic groups, and continuing decentralization of economic activities from the traditional CBD are a few examples of changes in lifestyles that will have a profound impact on the future form, function, and structure of urban and regional systems. Undoubtedly, these changes, along with changes in the demographic and social patterns, affect urban locational behavior and travel demand in urban areas.

Changes in technology also have profound impacts both on lifestyles as well as the functions and structures of urban and regional systems. Several computer companies, including IBM and Kurzweil Applied Intelligence Inc. are developing new computers known as 'talkwriters' that understand spoken words and print them on a screen. Having an electronic map on the dashboard of a car is no longer technically improbable, but is only unpopular because of its cost. Some emergency vehicles in California have devices on their dashboards showing the drivers how to get to their destinations. Currently, the device costs around $ 3,000. Recent developments in superconductivity may speed up the development of trains that go 250 mph or more, which is as fast as some planes. High-speed trains called 'MAGLEVS' which levitate on a cushion of magnetism, are being developed in Germany and Japan (USA Today, December 28, 1988).

Easy accessibility to telecommunication facilities, including microcomputers, facsimile machines, and picture-phones may also affect urban locational and travel behaviors. Some travelling may be substituted by using one or more of these telecommunication means. People who do not need frequent face-to-face contacts in their production activity may choose to locate further from the CBD where traditional modes of transportation is not easily accessible. Based on the data from the Nationwide Personal Transportation Study (NPTS) conducted by the U.S. Department of Transportation in 1977 and 1984, Kumar (1989) showed that commuting distances of workers in information-dominated industries in metropolitan areas have been declining. This suggests that some of the work trips might have been substituted by telecommuting.

In fact, the last two decades of rapid advances in information and computer technology have led some analysts to claim that mankind is now entering into an 'information society' (Bell, 1979; Rogers, 1986). With the growing use of telecommunication systems, it is becoming possible for different productive activities such as manufacturing, distribution, management, accounting, and inventory control to be located at different places and yet are operated efficiently as a single enterprise (Kumar, 1989). This emergency trend may be represented by new forms of economic agglomerations in a noncontiguous manner. Coopers and Lybrand estimated that electronic messages will soar from \$ 2.4 billion in 1988 to \$ 8 billion in 1991 (USA Today, December 28, 1988).

Based on these changes, which have significantly affected the shaping of urban and regional functions and structures, this discussion will now assess emerging research issues in transportation in the three categories:

1. a supply side view of transportation systems;
2. a demand side view of transportation and spatial interaction; and
3. a technological development view of intelligent information systems on transportation.

3. Transportation Systems: A Supply Side View

There are several excellent works available that provide the basic framework for understanding the supply side of transportation services, including Morlok (1976), Manheim (1980), and Florian and Gaudry (1980). While there are various views of the supply side of transportation, Turnquist (1985) summarized that the supply side of transportation involves both the technology of providing services and the management actions of the suppliers.

3.1 Transportation system characteristics and operations

An important point to note here is that we have focused on person trip movements for the past thirty years and have paid little attention to understanding the urban goods movement. Likewise, only a few authors paid attention to the problems of vehicle fleet sizing and empty vehicle distribution for trucks and rail cars (Jordan and Turnquist, 1983; Haghani and Daskin, 1984; Jordan and Burns, 1984; Powell, Sheffi, and Thiriez, 1984). The problem of vehicle size is especially acute in developing countries, where overloading practice is sometimes beyond imagination.

Implications of such overloading practice in transportation planning is that the tonnage shipped and the maximum capacity of vehicles passing through certain checkpoints diverges widely. At the same time, network capacity investment planning based on the number of vehicles forecasted (not considering vehicle fleet size which accounts for the overloading) would result in smaller amounts of capacity increase. In addition, overloading not only harms vehicles, but also damages the road systems. In fact, improvement of road conditions has become a major concern for many developing countries in the 1980s. In many developing countries, roads have deteriorated in the past decade due to their carrying heavier traffic loads than were originally intended for them, and funds have been allocated for expanding rather than maintaining the network (World Bank, 1983).

3.2 Transportation logistic system

Transportation logistics can be defined as the study of how manufactured goods overcome time and distance in the process of being transformed from raw materials to consumed products (Daskin, 1985). Logistics is an important component of the economy, as can be seen in the U.S., where logistics costs have been estimated at between 15 and 23 percent of the GNP, depending on whether or not service industries are included in the GNP.

In many logistical studies, the emphasis has been on the inter-action of shippers and carriers and their decision-making processes. The possibility of modelling the interaction of shippers and carriers in bilevel programming frameworks has been suggested and new findings in algorithms for solving a large-scale bilevel programming problems are reported in Suh (1989) and Kim (1990).

Expert systems and neural systems, two branches of artificial intelligence, have emerged as promising tools for developing integrated models of several logistical functions such as location, fleet selection, inventory, production, marketing, and vehicle routing. An interesting application of an expert system is to search for the optimal routing interactively and to solve routing problems in real time by combining an expert system with an optimization routing model (Golden and Baker, 1985). A detailed explanation of the possibility of combining expert systems with other existing tools can be found in Han and Kim (1989) and will be introduced in the latter part of this paper.

3.3 Transportation network and facilities

Research opportunities on transportation facilities can be highlighted by the fact that about 44 percent of 565,000 bridges on the U.S. street and highway system are considered deficient. There are about 3.9 million miles of streets and highways and about 266,000 miles of railroad lines in which deficiencies have been developing at a rate faster than funds have become available to maintain it. Current estimates of costs to repair, rebuild, and upgrade public works in which transportation facilities of all kinds account for a major portion of these needs range from one trillion to three trillion dollars over the next twenty years (Moavenzadeh, 1985). Research issues and problem areas in transportation facilities are well analyzed in both papers by Moavenzadeh (1985) and Garrison (1985).

4. Transportation and Spatial Interaction: A Demand Side View

Among the urban and regional systems models developed thus far, only a few models have explicitly addressed some of the important characteristics of urban areas identified above. Mills (1982) first introduced a model that explicitly analyzes the intensive use of urban land employing a three-dimensional input-output model. He also introduced a method to analyze transportation congestion first conceptually (Mills, 1972) and then numerically (Mills, 1976) using the integer programming technique. The transportation networks discussed in Mills' models, however, are abstract forms of real-world transportation networks. The form of public-private interaction in Mills' models is represented in an optimization framework that permits calculation of an optimum allocation of both public and private resources. Furthermore, Mills' models demonstrated that competitive markets can sustain an optimum allocation of resources if the public sector provides its services in an optimum fashion.

A method for determining the transportation service requirement for both passengers and freight combined was first introduced by Leontief and Strout (1963) in a regional context and later expanded by Wilson (1970a, 1970b). In these models, Wilson used an input-output structure for the production and an entropy maximization for the transportation sector.

Erlander (1977) introduced the notion of 'dispersion factors', which can now simulate the fact that a city has been incrementally built and that the present form and structure of a city, have deviated from its optimum form and structure. He used entropy as a constraint instead of maximizing it, compared to Wilson (1970a) who used it in the objective functions. Kim (1989) introduced an extended model with a nonlinear structure by combining Wilson's commodity flow model, Erlander's dispersion factors, and Mills' urban activity model.

While past and current research on combined models of urban activities has been limited, there has been a considerable amount of related research that provides a basis for building combined urban activity models and developing efficient solution algorithms.

Of the existing urban systems models that are formulated to guide public sector planning for future urban growth, most are inadequately articulated to permit determination of whether market solutions are efficient or, if not, what kind of public intervention would be best to ensure efficiency. Such models derive the urban spatial patterns from conditions of supply and demand, profit and utility maximization, and market equilibrium without optimality criteria.

Improvements have been made in models by Mills (1972, 1976) and Kim (1978, 1979), and Moore and Wiggins (1990) and Moore and Gordon (1990) in handling public-private interaction. Even in these models, however, a single objective criterion was assumed to represent society's welfare despite the fact that the objectives of the public and private sectors are frequently different, if not conflicting. While the need for the valuative criteria is prevailing for the assessment of public policies, the applications of a single valuative criterion is increasingly complex. After all, one of the critical questions in a mixed economic system is how much the public sector should intervene in order to enhance the market efficiency. Simultaneous representation of different and/or conflicting objectives in building urban and regional systems models should be the subject of continuing research.

Although the rational decision-making models have provided useful recommendations for public policy, it is important to recognize that there are other useful perspectives on decision making, particularly on public decision-making behavior. Linstone (1984) argues that alternative perspectives to the rational decision-making procedure exist, that viewing problems through several perspectives is like viewing an object through several different lenses, and that multiple objectives can clarify the analysis of any area of policy planning. Allison (1971) introduced the notion of multiple perspectives in policy making; and Steinbrunner (1974) applied the same basic concept to an analysis of planning a proposed multinational force for NATO. Nijkamp (1977) provides comprehensive surveys of multiobjective programming methods.

Wachs (1985) recognized that one of the major attributes of organizational perspectives in decision making is that there are multiple decision-making units, each is pursuing its own goals, and each interested in enhancing its status. As described previously, there are at least two actors making decisions affecting urban systems in a mixed economic system: the public and private sectors. Within public sectors, there are many different units making decisions that affect urban systems: the federal government, the state governments, and various local units. The foregoing analyses strongly suggest the formulation of urban systems models in a multilevel programming framework. A special case where only two different decision units are involved is called a bilevel programming problem, the nature and structure of which are introduced in Suh and Kim (1989).

While development of efficient solution algorithms should be the subject of continuing research (see Kim, 1990, for a detailed solution algorithm), the multilevel programming approach has many features that are not found in other techniques.

1. Different, if not conflicting, objectives or two or more decision makers and/or different units can be simultaneously analyzed. It implies that the public-private interaction explicitly can be evaluated.
2. Multiple valuative criteria can be explicitly included in a model, compared to a single valuative criterion in conventional programming approaches. In this view, a multicriteria decision-making method can provide realistic policy recommendations.
3. The multilevel programming approach is a conflict resolution method because the approach can accommodate conflicting objectives. Many conflicting public objectives, such as economic development and environmental preservation in an urban area, can now be evaluated in terms of broader societal goals.

Further research on bilevel and multilevel programming approaches and their applications to urban and regional systems analysis is one area in which research promises major payoffs.

5. Transportation and Artificial Intelligence

The salient characteristic of the computer, its enormous capacity and speed in storing, accessing, and processing data, has made it an indispensable tool for urban decision makers who deal with varied information in their problem-solving tasks.

Recent advances in artificial intelligence (AI) technology raise interests in how existing urban systems modeling techniques might benefit from AI. Artificial intelligence is not a technology that solves urban problems by itself. Rather, it is a newly emerging and promising technology that can be incorporated or integrated into the existing urban systems models to provide more intelligent and effective solutions to urban problems. Ongoing AI research shows the possibilities of developing intelligent computer systems that are able to reason and learn by themselves on a certain domain as time goes by. Such an **adaptive** learning characteristic of the intelligent computer systems can be fully exploited in order to incorporate the nature of the dynamic changes in urban areas over a period of time. As decision makers' tasks become more complex, the role of computers and computerized information systems becomes increasingly important to urban systems analysis (Kim, Wiggins and Wright, 1989). In the face of uncertainty and dynamic changes in urban society, the rational decision-making model alone may not be adequate for responding to continuously challenging urban problems. Indeed, the model may have to be supplemented by additional perspectives based on the following attributes of decision makers' behavior in an organization (Wachs, 1985):

1. Short-term consequences of decisions are emphasized over long-term consequences.
2. Innovation is less important than organizational security and enhancement.

3. Preferred policies are those most easily absorbed into an organization's on-going values and programs.
4. Greater reliance on relationships than on formalized technical information.
5. Decisions are preferred on the basis of the extent to which they enhance one's role in an organization or one's standing in a profession.
6. Intuition and rules of thumb are at least as important as analysis.
7. Memory and personal experience are more important than technical analysis.
8. Decision processes focus on simple hypotheses and a few important alternatives.

Few would dispute the importance of these decision-making attributes in developing alternative strategies for urban problem solving. While the conventional spatial systems modeling techniques cannot handle these important attributes, expert systems and neural systems may be able to do so. While it is beyond the scope of this paper to describe those intelligent computer systems in detail, existing urban systems models can be attached to artificial intelligent systems in which rule-based reasoning, qualitative simulation and new knowledge elicitation and updating through self-modifications are tied into quantitative systems analysis.

6. Intelligent Computer Systems

6.1 Knowledge-based expert systems

In the decision-making process, intelligence may be more useful to planners than data or information. According to Catanes (1979) and Harris (1987), intelligence is defined as **meaningful information derived from catching the essential factors from complex data to provide a guide to action.**

In general, the expert system is defined as 'an intelligent computer system that uses a representation of human expertise in a specialty domain in order to perform functions similar to those normally performed by a human expert in that domain' (Goodall, 1985). The essence of expert systems is that they attempt to incorporate the judgement, experience, rules-of-thumb, and intuition of human experts into problem solving.

The efforts made by several authors to distinguish expert systems from conventional programming are useful in understanding the nature of expert systems. According to Waterman (1986), 'an algorithmic method of conventional programming is designed to produce optimal solutions, whereas the heuristic method of expert systems produces an acceptable solution most of the time.' Thus, the problems suitable for expert system development must require heuristic solutions rather than algorithmic solutions. Such problems in spatial systems analysis may include finding acceptable alternatives while considering those attributes described above. Or it may include settling a lawsuit in which heuristics and rules-of-thumb can effectively limit the search for solutions.

As depicted in Figure 1, combining expert systems and urban systems models requires (1) a user interface, which allows the user to communicate with the system, providing necessary data to the system; (2) an inference engine, which solves given problems using input data from the user and knowledge from the knowledge base, through its own reasoning methods; (3) a knowledge base and qualitative analysis that contain the knowledge obtained from a domain expert, including facts, beliefs, rules-of-thumb, and other judgmental factors of the human expert; and (4) a model base containing urban systems models.

The inference engine uses a rule-based paradigm to process domain knowledge that is predominantly declarative. Model base is made up of the conventional systems models that are quantitative in nature. In addition to the concept of combining expert system with operations research models, there are other potential combinations that merit further attention from regional scientists' further attention.

1. Combining Expert systems (ES) and database management systems (DBMS) is one area of research whose time has come. The idea of combining ES and DBMS has received a great deal of attention recently, creating

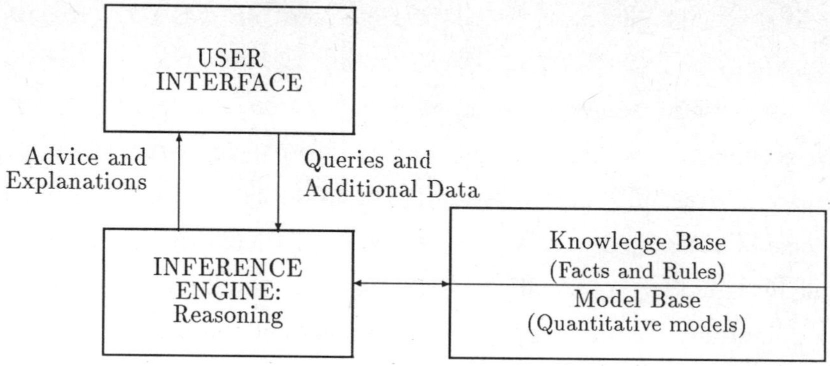

Figure 1 Combining Expert System and Urban Model Systems

another hybrid information system called the 'expert database system' (EDBS). The marriage of ES and DBMS takes two fundamental forms. The first type of coupling is with 'an expert' or 'intelligent' interface to a standard DBMS or an 'expert' query optimizer with a standard DBMS. The second type of coupling includes a standard expert system integrated with a large relational database of facts (Brachman,
1987).

The functions of an intelligent interface may include formulating efficient queries for a user by incorporating the knowledge of a human expert in the domain of a database, or interpreting data retrieved from a database and eliminating any inconsistencies (for instance, unifying the different units of measurement used in the database). The functions of the database system in the second type of coupling include the direct use of variables in the database system of the rules in the expert system. In fact, the inability of an ES to access or utilize a database already developed has been one of the major deficiencies of ES.

2. A second field involves combining expert systems (ES) and decision support systems (DSS). The idea of integrating ES into DSS to create more powerful and useful computer-based systems has been given much attention recently, creating a new terminology, 'expert decision support system' (EDSS), or 'intelligent decision support system.' The possible contributions of ES in EDSS include (1) helping users in selecting models, (2) providing judgmental elements in models, (3) simplifying building simulation models, (4) enabling friendlier interface, and (5) providing explanation capability (Turban and Watkins, 1986).

Most of all, the ES can play an important role in EDSS with model selection and building. As Strauch (1974) points out, the process of problem analysis (or modeling) usually involves three interrelated components: formulation of the formal problem, mathematical analysis, and interpretation of the results. While the mathematical analysis is handled well with DSS, the formulation requires the subjective knowledge of the user. Further, the interpretation requires the personal judgement of decision makers. The coupling of ES and DSS in this case is based on the assumption that subjective knowledge and personal judgment can be best made by experts.

An example of the use of ES for this purpose is the Advisory System for Ground Water Quality Assessment (Armstrong, 1987), which maintains a knowledge base in addition to the model base to assist decision makers in determining proper sample patterns and size given a specified confidence level and evaluating samples obtained. Wood and Wright (1987) also show an example of adding a rule-based system to the simulation model for storm water management for the purpose of aiding users with model calibration and result interpretation.

The intelligent interface provided by ES for the modeling tasks in DSS may stimulate planners to employ mathematical models more frequently and easily in their problem-solving processes. As often criticized by planners, modeling components tend to be treated as a black box; they are inadequate in recognizing the need for judgments by the users. Further, they often

Table 1 Taxonomy of Decision Problems

Type of knowledge needed for problem solving	Nature of problems	
	Structured or routine	Unstructured or semistructured
Quantitative reasoning: numeric computation	Database management systems	Decision support systems
Qualitative reasoning: intuition and rules of thumb	Expert systems	Human experts

conceal implicit judgments and assumptions by the users (Langendorf, 1985).

The study by Wellman (1986) that attempts to connect ES to mathematical modeling is worth much attention. He reports a rule-based system that generates parameters for the user to make the mathematical models easy to use. In that system, ES serves as an extra layer between the model and the user, translating qualitative criteria into the numeric input and also translating the model's numeric output to qualitative concepts that are more intuitive and informative to the user (Wellman, 1986). This approach certainly is a great improvement over the unaided use of modeling algorithms, encouraging easier use of quantitative modeling to support many planning decisions.

Although some authors see expert systems as special instances of decision support systems (DSS), many authors distinguish ES from DSS because of the former's unique structure, technology and ability to solve different types of problems (see, for instance, Kroeber and Watson, 1987; Liang, 1988). Table 1 summarizes major differences between ES and DSS in terms of types of computer processing involved and types of decision problems solved.

In DSS, the computer is used to store data and various decision models and the user interacts with the computer through the user-interface, providing the computer with judgments. In ES, the computer stores all expert knowledge, including judgments, into a program and recommends a solution by using appropriate analyses and its own programmed logic.

3. Combining ES and geographic information systems (GIS) is another promising area of research. The major function of GIS is to store and manipulate spatial data for cartographic modeling. The basic need for coupling ES with GIS stems from the fact that many tasks involved in cartographic modeling require expertise on the part of the user in the particular subject matter, and that expertise can be provided by experts or expert systems (Kim, Wiggins and Wright, 1989).

For example, the task of identifying suitable sites for a particular land use can be effectively assisted by GIS, which easily identifies the area with desired or undesired characteristics through several overlay manipulations. Another use of expert systems in GIS may be to provide a useful and more intelligent interface to the user. ES may be designed to help users to devise efficient operating procedures for cartographic modeling and interpret the results of spatial analysis. ES can also be used in GIS for uncertainty reasoning and to resolve inconsistent and contradictory information obtained from GIS. Among the problems of integrating ES with GIS, Robinson, Frank, and Blaze (1986) point out the limited capability of current ES shells and the lack of formalism typical in geography.

6.2 Machine learning and neural systems

As frequently observed, the greatest bottleneck in developing expert systems lies in knowledge acquisition process (see Feigenbaum, 1977). Probably the future AI research should be directed towards the improved automation of the acquisition of operational knowledge either (1) from expert-supplied data, or (2) from raw observational data or (3) from data generated exhaustively from logical specification and simulations.

One obvious shortcoming of expert systems is that they cannot respond to changing conditions. Expert systems may be able to duplicate human expertise at a given moment - like taking a snapshot of the expert's knowledge - but they cannot respond to new developments and changes in a domain. Thus the ability to build expert systems that can learn by themselves is a necessary and natural extension of the existing technology (Salzberg, 1988). Learning itself is a very general term denoting the way in which people or computers increase their knowledge domain and improve their skill. From the very beginning of AI, researchers have sought to understand the process of learning and create computer programs that are able to learn (Cohen and Feigenbaum, 1989). As we can see in Figure 2, both the feedback and inputs contribute to the learning process by referring to knowledge that has already been learned.

In classical AI field, experts are used to supply their domain knowledge to give the computer the basis for appropriate answers for a specific domain. On the other hand, a neural system[1], another branch of AI learns directly by interacting with the domain. It may not need to have any expert knowledge about domain. Given enough time and experience or training, the neural system may be designed to learn most of the things about domain, and ultimately it will be able to learn what is presently not known by any of the experts.

Neural network can be defined as computing systems made up of a number of simple, highly interconnected processing elements, which processes information by its dynamic state response to external input. Unlike conventional serial systems, the unique nature of neural networks is neither sequential nor deterministic, it responds, in parallel, to inputs presented to it (see Caudill, 1989).

How can we apply these artificial intelligent models for analyzing issues in transportation and spatial systems? Urban and regional systems are complex since spatial forms, functions and forms are results of interaction among individuals who act independently, pursuing their own interests. Great many scholars in Regional Science and other related disciplines have contributed much to our understanding of how a spatial system works, particularly in mixed economic systems. There are also many scholars who

---

[1] Some scholars call it neural networks, others call connectionist models, parallel distributed processing.

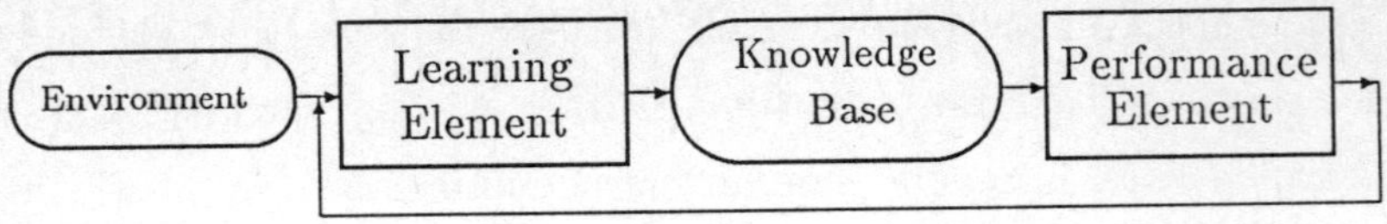

Figure 2 Simple Learning Model; Cohen and Feigenbaum (1989)

Note: Ovals denote declarative bodies of information and boxes denote procedures

shed light on the issue of formulating public policies and plans that enhance the welfare of the entire system.

And yet, the dynamic nature of urban and regional systems mandates that public policies and plans to be revised due to the constantly changing internal and external factors. These factors include political changes through elections, internal changes in party platforms, and changes in citizen's preference related to such issues as tax rates, revenue allocation, and priority of government spending, all of which affect public policies and plans.

Thus, while we may be able to explain a great deal of causes and effects of spatial phenomena prevailing in modern metropolises, there are still many unanswered questions and unsolved issues due to the dynamic, uncertain, and stochastic nature of interactions among activities inside the urban systems. There are still many unclear linkages between causes and effects due to multiplying impacts of each activating variable. Furthermore, we know little about evaluating public policies with respect to their impacts on the society. We also know little about forecasting the future patterns of urban form, function and structure. And yet, we frequently have to forecast for deriving public policies.

Learning systems in artificial intelligence generally concerns with accurate interpretation of broader phenomena, possibly all variables involved, and attempts to derive abstraction and generalization in such a way that is more economical than the conventional techniques. Organized like human brain cells, neural systems, for example, are designed to learn on their own and to make judgements the way human beings do.

In dealing with many unanswered and unsolved questions in spatial systems, learning systems such as machine learning and neural systems may shed light on finding the missing links between causes (say a new subway system in Los Angeles) and effects (say urban form, structure and function in Los Angeles), while considering all possible variables involved in a given metropolis. In this example, the missing links include subway fare, gasoline price, level of dispersion of socio-economic activities, land values, employments, migration, etc.

Even though artificial learning systems and neural systems are gaining their application capabilities in engineering (Shea and Lin, 1989) and commerce (Liang, 1989), they are still in their infant stage, particularly in the application to analyzing social science issues. Nevertheless, the fact that a piece of neural network software is being sold in the market less than $ 100 is an indication that users will soon come up with new application areas possibly including transportation planning and engineering, regional science and social science in general.

7. Summary

Telecommunication systems, computer communication networks, and intelligent information systems have been growing at an explosive rate in the past decade. And many indications show that the same trend will continue into the future. Gavish (1990) explains this phenomenal growth by three major factors: 1) development of related technology such as fiber-optics, local area networks, digital and fast-switching, microelectronics, and satellite technology; 2) deregulation of the telecommunication industry in the U.S. and other countries, which enabled new entrants to the market, contributing to a higher innovation rate, new product and service introductions, and increased competition; and 3) globalization of the market place and the need to coordinate manufacturing, marketing, services, and financial activities over longer distances which have contributed to price reductions.

Telecommunication networks are having a profound impact on industrial structures which, in turn, effect transportation patterns, and in the long run, on the spatial interaction of socio-economic activities. Driven by these forces, urban form, structure and functions will inevitably be affected in the future. At the same time, the rapid development of intelligent information systems, artificial intelligence including expert systems, machine learning, and neural systems can shed light on explaining unknown linkages between causes and effects in spatial phenomena.

In the light of these new changes caused by the rapid developments of telecommunication and artificial intelligence, an attempt has been made in this paper to suggest a research agenda for analyzing future transportation and spatial interaction. Brief descriptions on research issues were discussed in the categories of 1) transportation systems, 2) transportation and spatial interaction, 3) transportation and artificial intelligence, and 4) intelligent computer systems including expert systems, machine learning, and neural systems. We do not claim that these are comprehensive nor complete lists of areas from which one can draw a research agenda for analyzing future issues in transportation and spatial interaction. Nevertheless, we hope that this paper opens a dialogue among scholars in assessing a research agenda in this critical field of Regional Science.

References

Allison, G., (1971), Essence of Decision: Explaining the Cuban Missile Crisis, Boston, Little, Brown.

Armstrong, M.P. (1987), A Rule-Based Advisory System for Ground Water Quality Assessment at Hazardous Waste Disposal Sites. Department of Geogrpahy and Computer Science, The University of Iowa.

Bell, D. (1979), Communications Technology for Better or for Worse, Harvard Business Review May-June, 20-42.

Boyce, D. (1985), Transportation Research: the State of the Art and Research Opportunities, Transportation Research 19A.

Brail, R.K. (1987), Microcomputers in Urban Planning and Management, New Jersey, Center for Urban Policy Research.

Catanese, A.J. (1979), Information for Planning, in: The Practice of Local Government Planning, Washington DC, International City Management Association.

Caudill, M. (1989), Using Neural Nets: Representing Knowledge, AI Expert, December.

Cohen, P.R. and E.A. Feigenbaum (1989), The Handbook of Artificial Intelligence: Vol III, Reading MA, Addison-Wesley Publishing Inc.

Daskin, M. (1985), Logistics: An Overview of the State of the Art and Perspectives on Future Research, Transportation Research 19A:383-398.

Erlander, S. (1977), Accessibility, Entropy and the Distribution and Assignment of Traffic, Transportation Research 11:149-153.

Feigenbaum, E. (1977), The Art of Artificial Intelligence: Themes and Case Studies of Knowledge Engineering, Stanford CA, Technical Report Department of Computer Science: STAN-CS-77-621.

Florian, M. and M. Gaudry (1980), A Conceptual Framework for the Supply Side in Transportation Systems, Transportation Research 14B:1-8.

Garrison, W.L. (1985), Basic Research on Transportation Facilities, Transportation Research 19A:511-518.

Gavish, B. (1990), Telecommunication: A New Area for Operations Research, OR/MS Today Feb. 62-64.

Golden, B. and E. Baker (1985), Future Directions in Logistics Research,

Transportation Research 19A:405-409.

Goodall, A. (1985), The Guide to Expert Systems, Oxford, England, Learned Information Ltd.

Haghani, A. and M. Daskin (1984), A Short-Term Rail Freight Optimization Model. Paper presented at the Dallas ORSA/TIMS Meeting.

Han, S. and T.J. Kim (1989), Can Expert Systems Help with Planning? Journal of American Planning Association 55(3):296-308, Summer.

Harris, B. (1987), Information is not Enough, URISA News 90.

Jordan, W. and L. Burns (1984), Truck Back-Hauling on Two Terminal Networks, Transportation Research 18B:487-503.

Jordan, W. and M. Turnquist (1983), A Stochastic Dynamic Model for Railroad Car Distribution, Transportation Science 17:123-145.

Kim, T. (1990), Advanced Transport and Spatial Systems Models: Application to Korea, New York, Springer-Verlag Publishers.

Kim, T. (1979), Alternative Transportation Modes in a Land Use Model: a General Equilibrium Approach, Journal Urban Economics 6:197-215.

Kim, T., L. Wiggins and J. Wright (1989), Expert Systems: Applications to Urban Planning, New York NY, Springer-Verlag.

Kim, T.J. (1978), A Model of Zoning for a Metropolis, Environment and Planning A 10:1035-1047.

Kim, T.J., L.L. Wiggins and J.R. Wright (1989), Expert Systems: Applications to Urban Planning, New York, Springer-Verlag Publishers.

Kroeber, D.W. and H.J. Watson (1987), Computer-Based Information Systems, New York, Macmillan Publishing Company.

Kumar, A. (1989), Impact of Technological Developments on Urban Form and Travel Behavior. Technical Report, School of Urban and Regional Planning, University of Southern California, Working Paper.

Langendorf, R. (1985), Computer and Decision Making, Journal of American Planning Association 422-433, Autumn.

Leontief, W. and A. Strout (1963), Multiregional Input-Output Analysis, in: T. Barna (ed.), Structural Interdependence and Economic Development, pp. 119-150, New York, Macmillan.

Liang, T. (1989), A Composite Approach to Inducing Knowledge for Expert Systems Design. Faculty Working Paper, Bureau of Economics and Business Research, University of Illinois at Urbana-Champaign, NO 89-1534.

Liang, T. (1988), Expert Systems as Decision Aids: Issues and Strategies, Journal of Information Systems 2(2):41-50.

Linstone, H.A. (ed.)(1984), Multiple Perspectives for Decision-Making: Bridging the Gap between Analysis and Action, New York, North Holland.

Manheim, M. (1980), Understanding Supply in Transportation Systems, Transportation Research 14A:119-136.

Mills, E.S. (1972), Markets and Efficient Resource Allocation in Urban Areas, Swedish Journal of Urban Economics 74:100-113.

Mills, E.S. (1976), Planning and Market Processes in Urban Models, in: R.E. Crieson (ed.), Public and Urban Economics: Essays in Honor of William Vikrey, Lexington MA, Heath.

Moavenzadeh, F. (1985), Research Needs in Transportation Facilities: Guideway Technology and Materials Research, Transportation Research 19A:497-509.

Moore II, J. and P. Gordon (1990), A Sequential Programming Approach to the Formation of Urban Subcenters. Submitted to the Journal of Socio-Economic Planning Sciences.

Moore II, L and J.E. Wiggins (1990), A Dynamic Mills Heritage Model with Replaceable Capital. Forthcoming in Papers of the Regional Sciences Association.

Morlok, E. (1976), Supply Function for Public Transport: Initial Concepts and Models, in: M. Florian (ed.), Traffic Equilibrium Methods pp. 322-267, Berlin, Springer-Verlag.

Nijkamp, P. (1977), Theory and Application of Environmental Economics, Amsterdam, Holland, North Holland.

Powell, W., Y. Sheffi and S. Thiriez (1984), The Dynamic Vehicle Allocation Problem with Uncertain Demands, in: Ninth International Symposium on Transportation and Traffic Theory, pp. 357-374, Utrecht, The Netherlands, VNU Science Press.

Richardson, H. (1988), Monocentric vs. Policentric Models: The Future of Urban Economics in Regional Science. Presidential address to the Western Regional Science Association Twenty-Seventh Annual Meeting, Napa Valley CA.

Robinson, V.B., A.U. Frank and M.A. Blaze (1986), Expert Systems Applied to Problems in Geographic Information Systems: Introduction, Review, and Prospects, Computer, Environment, and Urban Systems 11(9):161-173.

Rogers, E. (1986), Communications Technology: The New Media in Society, Macmillan, Inc., The Free Press.

Salzberg, S. (1988), Machine Learning Moves Out of the Lab, AI Expert 44-52, Feb.

Schofer, J. and D. Boyce (1985), Conference Summary and Conclusions, Transportation Research 19A:351-354.

Shea, P. and V. Lin (1989), Detection of Explosives in Checked Airline Baggage Using an Artificial Neural System, in: International Joint Conference on Neural Networks, Vol. II, pp. 31-34, Washington DC, Institute of Electrical and Electronic Engineering and International Neural Network Society, June 18-22.

Steinbrunner, J. (1974), The Cybernetic Theory of Decision, Princeton N.J., Princeton University Press.

Strauch, R. (1974), A Critical Assessment of Quantitative Methodology as a Political Analysis Tool, Santa Monica, Rand.

Suh, S. (1989), Implementation and Evaluation of Nonlinear Bilevel Programming Model of Equilibrium Network Design Problem, PhD thesis, University of Illinois at Urbana-Champaign.

Suh, S. and T. Kim (1989), Solving a Nonlinear Bilevel Programming Model of Equilibrium Network Design Problem for Korea, Papers of the Regional Science Association: Stockholm Volume 66:47-59.

Turban, E. and P.R. Watkins (1986), Integrating Expert Systems and Decision Support Systems, MIS Quarterly 231-245, June.

Turnquist, M. (1985), Research Opportunities in Transportation System Characteristics and Operations, Transportation Research 19A:357-366.

Wachs, M. (1985), Planning, Organizations and Decision-Making: A Research Agenda, Transportation Research 19A(5/6):521-531.

Waterman, D. (1986), A Guide to Expert Systems, Massachusetts, Addison-Wesley Publishing Company.

Wellman, M.P. (1986), Reasoning about Assumptions Underlying Mathematical Models. in: J. Kowalik (ed.), Coupling Symbolic and Numerical Computing in

Expert Systems, New York, Elsevier Science Publishers B.V.

Wilson, A.g. (1970), Entropy in Urban and Regional Modelling, London, Pion Ltd.

Wilson, A.g. (1970), Interrgional Commodity Flows: Entropy Maximizing Approaches, Geographical Analysis 2:255-282.

Wood, D.M. and J.R. Wright (1987), Calibrating Complex Simulation Models Using Rules-Based Inferencing, Working Paper, School of Civil Engineering, Purdue University.

World Bank (1983), World Development Report, Oxford UK, Oxford University Press.

Chapter 7

URBAN LAND USE AND THE LOCATION OF POLLUTING PLANTS

M. Spiegel
U. Spiegel

1. Introduction

In the early 1970s numerous articles were written to analyze urban land use. Some interesting examples include Mills and Mackinnon (1973), Oron Pines Sheshinski (1973), Alonso (1969), Dixit (1973), Mills (1972), Mirlees (1972), and Solow (1972, 1973). The models considered in all of these articles have used the following simplifying assumptions:

a. The city is monocentric and all economic activity, i.e., production and marketing, is carried out in the Central Business District (CBD) where all job opportunities are located.

b. The transportation system in the city is radial and equally dense in every direction.

c. The only travel cost is the community cost of workers between their residence and their place of work in the CBD.

d. No local public goods or bads are in evidence nor are there any neighborhood externalities. Therefore, the only spatial characteristic that matters to households is the distance from the CBD.

The common implication of the above literature is that the bid rent is a decreasing function of the distance from the CBD. Masahisa Fujita (1986) considers neighborhood externalities and their effects on residential land use. Two models of externalities are discussed in his article 'Crowding Models and Racial Models'. Others also show how these kinds of externalities affect city boundaries and the borders between different residential areas. See Rose Ackerman (1975, 1977) and Bailey (1959) among others.

This paper differs from the aforementioned literature in the nature of the externalities considered. We concentrate on the detrimental externalities created as part of the production process. We discuss how detrimental externalities affect the social decision of where to locate the polluting plants: on the outskirts or in the CBD. As such it differs from Stull (1974) who studies externalities in production and assumes that 'Manufacturing firms are constrained to locate near the urban center while households are constrained to outlying locations.'

The structure of the paper is as follows: in Section 2 we consider whether to let polluting plants agglomerate in the CBD or force them out to the city outskirts. Section 3 presents the optimization problem for determining the most desirable location of polluting plants. Section 4 discusses the implied equilibrium bid rent function.

2. Pollution and Market Failure

Consider a monocentric circular city with a spaceless CBD. In order to minimize total transportation cost of buyers, all marketing activity is being carried out in the CBD. We assume that output is produced in the city by many small independent producers. As part of the production process, detrimental externalities are also created. The amount of pollution (such as smoke, noise, dust, etc.) is positively related to the volume of output. The amount of pollution received by a given consumer is negatively related to the distance between the pollution sources (the plant location) and the location of that consumer.

A market failure is defined as the case where the private profit maximization of plant owners will lead them to agglomerate their spaceless plants in the CBD, while the socially optimal solution calls for locating plants on the circumference of the circular city. Therefore, in the case of a market failure, the social cost of producing the given output in the CBD is larger than the social cost of producing the same level of output on the outskirts of the city.

When all plants are located in the CBD, the total social cost of production and transportation is given by:

(1) TC(CBD) = Factors costs + Pollution costs + Workers' commuting

costs.

When all plants are located on the outskirts of the city, the social cost of production and transportation is given by:

(2) TC(outskirts) = Factors costs + Pollution costs + Workers' commuting costs + Transportation costs of outputs from plants to CBD + Interplant transportation costs of intermediate goods.

We assume that the factors costs and the worker's commuting costs are the same, independent of the plant's location.[1]

Keeping output constant, the differences in the social costs due to a change in a plant's location results from: First, the difference in the 'pollution cost'. We define pollution as the cost of cleaning or avoiding the pollution. When plants are located on the outskirts, the cost of pollution for the residents is, at most, one half of the pollution cost when the plants are located in the CBD. This is due to the assumption that diffusion of pollution is such that each of the iso pollution curves is concentric. Once plants are located on the outskirts, more than one half of the pollution lands on non-residential sites, and pollution cost is lowered by more than 50 percent. Second, when plants are in the CBD, no transportation of the final goods from plants to market takes place. This cost is positive when plants are on the outskirts. Third, if the economies of agglomeration of plants (of being close to each other) is significant, there is an additional loss from moving plants from the CBD to the outskirts as the distance between plants will rise.

The market failure is the case where the market solution (of the independent producers) does not coincide with the social cost minimizing solution. Specifically, market failure is the case where the decrease in pollution cost outweighs the increased transportation cost and the loss of economies of agglomeration (due to the increased distance between plants).

The social planner would decide to locate the plants on the outskirts of the city if half of the pollution cost exceeds the sum of the following:

(3) $R_2 \circ h \circ Y + g(E,R_2)$

where R_2 is the distance of the plants that are located on the outskirts from the CBD measured in miles, h is the cost for each unit of output Y per mile of transportation, Y is the units of output which are transported from the outskirts to the CBD, and $g(E,R_2)$ represents the additional loss of moving plants R_2 miles from the CBD (to the outskirts) due to losing the economies of agglomeration. In the next section, we refer to the case where the plants are not located in the CBD according to the decision of the social planner, but still can be located within the boundaries of the city or even further away from the outskirts. We solve for the optimal plant location in the case of detrimental externalities.

3. Market Failure and the Optimal Plant Location

We assume that the size of the residential area is given. Therefore, in case of market failure, the problem faced by the social planner is to define the optimal plant location, that is, to decide what is the optimal distance from the city limits the polluting plants should be located. This distance defines the location for which the social costs of producing the given level of output is minimized. By moving the polluting plants away from the residential area, a buffer zone of the width $Z>0$ miles is created. We assume that all land outside the city limit is used for agriculture only, and the pollution damages for that use is negligible. The advantage in creating a buffer zone, Z, is that a significant portion of the pollution emitted by the plants lands on agricultural grounds and not on residential neighborhoods, thereby reducing the damage and pollution cost. On the other hand, locating the plant Z miles beyond the residential boundary, increases the transportation costs of the final goods from the plants to the CBD, increases workers' commuting costs from the residential area to the plants, and increases production costs due to not having complementary business nearby (diminished economies of agglomeration). In order to determine the optimal distance, we assume the linear pollution supply function

(4) $$P(x) = \begin{cases} 2(A - Bx) & \text{for } x \le A/B \\ 0 & \text{for } x > A/B. \end{cases}$$

where A is the highest level of pollution consumed at the source and B indicates the rate of pollution reduction as a function of distance from the source. The total amount of pollution created by a single plant is:

(5) $$TP = \int_0^{A/B} P(x)dx = 2Ax - Bx^2 \Big|_0^{A/B} = \frac{A^2}{B}$$

When the radius of the residential neighborhood is larger than A/B, it is clear that the plants located in the CBD will dump all their pollution on residential land. If the plants are located on the circumference of the city, at most half of the TP will land on the residential area.

If plants are moved Z miles into the agricultural land and away from the residential area where $0 < Z < A/B$, the total pollution that lands on agricultural ground and thereby avoids the residential neighborhood is:

(6) $$SP = \int_0^{Z} (A - B)dz = AZ - \frac{BZ^2}{2}.$$

This is the economic benefit from creating buffer zone Z. The problem of the social planner is to define the socially optimal width of the buffer zone.

The net social benefit of a zone of size Z is W(Z) which is:

(7) $$W(Z)=[C1(AZ - (B/2)Z^2)]-[2t\circ Z\circ n\circ N]-[Z\circ h\circ Y + g(E,R_3)-g(E,R_2)].$$

The term in the first bracket represents the benefit to the residential area which arises from plants being Z miles away from the city limits. This term indicates the reduction in the pollution cleaning cost resulting from moving the polluting source away from the residential area by Z miles where C1 is the cost of cleaning of each unit of pollution. The term in the second bracket represents the worker's commuting costs of the additional 2Z miles of n workers N times during the production period where t is the travel cost per mile. These costs are raised when the plant is moved further away from the outskirts by Z miles. The term in the third bracket has two components: the first term is $Z \circ h \circ Y$, which measures the additional transportation of the output from the city limits Z miles away, and the other two terms measure the additional cost of interplant transportation cost and efficiency loss due to the loss of economies of agglomeration, which result when plants move from R_2 to $R_3 = (R_2 + Z)$ miles from the CBD. The first order condition of this problem implies the optimal distance (from city limits) Z^*:

(8) $$Z^* = \{1/B \circ C1\} \{C1 \circ A - [2 \circ t \circ n \circ N + h \circ Y + g_z(E,R_3)]\}.$$

4. The Rent as a Function of Location from the CBD

Our model adopts the basic Muth type model and its basic assumption about the circular city and the CBD. All consumers are assumed to travel to the CBD to purchase their consumption goods. We differ from the Muth model assuming that the production activity, which is the source of the detrimental externality, is done outside the residential zone. For the sake of simplicity, we assume that all polluting plants are evenly distributed on the circumference of the circular city. We also assume that the residential zone and the production site are separated by a buffer of nonresidential (agricultural) land. As we assume that all consumers are employed in the production, and since production and marketing activities are at different locations, the consumer has to commute N times from his home to his place of work and M times from his home to his shopping area in the CBD. In addition, we assume that the total transportation cost depends linearly on the distance travelled.[2]

In the model all consumers are identical in their income of y dollars and have the following identical utility function, U(C,S). This function is quasi concave and strictly increasing where C represents the composite consumption good with a numeraire price of 1, and S represents the area consumed lot size of the house.

Consumer's income y is spent on consuming C dollars and S units of land at a distance x from the CBD at a price r(x), travelling cost of T(x) and cleaning up the pollution cost for each consumer cl(x). Since the expenses T(x) and cl(x) are constant per location the net income per location x is:

(9) $$Y - T(x) - cl(x).$$

The budget constraint faced by a consumer is:

(10) $$C + r(x) \circ S = y - T(x) - cl(x).$$

The optimization problem of each household is twofold:

(A) for each location solve:

(11) $$\max_{C,S} \left\{ U(C,S) \quad C + r(x) \circ S = y - T(x) - cl(x) \right\}.$$

(B) Given the set of solutions to (11) of C(x), S(x), the consumer will choose the location which solves:

(12) $$\max_{x} U[C(x), S(x)].$$

Following the Von Thunen model of land use, we assume that if the utility of a certain location x is higher than that of other locations, competition among consumers will bid the rent of that location such that utility of location will be equalized across all locations. We apply the same concept of bid rent here. This is the maximum rent per unit of land a household would be willing to pay for residing at location x, while enjoying a fixed level of utility U. Formally, the bid rent (x,U) can be defined as:

(13) $$\rho(x,U) = \underset{C,S}{\text{Max}} \left\{ \frac{Y - T(x) - Cl(x)}{S} \quad U(C,S) = U^* \right\}.$$

Denote S(x,U) as the bid max lot size, and $S^*(r,U)$ as the compensated demand for land at distance x. Holding utility constant at U^*, $S^*(x,U)$ is the solution for a cost minimization problem:

(14) $$\text{Min} \left\{ C + r(x) \circ S \quad U(C,S) = U^* \right\}$$

Since (14) is the dual problem of utility maximization under the budget constraint ((11) above), it holds identically that:

(15) $$S(x,U) = S(\rho(x,U),U).$$

Through an application of the envelope theorem to (13) we have:

(16) $$\frac{\partial \rho(x,U)}{\partial x} = - \frac{[T'(x) + Cl'(x)]}{S(r,U)} \equiv - \frac{TC'(x)}{S(r,U)}.$$

where TC is the total cost (of travelling and pollution cleaning) per location. Condition (16) implies that the bid rent will change in an inverse relation to the change in the total cost per location. Hence, from (15) we get:

(17) $$\frac{\partial S}{\partial x} = \frac{\partial S}{\partial r} \circ \frac{\partial \rho(x,U)}{\partial x}.$$

In the traditional model of location, where all economic activity is carried out in the CBD, the signs of (16) and (17) are negative and positive, respectively. That is, the rent will decrease as the distance from the CBD increases, and the lot size consumed will go up.

Our approach to the division of economic activity into two centers, production on the outskirts of the city and marketing in the CBD, indicates that the signs of (16) and (17) are not necessarily the same. In order to identify the sign of the slope of the bid rent function, we have to discuss the T(x) and c1(x) functions.

The Travelling Cost, T(x)

We assumed above that total transportation cost depends linearly on the distance travelled by the consumer to the production and marketing places. We define travelling cost per mile as t, therefore the total transportation costs of the consumer located at the distance x miles from the CBD is:

$$T(x) = t \circ x \circ M + t(R_3 - x)N \qquad \text{for } x > 0 \text{ and } N > M > 0. \tag{18}$$

where R_3 is the distance of the plants from the CBD. Therefore, we can define R_3 as: $R_3 = R_2 + Z$ where R_2 is the limit of the residential area and Z is the buffer zone.

The Cleaning Cost, cl(x)

According to our simplified assumption in the previous section, the supply of pollution diminishes at a constant rate with the distance from the plants. If the plants are located at a distance of R_3 miles from the CBD and at Z miles from the outskirts of the residence area which is R_2 miles away from the CBD, we can define the cleaning cost at location x from the CBD, cl(x), as follows:

$$Cl(x) = \begin{cases} 0 & x < R_1 \\ Cl \circ b(x - R_1) & R_1 \le x \le R_2 \end{cases} \tag{19}$$

Explanation

Given that the polluting plants are located at R_2, R_1 is the distance from the CBD where the supply of pollution approaches zero. We write the supply of pollution per resident location as follows[3]:

$$p(D) = a - b \circ D. \tag{20}$$

where D is the distance from the plants. At D = a/b from the plants the supply of pollution approaches zero. This point is located R_1 miles away from the CBD or $R_3 = D + R_1$. (See Figure 1). As C1 is the cleaning cost per unit of pollution, we obtain equation (19) above.

We further assume D > Z, which means that $R_2 > R_1$, therefore at distance x from the CBD where $R_2 > x > R_1$, cleaning costs are positive. Since the location dependent cost is:

$$(21) \qquad TC(x) = T(x) + cl(x).$$

we get:

$$(22) \qquad TC'(x) = T'(x) + cl'(x).$$

As N > M, then for any x > 0, T'(x) < 0. Therefore, the sign of Tc'(x) will have the following values:

$$(23) \qquad TC'(x) = \begin{cases} T'(x) = t(M-N)<0 & \text{for } x < R_1 \\ T'(x)+cl'(x) = t(M-N)+Cl\circ b \gtreqless 0 & \text{for } x \geq R_1 \end{cases}$$

For location x, where $0 < x < R_1$, which is not polluted, moving away from the CBD will lower the transportation costs (T(x) < 0), and as a result, the bid

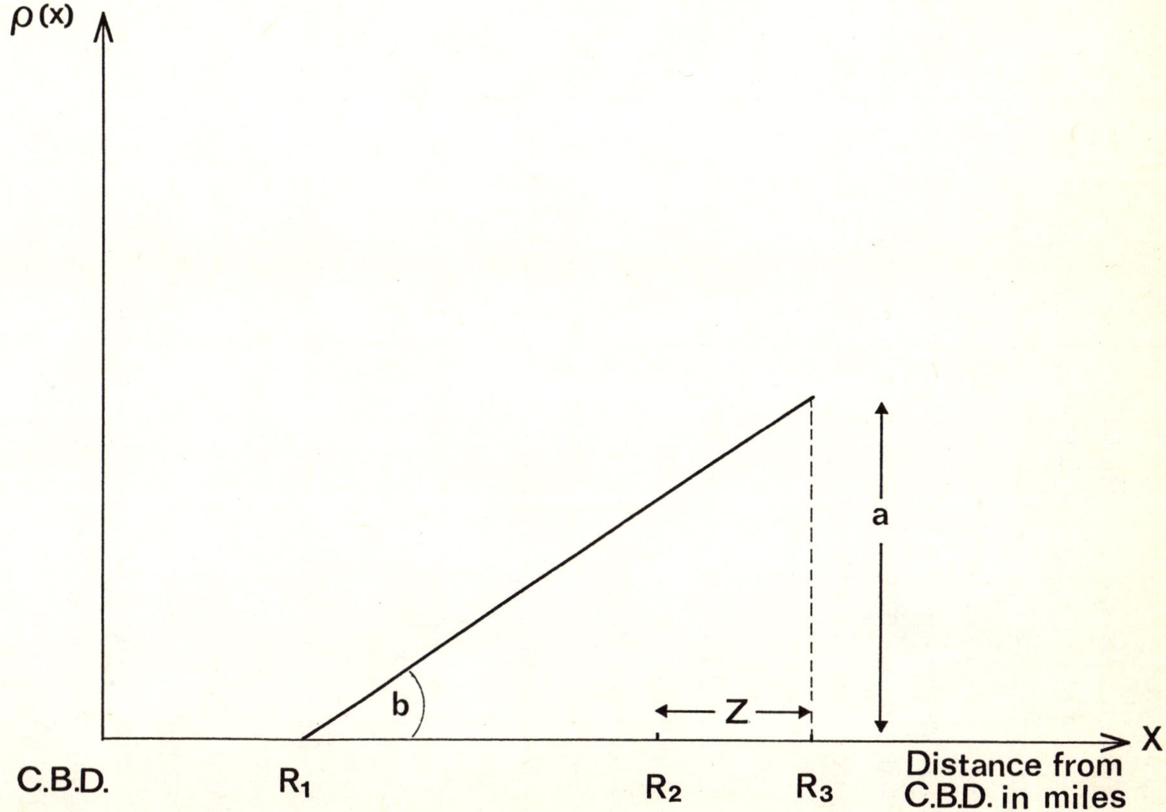

Figure 1. Pollution Supply Curve

rent function p(x,U) will rise continuously. For location x, where $R_1 \leq x \leq R_2$, in the polluted areas, one of the following possibilities might arise:

a. If $C1_o b > t(N-M)$, i.e., the marginal pollution cost is larger than the marginal saving of traveling costs (see Figure 2), then $TC'(x)>0$ and the bid rent function will decline with x in this area.

b. If $c1_o b = t(M-N)$, then the location total cost function, TC, is a constant and therefore $TC'(x) = 0$. As a result, the bid rent function increases with distance x from the CBD for $0 < x < R_1$ and is constant for $R_1 \leq x \leq R_2$.

c. If $C1_o b < t(N-M)$, then $TC'(x) < 0$ and the bid rent function rises continuously with distance x from the CBD over all $0 < x < R_2$. The rate of increase of the bid rent is higher for locations close to the CBD where $x < R_1$, and is lower for $R_1 \leq x \leq R_2$. (See Figure 3).

Under the assumption of normality of the demand for land, S, we get from (17) that the quantity of space used as a function of the distance from the CBD is inversely related to the slope of the bid rent function. As the bid rent increases with distance from the CBD, lot size decreases.

4. Conclusion

We have developed a simple model which considers the optimal spatial location of polluting plants in or outside the residential neighborhood. The different market forces in our model are represented by consumers who dislike the pollution, by producers who want to be close to their output markets located in the CBD and by workers who want to minimize their travelling costs to and from their place of work.

The specific market situation considered in the model is of a market failure. In that case, the intervention of the legislator in setting zoning regulations is required in order to obtain the socially desired outcome.

Once the location of the polluting plants has been determined, the derivation of the bid rent for land use implies that land prices are not necessarily monotonic with the distance form the CBD. Under certain circumstances the bid rent price might be rising with the distance from the CBD. This suggests that the location of polluting plants should be introduced

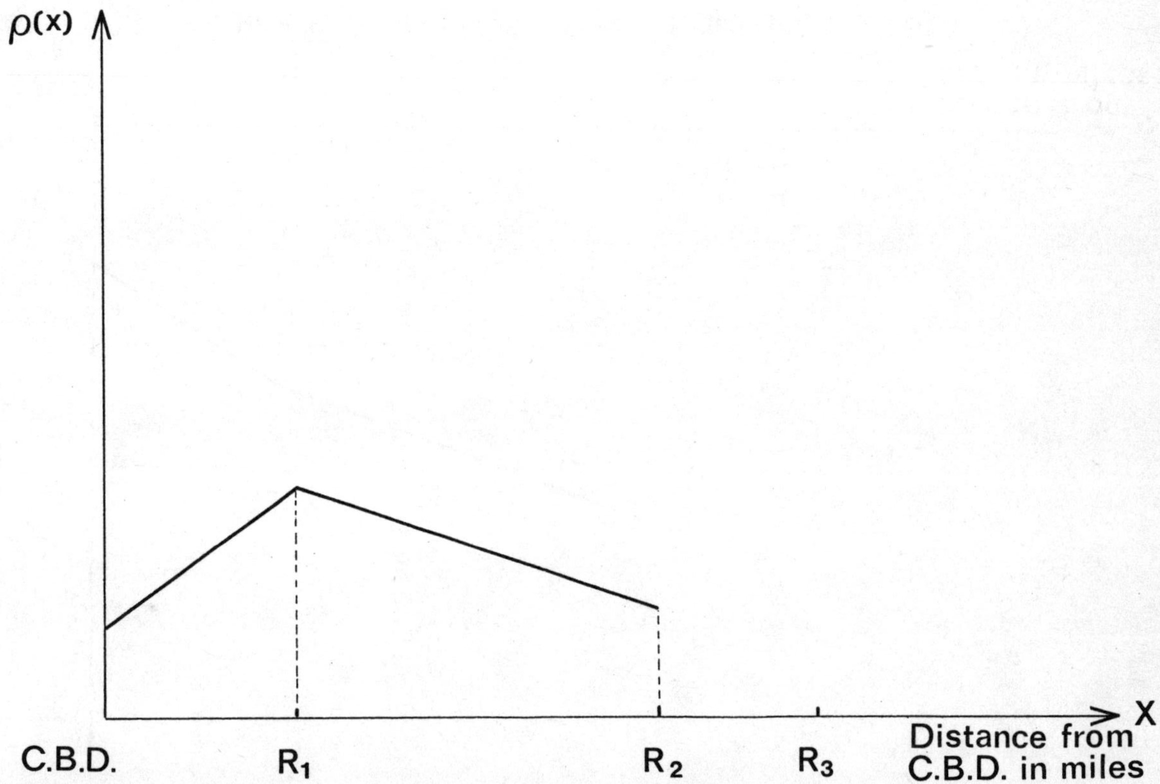

Figure 2. The Bid Rent Function when the Marginal Pollution Cost ($C1_ob$) is larger than the Marginal Saving on Transportation [t(n-m)].

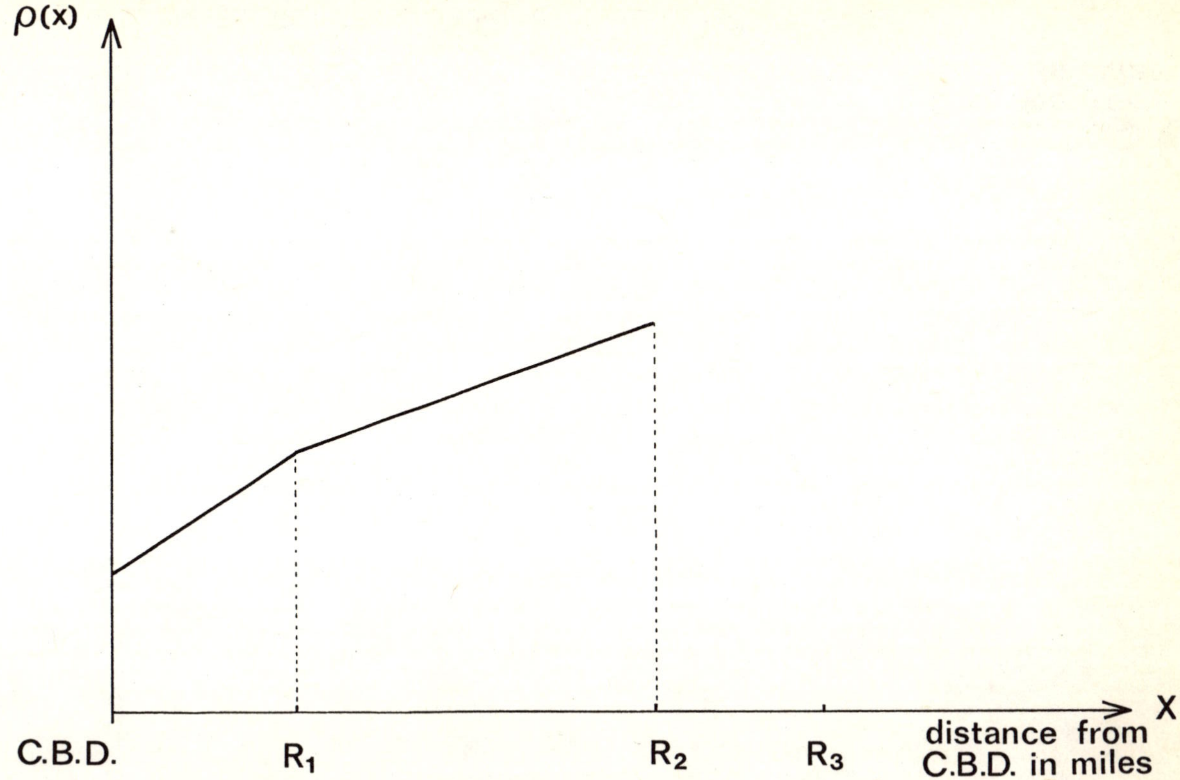

Figure 3. The Bid Rent Function when the Marginal Pollution Cost [$C1_ob$] is Smaller than the Marginal Saving on Transportation [t(n-m)].

explicitly into the empirical analysis of urban planning and consumers housing decisions and the implied travelling habits.

Endnotes

1. We assume that the density of workers' residence is constant, which implies that the total travelling costs of the workers are independent of plant locations.

2. If, for example, N = M, the total transportation costs are independent of location, i.e., all consumers will travel the same distance independent of the location of their residence. However, we use a more reasonable assumption that N > M so consumers who are located near the CBD would have higher travelling costs.

3. Note that equation (4) relates to the total pollution at a given distance from a single plant and not to the total pollution per resident from all sources.

References

Alonso, W. (1964), Location and Land Use, Cambridge MA, Harvard University Press.

Bailey, M.J. (1959), Note on the Economics of Residential Zoning and Urban Renewal, Land Economics 35:288-292.

Dixit, A. (1973), The Optimum Factory Town, Bell Journal of Economics and Management Science 4:637-651.

Fujita, M. (1986), Urban Land Use Theory, in: J.J. Gabszewisz et al. (eds.), Location Theory, New York, Harwood Academic Publishers.

Mills, E.S. (1972), Studies in the Structure of the Urban Economy, Baltimore, Johns Hopkins.

Mills, E.S. and J. MacKinnon (1973), Notes on the New Urban Economics, Bell Journal of Economics and Management Science 4:593-601.

Mirrlees, J.A. (1972), The Optimum Town, The Swedish Journal of Economics 74:114-135.

Muth, R. (1960), Cities and Housing, University of Chicago Press.

Oron, Y., D. Pines and E. Sheshinski (1973), Optimum vs. Equilibrium Land Use Pattern and Congestion Toll, Bell Journal of Economics and Management Science, 4:619-636.

Rose Ackerman, S. (1975), Racism and Urban Structure, Journal of Urban Economics 2:85-103.

Rose Ackerman, S. (1977), The Political Economy of a Racist Housing Market, Journal of Urban Economics 4:150-169.

Solow, R.M. (1972), Congestion, Density and the Use of Land in Transportation, The Swedish Journal of Economics 74:161-173.

Solow, R.M. (1973), Congestion Cost and the Use of Land for Streets, Bell Journal of Economics and Management Science 4:602-618.

Stull, W.J. (1974), Land Use and Zoning in an Urban Economy, The American Economic Review 64:337-347.

Part B. SPATIAL IMPACTS OF NEW TECHNOLOGY

Chapter 8

HIGH TECHNOLOGY INDUSTRIES - A PANACEA FOR PERIPHERAL REGIONS? THE EVIDENCE FROM ISRAEL

D. Shefer

1. Introduction

Industrial development is perceived as a vehicle for the regional economic growth of peripheral regions. In the past two decades, the high-tech industrial complex has undergone tremendous expansion worldwide, stimulating a new wave of industrial growth. Historically, dominant industrial processes and products have gone through phases; for example, the textile industry predominated in the last century; coal and steel rose to preeminence at the beginning of this century; when these stabilized, the car and durable consumer goods industries took off. Each development cycle or 'Kondratieff long wave', exhibiting growth, dominance and, eventually, contraction, has had a profound influence on the social and political environment and on labor-market relations in industrial agglomeration centers (Rothwell, 1982; Scott, 1987).

2. Survey of the Literature

The literature concerned with locational decisions of high-tech industries has varied in emphasis. In most cases, however, studies of high-tech industrial complexes around the world have found that certain regional attributes are attractive for manufacturing branch-location while other criteria are employed for services. Headquarters are usually established either in large urban centers or in medium-sized towns at the edges of urban centers (suburbs) which offer a high quality of life and the proximity of university 'brain centers' (Dorfman, 1983; Farley & Glickman, 1986; Markusen, 1983; Dieperink & Nijkamp, 1987). Planners interested in developing high-tech industries in outlying regions must take careful note of the high agglomeration tendencies they exhibit and also be warned of the exaggerated expectations often accorded, most notably by politicians, to high-tech industries; (Applebaum, 1983; Boddy and Levering, 1986; Myers, 1984).

Attempts to attract high-tech industries specifically to those peripheral regions that seem obviously disadvantaged by their distance from urban centers may encounter unsolvable problems. Yet, many planners maintain an optimism that high-tech industrial development is possible anywhere because of expanding communications technology which continually increases the 'footlooseness' of industry and the freedom of not being tied down to a specific source of raw materials or market, a concept that was explored over 20 years ago by Klaassen, 1967.

The valued image of 'high-tech' industry is derived mainly from the well-paid, prestigious jobs in research and development. Industries, however, are comprised of many activities, each fulfilling a different function. Production, which demands mostly semi-skilled and unskilled manual labor, is a more footloose activity than is research and development, and hence is more likely to be moved to peripheral areas. Such differentiation in location has been explained in reference to the 'product life cycle' theory, (among others, see Vernon, 1966, 1979; Hekman, 1980; and Norton and Rees, 1979). This hypothesis presents an historical analysis of the functional stages that a product undergoes before it reaches the consumer; namely: research and development, prototype production, mass production, marketing, sales, etc.

Initially, there is a period in which product development undergoes rapid changes, and a variety of services and experimental inputs is constantly demanded. During this stage, the producer can remain competitive only by locating in a centre that provides quick access to all these necessary inputs and services. As the product matures, the production process and the product itself become more standardized, and it becomes easier to transfer production operations to a location where labor is cheaper than in the center. A firm's flexibility in conducting its research and development, marketing and most administration from its home base while moving branch plants to lower-labor cost locations for mass production is critical. This flexibility enables firms, especially large corporations that are well-endowed to take advantage of national and world-wide opportunities, to expand their markets and to stay ahead of the competition (Moriarty, 1983; Norton and Rees, 1979; Scott, 1987; Gillespie et al., 1986).

The search for different attributes to satisfy different locational needs has resulted in an unbalanced division of industry functions between core and peripheral areas. There is a tendency for large companies to establish only production work, providing mostly low-paid jobs (and a few administrative positions) in peripheral areas. The expected improvements in the regional economy and in local labor opportunities have often been disappointing, since production branches have been found to create very little, if any, spin-off effect to increase the local multiplier (see Norton and Rees, 1979; Moriarty, 1983; Townroe, 1971). Multi-national corporations that move production operations around the world may be sought by developing countries, but these companies do not offer much more opportunity or job security to the local population than do traditional industries. According to Sayer, 1986 and Gillespie et al. 1986, there are even signs that a decrease in the expected profitability lifespan of products reinforces the locational advantage of core regions for production work, as the cost of transferring to distant locations is justified only by very long production runs. In order to analyze the forces behind changing locational trends, it is essential to understand the considerations involved in corporate decision-making, which is motivated by profit and is not primarily concerned with local or regional economic growth (see Vernon, 1979; Razin, 1984a; Malecki, 1986; Scott, 1987).

Given the 'rational' economic considerations that guide corporate policies, some studies have warned of the likelihood of peripheral areas developing 'branch-plant economies.' As described by Watts, (1981), these are regional economies based on production plants owned by non-local resident investors who locate their high-level jobs in their headquarter towns. Branch-plant economies may boost employment; in the long run, however, they do not develop and diversify into strong, mixed economies that would create the right environment for innovation. An environment that 'breeds' innovation has been found by Pottier, 1985; Dorfman, 1983; Hekman, 1980; and Dieperink and Nijkamp, 1987, among others, to be vital for succeeding in the world of high-tech industry. Researchers like Townroe, 1975; Coffey and Polese, 1985; O'Farrell, 1986; and Razin, 1984b; 1988, have concluded that pursuing local-based (endogenous) industrial development is perhaps the only way to build a more balanced economy, one that is less vulnerable to plant closures and, perhaps more importantly, that utilizes local inputs, fosters local links and encourages local entrepreneurship and innovation. Examples of such research and development centers are Boston, Northern California's Silicon Valley and Cambridge, England; (Markusen, 1983; Dorfman, 1983; Segal and Quince, 1987).

The tremendous advances in communications technology in the last decade have sparked enthusiasm for the theory that physical distance can be reduced in importance and transportation replaced to some extent with the so-called New Information Technologies (NIT); (Giaoutzi and Nijkamp; 1988). These technologies are undoubtedly going to influence the way business will be conducted in the future, but exactly how much and in which direction are still subjects for debate. Perhaps, as Salomon, (1988) has pointed out, NIT's potential effect will be proportional to a country's size, such that the benefits that are felt in peripheral towns, previously thought ill-suited for certain industries in a large country, will be greater than those felt in a small country like Israel, where the shorter distances allow face-to-face contact to remain the preferred way of doing business. An evaluation of the influence of NIT on the peripheral regions' potential share of the 'high-tech pie' must also examine the importance attached by firms to communications quality in making their locational decisions.

3. Population Dispersal and Urban Development Policies in Israel

In Israel, both the local authorities of peripheral towns and the national government have assigned high priority to the promotion of 'high-tech,' or 'knowledge-intensive,' industries in recognition of their economic potential and because of their association with prestigious employment opportunities and regional social and economic growth. High-tech industries are attractive too, because they are thought of as 'clean industry,' emitting small amounts of air pollutants into the air and generating less solid waste than did the older, traditional manufacturing industries.

Industrial development in peripheral areas in Israel, such as the Northern Galilee and the Southern Negev (see Map 1 and Table 1), was related to the government's population dispersal policy initiated shortly after the founding of the State in 1948. Concomitantly, the Law for the Encouragement of Capital Investment and other subsidy programs were initiated by the central government of Israel. Under these programs, the country was divided into development zones designating the priority of receiving various degrees of financial assistance for selected industrial sectors (see Map 1). Although these economic incentive programs proved to be marginally effective their total overall effect on the locational decision of major industrial activities were found to be very negligible, (see Shefer and Frenkel, 1986; and 1987). The aim of the population dispersal policy was to relieve population pressure on the coastal plain, where the majority of the country's residents was concentrated. The population dispersal policy manifested itself in an urbanization policy that was based on the creation of a hierarchy of urban centres resembling the form advanced by Christaller's central place theory and the theoretical spatial patterns of Losch. To fill in the gap in mid-size cities in this hierarchy, new urban centres, or development towns, were created. These, however, preceded the provision of adequate industrial employment (Shachar, 1971).

More than 30 development towns have been designated since the establishment of the State. Most of these towns were constructed by the mid-fifties, and the last two by the mid-1960's. The majority of the development towns were established on entirely new sites; a small number was designed for sites of existing small settlements (see Map 2).

Criticism has been levelled against this large-scale dispersal program on several accounts; primarily, critique has centered on the diffusion of effort over too large a number of new urban centres, which reduced the impact that

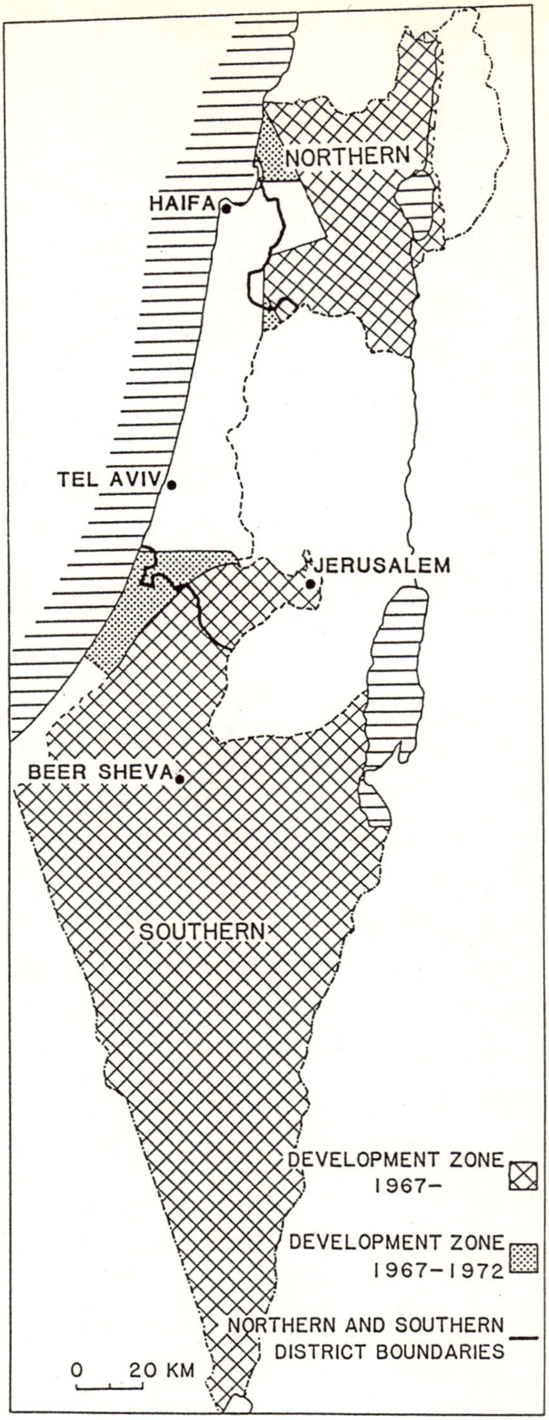

Map 1. Location of Israel's Three Major Districts and Largest Urban Area

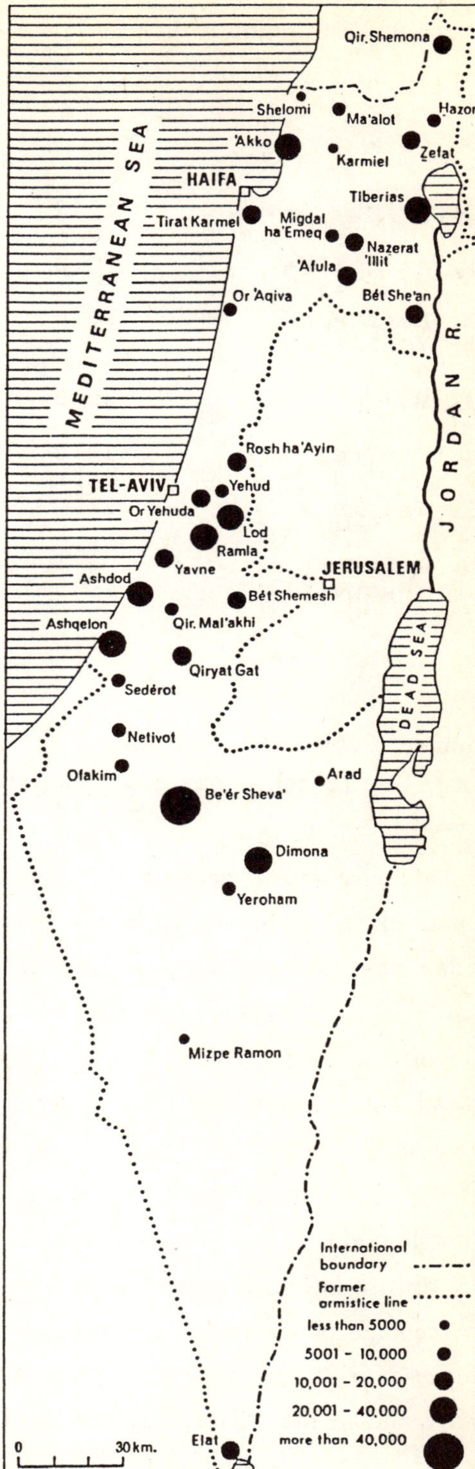

Map 2. Israel's Development Towns
Source: Shachar (1971, p. 366)

Table 1. Distribution of Israel's Population by District
(in percentage)

District	1948	1961	1972	1983
1. Jerusalem	10.2	8.8	11.0	11.7
2. Northern (Galilee)	16.8	15.5	15.0	16.2
3. Haifa	20.5	17.0	15.4	14.2
4. Central	14.3	18.7	18.5	20.6
5. Tel Aviv	34.7	32.1	28.8	24.8
6. Southern (Negev)	2.5	7.9	11.3	11.8

Source: Central Bureau of Statistics, 'Census of Population and Housing,' 1948, 1961, 1972, 1983, Jerusalem

could have been achieved if the same effort and resources had been concentrated in fewer, but well-selected centres. Indeed, even today, 35 years after most of these towns were established, a large number of them are still small in size and suffer from stagnation and a lack of employment opportunities. Socio-economic problems continue to plague many of the development towns, especially during periods of economic slowdown and recession.

In spite of the heroic efforts that were invested to effect the spatial distribution of population in Israel, no major change in percentage distribution by district has taken place, as can be seen from Table 1. It is true that the percentage of the population residing in the Tel Aviv district fell from 34.7% in 1948 to 24.8% in 1983; however, the combined percentage of those in Tel Aviv and in the adjacent central district decreased only marginally, from 49%

in 1948 to 45.4% in 1983. Furthermore, it is clear that the Tel Aviv Metropolitan area is gradually expanding in area, undergoing a process that may also be observed in a large number of viable metropolitan areas throughout the world.

Industrial development in the development towns has been hindered for a number of historical reasons. First, the Israeli political/governmental structure is very centralized, and a large portion of development town industry since the 1950s has depended on either central government initiative or direct and indirect government incentives. Although the various governments' intentions have been genuine, the bureaucracy involved in enforcing policies has often produced significant problems resulting in impediments to growth in peripheral areas. As in most Western countries, major industry in Israel, too, has tended to agglomerate in and around the major cities (particularly Tel Aviv and Haifa), where easy and convenient proximity to government offices, a large pool of skilled labour, information services, and transportation are all available.

Secondly, the development towns were inhabited by a newly arrived, usually socially and economic weak population that was ill-equipped to compete with those who had the advantages offered by the large central cities. Although government policy succeeded in stimulating population growth in these towns, there was a lack of simultaneous economic growth essential for establishing a firm economic base, especially those development towns located in peripheral regions (Razin, 1988). As a result, reviving the economic base is a challenge now facing the new generation of Israeli-born development-town residents; and this resuscitation must succeed quickly in order to halt the exodus of that generation - particularly its strong, educated element - from the towns.

The first section of this paper reviews the literature on high technology industrial location and analyzed the forces that have influenced the form of the high-tech landscape. The second section will present findings from a recent study by Shefer and Frenkel (1986), in which 53 randomly selected high-tech industrial firms in Israel were interviewed in regard to their critical locational determinance. The peripheral regions' share in this industries will also be assessed.

Since no consensus on an exact definition of 'high-tech industry' exists as yet in the literature, this paper will utilize five widely-accepted industrial categories, based on their high proportion of research and development activity and on the types of products produced by these industries. The categories are as follows: (1) pharmaceuticals, (2) communications equipment, (3) electronics and computers, (4) optics, and (5) precision instruments.

4. Israel and the Push for High-Tech Growth

Boosted by increases in military spending and the government's shift in the 1960s toward encouragement of export industries, high-tech industries in Israel have grown in the last 15-20 years. Investment has come from the government, overseas investors, the Histadrut (the General Labor Federation), and private corporations. In order to create prestigious employment opportunities, the strategy advanced by the government to build up high-tech industries in peripheral regions has resembled past industrial policy: pursuing external investors, either from Israeli core areas or from abroad; in some cases, as Razin, (1988) has pointed out, the Government has ended up paying 85% of the investment needed.

The peripheral share in the new, dynamic high-tech industries has, however, been minimal because, regardless of location, all firms in Israel that export a certain percentage of their product are eligible for government incentives. Therefore, most high-tech firms have located near agglomeration centres in order to enjoy their advantages in terms of skilled labor and services. The most concerted attempt to merge high-tech industrial growth with population distribution has been the creation of a 'science and industry region' in the Galilee, (see Map 1). The region's goals are to incorporate such principles as those enunciated in the plan put forth by the Jewish Agency, (1983): the accommodation of technology for social as well as economic uses, social pluralism, integration of settlement types, a high and unique standard of living and, of course, employment based on high-tech development. Some elements of the plan are already in evidence in the region. Yet, a high-aiming policy such as establishing R&D in peripheral regions faces strong odds; it is not even known whether the goals are potentially attainable. In order to determine whether, and how, high-tech industry can be attracted to Israel's peripheral regions, like the Galilee in the north or the Negev in the south, planners must translate these goals into operational terms. The initial factors that are important to companies when choosing their preferred locations must be identified, and then public policies devised that can be used as instruments for creating the necessary business environment.

5. Data Source

In 1986, Shefer and Frenkel conducted an empirical survey of 53 randomly selected Israeli high-tech firms in order to investigate the impact of advanced means of communications on the location and operation of such firms in Israel. Tables 2 and 3 give the distribution of these firms by, respectively, geographical area and industry type. The study gathered information on the spatial distribution of the firm's operations, searched for the factors involved in their location and operation decisions and, to some extent, measured firms' satisfaction with their choice. The methodology employed involved a detailed personal interview, through a questionnaire, of an executive or high-level officer in each of the firms sampled. Questions dealt with company ownership, locational decision factors, use of communication means, annual company turnover, government financial aid received (grants and/or loans), employee types and satisfaction or dissatisfaction with the company's chosen locations.

Distinct differences in character were found between peripheral and core firms. More specifically, the location requirements of high-tech firms belonging to a certain category (e.g., the optics industry), were found to be much less flexible than those of another (e.g., the communications equipment industries); uneven distribution was found in the types of activity (i.e., research and development vs. production) performed by firms belonging to the same industrial category; and a correlation was established between location and

Table 2. Distribution of Sample of High-Tech Firms by Geographical Area

Geographical Area	Number	Percentage
Metropolitan	17	32.1
Urban Fringe	15	28.3
Development Town	14	26.4
Kibbutz	7	13.2
	-----	-----
Total	5	100.0

Table 3. Distribution of the Sample of High-Tech Firms by Industry

Industry Type	Number	Percentage
Pharmaceuticals	8	15.1
Communications Equipment	11	20.8
Electronics and Computers	26	49.0
Optics	8	15.1
	---	-----
Total	53	100.0

ownership type (private vs. government). Accordingly, a development area firm may be characterized as being primarily large, government- or public-owned, engaged in mass production and belonging to the communications equipment industry.

A core firm can be characterized as being medium- or small-sized, privately-owned, engaged primarily in research and development and belonging to the pharmaceutical, computer or optics industries.

6. Findings

The survey responses revealed that three main variables influenced site choice:

1. availability of a large skilled labor pool: scientists, engineers, administrators;
2. governmental financial assistance;
3. convenience for management.

Communications considerations were not among the highest priorities mentioned, but good communications are critical to the functioning of industries both in the core and in the peripheries. Some 77% of the firms surveyed had internal computers; this figure jumped to 86% in development areas. Expenditures on communications (excluding transportation), however, made up only 2-3% of the average firm's overhead, so that communications use was seen to be fairly fixed and inelastic. Newer firms proved to be bigger consumers of advanced means of communications - facsimile, electronic mail, etc. - than older firms. This tendency held for all three of the mutually exclusive locational divisions into which the study was divided (metropolitan center, urban fringe, development area). (For a more in-depth discussion on the effect of communications on the location of high-tech industries, see Shefer, 1988).

The results of the data analysis is reinforced findings documented in the literature regarding locational decisions of high-tech industries; e.g., peripheral locations were chosen for different functions than were metropolitan and suburban locations. The factors considered important by firms in development towns were government financial aid, the possibility of future expansion, physical attributes and ideological reasons; whereas in the metropolitan areas, the criteria for location were based more on the existence of a large pool of highly skilled workers and main-market proximity. The urban fringe firms stressed the importance of locating near universities, scientific laboratories and services, as well as convenience for management and other cultural and environmental amenities.

Table 4 shows that, as expected with production branches, the median size of development-area firms was significantly larger than that of firms located in the metropolitan center and fringe.

The dominance of production over all other types of activity in development areas is shown in Table 5. It might be noted that although production occupies the highest percentage of workers in all areas, their distribution is much more highly skewed in development areas and more balanced by research and development, administration and services in metropolitan areas and the urban fringe.

Table 4. Median Firm Size (According to Number of Employees) by Geographical Area

Area	Median No. of Employees
Metropolitan	55
Urban Fringe	40
Development Town	93
Kibbutzim	45

Total Sample	55

Table 5. Distribution of Employee Activity by Geographical Area (in percentages)

Activity	Metro-politan	Urban Fringe	Development Area	Kibbutz
Administration	14.5	13.0	13.1	17.8
Production (skilled)	27.3	29.4	44.2	35.6
Production (unskilled)	18.0	24.8	32.8	29.1
Services and Maintenance	5.3	2.2	4.0	6.5
Sales	20.3	3.2	1.9	7.1
R & D	14.6	26.8	4.0	2.5
Other	0.0	0.6	0.0	1.4
Total	100.0	100.0	100.0	100.0

Table 6. Government Financial Aid as % of Total Set-Up Cost by Geographical Area

Area	Total	Grants	Loans
Metropolitan	46	13	36
Urban Fringe	40	5	36
Development Area	59	22	37
Kibbutzim	38	13	25

Another important finding of the survey was that government financial aid was not clearly directed. Although grants in development areas were higher than in other areas, the loan policy was seen to be indifferent (see Table 6). This explains why grant incentives were not significant enough to encourage hesitant firms to locate in a development town. The survey results thus point to an urgent need for the government to show an unambiguous, more meaningful preference for development areas in its industrial financial-aid policy.

7. Conclusions

Studies in many countries have shown that pursuing high-tech industrial development in peripheral regions has often resulted only in the establishment of production branch plants that have few links with local suppliers or services; (Townroe, 1975; Felsenstein, 1986; Malecki, 1986; Scott, 1987; Nijkamp, 1988). Although distances between metropolitan and development areas in Israel are relatively small, the high-tech industries included in our survey followed the same pattern. Firms involved in significant amounts of research and development have chosen to locate in areas of agglomeration, whereas the peripheries have attracted large production activities.

Because of Israel's small size, improvements in communications can make distances almost negligible for some purposes. In the empirical survey, communications needs were not cited as being among the most important considerations in locational decision-making; on average, only 2-3% of the firm's total expenditure were devoted to these needs. Since, however, newer firms were seen to use more advanced means of communication - facsimile, inter-computer communication, electronic mail, etc. - than did older firms, it is reasonable to suggest that the existence of a high-quality communications infrastructure and related services is a necessary, though not sufficient, factor for firms considering a peripheral or development area location.

The government must show a commitment to its stated objectives of 'science and industry' development in the Galilee (and form a parallel plan in the Negev). It must genuinely and ardently direct its policies to distribute not only branch plants but also the high-level activities of high-tech industries to peripheral regions. Local services and suppliers must be built up and a solid clientele attracted to them in order to boost regional economies by increasing the local multiplier. To achieve this goal, Government policy must strive to endow peripheral regions with some of the attributes available in core regions.

The presence of a skilled labor force is crucial in the locational decisions of firms. The skilled employees' demands for urban cultural and educational amenities must also be taken into account. In addition, programs for labor training and business service development must be encouraged by government policy to enable peripheral regions to raise their endogenous levels of innovation capacity. The pursuit of investment by external industries must not be a goal in itself; rather, peripheral regions should strive to attract only those investments that will act as catalysts to local entrepreneurship in order to bolster the long-term economic vitality of these regions.

References

Applebaum, E. (1983), Winners and Losers in the High-Tech Workplace, Challenge Sept./Oct., 52-55.

Boddy, M. and J. Levering (1986), High Technology Industry in the Bristol Corridor: The Aerospace/Defence Nexus, Regional Studies 20:217-231.

Coffey, W. and M. Polese (1985), Local Development: Conceptual Bases and Policy Implications, Regional Studies 19:85-93.

Dieperink, H. and P. Nijkamp (1987), The Spatial Dispersion of Industrial Innovation, Unpublished Working Paper, The Free University, Amsterdam.

Dorfman, N (1983), Route 128: The Development of a Regional High-Technology Economy, Research Policy 12:299-316.

Farley, J. and N.J. Glickman (1986), R&D as an Economic Development Strategy, American Planning Association Journal 52:407-418.

Felsenstein, D. (1986), The Spatial Organization of High Technology Industries in Israel. The Hebrew University, Jerusalem (unpublished thesis).

Giaoutzi, M. and P. Nijkamp (eds.) (1988), Informatics and Regional Development, Aldershot, Gower.

Gillespie, A., H. Williams and A. Thwaites (1986), Competition Internationalisation and the Regions, paper presented at Anglo-American Workshop on the Location and Growth of High Technology Industry, Cambridge, England.

Hekman, J.S. (1980), The Product Cycle and New England Textiles, Quarterly Journal of Economics 94:697-717.

Jewish Agency, Northern District (M. Mazan and T. Dochan) (1983), A Science and Industry District: Preliminary Ideas, Haifa, Jewish Agency.

Klaassen, L.H. (1967), Methods of Selecting Industries for Depressed Areas, Paris, Organization for Economic Cooperation and Development.

Malecki, E.J. (1986), Technological Imperatives and Modern Corporate Strategy, in: A.J. Scott and M. Storper (eds.), Production, Work, Territory: The Geographical Anatomy of Industrial Capitalism, Boston, Unwin and Allen.

Markusen, A.R. (1983), High-Tech Jobs, Markets and Economic Development Prospects: Evidence from California, Built Environment 9,1:18-28.

Moriarty, B. (1983), Hierarchies of Cities and the Spatial Filtering of

Industrial Development, Papers of the Regional Science Association 53:59-82.

Myers, D. (1984), Sliding Down the Ladder: Citizens' View of Austin's Quality of Life. Austin: Community and Regional Planning, School of Architecture, University of Texas, Austin.

Nijkamp, P. (1988), The Strategic Role of High-Tech Firms in Peripheral and Border Areas, Unpublished Working Paper, The Free University, Amsterdam.

Norton, D. and J. Rees (1979), The Product Cycle and the Spatial Decentralization of American Manufacture, Regional Studies 13:141-152.

O'Farrell, P (1986), Entrepreneurship & Regional Development: Some Conceptual Issues, Regional Studies 20:565-584.

Pottier, C. (1985), The Adaptation of Regional Industrial Structures to Technical Changes, Papers of the Regional Science Association 58:59-72.

Razin, E. (1984a), Location of Industrial Firms in Israel, Jerusalem, The Jerusalem Institute for Research on Israel (Hebrew).

Razin, E. (1984b), Location of Management and Production Branches of Industrial Firms in Israel and its Implication for the Development of Development Towns, Jerusalem, The Jerusalem Institute for Research on Israel (Hebrew).

Razin, E. (1988), Stages in the Development of Israel's Spatial Industrialization Policy. Paper presented at Dutch-Israeli Symposium on Locational Considerations of Industrial Organizations - Conditions & Effects; Herzliya, Israel.

Rothwell, R. (1982), The Role of Technology in Industrial Change: Implications for Regional Policy, Regional Studies 16,5:361-369.

Salomon, I. (1988), Transportation-Telecommunication Relationships and Regional Development, in: M. Giaoutzi and P. Nijkamp (eds.), Informatics and Regional Development, pp. 90-102, Aldershot, Avebury.

Sayer, A. (1986), Industrial Location on a World Scale: The Case of the Semi-Conductor Industry, in: A.J. Scott and M. Storper (eds.), Production, Work, Territory: The Geographical Anatomy of Industrial Capitalism, Boston, Unwin and Allen.

Segal, Quince, Wicksteed (1985), The Cambridge Phenomenon: The Growth of High-Technology Industry in a University Town, Norfolk, England, The Thetford Press.

Scott, A.J. (1987), The Semiconductor Industry in South-East Asia:

Organization, Location and the International Division of Labor, Regional Studies 21:143-160.

Shachar, A. (1971), Israel's Development Towns: Evaluation of National Urbanization Policy, Journal of the American Institute of Planners November, 362-372.

Shefer, D. (1988), The Effect of Various Means of Communication on the Operation and Location of High-Technology Industries, in: M. Giaoutzi and P. Nijkamp (eds.), Informatics and Regional Development, pp. 166-181, Aldershot, Avebury.

Shefer, D. and A. Frenkel (1987), Generating Alternatives for the Socio-Economic Development of Development Towns in Israel, Haifa, Israel, Samuel Neuman Institute for Advanced Study. Technion-Israel Institute of Technology.

Shefer, D. and A. Frenkel (1986), The Impact of Communications on High-Tech Industrial Development in Israel: An Empirical Study, Haifa, Israel, Samuel Neuman Institute for Advanced Study, Technion-Israel Institute of Technology.

Townroe, P.M. (1971), Industrial Location Decisions, University of Birmingham, Center for Urban and Regional Studies.

Townroe, P.M. (1975), Branch Plants and Regional Development, Town Planning Review 46,1:47-62.

Vernon, R. (1966), International Investment and International Trade in the Product Life Cycle, Quarterly Journal of Economics 80:190-207.

Vernon, R. (1979), The Product Cycle Hypothesis in a New International Environment, Oxford Bulletin of Economics and Statistics 41:255-267.

Watts, H.D. (1981), The Branch Plant Economy, London, Longman.

Chapter 9

THE IMPACT OF NEW INFORMATION TECHNOLOGIES ON SPATIAL INEQUALITIES

M. Giaoutzi
A. Stratigea

1. Introduction

New Information Technology (NIT) may be regarded as a pervasive, radical and innovative technology, which may potentially trigger off a systemic transformation in our economies.

As more and more economic activities become concerned with the generation, processing and exchange of information, technical developments which affect this activity are of potentially far-reaching significance.

The geographical impact of NIT is clearly strongly dependent on its degree of pervasiveness on the two following dimensions: sectors (e.g., small-scale vs. multinational firms), and regions (e.g., core vs. peripheral regions).

Under the new technological regime where economic activities are increasingly dependent upon information availability and use of reliable and up-to-date information systems, it seems plausible that core areas are in an advantageous position as far as access to these networks is concerned. But since new technology is a source of both **constraints** and **opportunities**, the question that is emerging is concerned with the new **prospects** and **threats** peripheral areas are facing as to their ability to overcome the gap separating them from core areas. More precisely, the issue at stake concerns the new role attributed to peripheral areas in the context of the new international division of labour and how they are going to cope with the competitive advantages of the core areas in the development process.

The aim of this paper will be to approach the above questions in the context of the new roles imposed on spatial structures (core vs. periphery) by the power relations formed by New Information Technology (NIT) networks, viz. the **structure** and **intensity** of these networks. The first part of the paper will provide some discussion on the foreseeable **prospects** and **threats** peripheral areas are facing in this new context.

2. Typology of Core and Peripheral Areas

Before we embark upon a theoretical discussion on the expected **prospects** and **threats** imposed on peripheral areas by the new logic of NIT developments it is necessary to put emphasis upon the diversification likely to be embedded in each inequality pattern.

Speaking of peripheries, we may well refer to previously industrialized (declining) areas, stagnating areas, underdeveloped areas, etc. Each of them has incorporated a diversified range of structural components, which in certain cases possess for example a number of fully developed networks, while in others any infrastructure is lacking which would be required for their development. In most of the cases their peripheral position is defined by their comparative performance as to the rest of the regions in each specific national context. In a global world market context though there has been a ranking as to their development potential which is placing them according to their competitive position in the world market. Prospects and threats in this context have a rather diversified qualitative texture. By generalizing one runs the risk to deprive certain regions from the region-specific description of their potential since the range and type of their constraints are of a different nature.

The above problems have been reflected in the efforts of many researchers to classify regions on the grounds of certain structural components they possess (see Swijngedouw, 1987 and Nijkamp, 1987).

In the context of analyzing either clusters with similar development patterns (e.g., economic indicators, social indicators, infrastructure indicators, etc.) or locational aspects related to innovative behaviour, the above typologies are important for the better framing and efficiency of regional policies.

The importance attached to technology in a regional development policy context has brought forward a range of new indicators related to factors such as:

- presence of professional and technical labour competence;
- threshold size of urban agglomerations present in the region which will allow for cultural activity and communication to be heightened;
- conditions that promote synergy or instability etc. (Oakey, 1985).

It is clear that the locational factors which are predominant in 'high-tech' industries and in the majority of the innovation-oriented regional policies are broader than those common in previous generations of regional policies. The importance of **information** and **instability** in the process of technological change itself brings this about.

In the discussion on socio-economic, cultural and political differences between **core** and **periphery** the efficiency-equity dilemma has played an important role. In an **international context** transnational trade flow policies and economic power blocks have permanently faced this dilemma. In a **regional-national context** though, regional urban policies also reflect the permanent struggle between an efficient use of scarce resources - usually taking place at the core - and the welfare discrepancies with respect to peripheral regions of a national system.

A most efficient way of classifying regions in order to better promote their development is an issue of importance to deal with. **Which type of policy is needed for which type of technology and in which region** reflects a question which reveals the growing awareness that **technology, economics** and **space** have an interwoven triangular relationship with close interactions between all three components (Nijkamp, 1987).

The later relation reveals the necessity for introducing a rationale for clustering cases which will be able to identify their relative position in an international context and place them later on in a more substantial context relevant for policy-making.

So the following matrix gives a rough classification of cases as these have been identified in the regional literature throughout the years employing the whole range of aspects approached by various theories.

	International	National	Local
Core	A	E	I
Center	B	F	J
Peripheral	C	G	K
Border	D	H	L

The use of such a framework enables decision making to proceed towards operationalizing the information required in each particular case and thus to identify more structured policies accordingly.

3. The Impact of NIT in the Development Process

Technological developments and more specifically (NIT) have grown dramatically over the past decades, a process which has shown the first clear signs of a transition phase or our industrialized societies. The economic situation which hit the Western economies in the seventies has evoked a new spirit in many countries as to their efforts to become more **efficient** and **competitive** by favouring the necessary conditions for an economic revitalization. The increasing importance attached to the role of NIT in the regional development context is mainly associated with the catalytic function of NIT for a set of structural transformations conveyed by the term 'Information Economy' (Goddard and Gillespie, 1987). The paramount role played by NIT in the light of the Information Economy perspective is mainly related to the fact that **knowledge** and **information-oriented** activities are being considered more and more as the 'strategic resources' **upon which competition and efficiency** are based. These strategic resources have nowadays become the necessary requirements for the survival and flourishing of (our) modern societies.

NIT may be characterized by its revolutionary impact on the regional development process, since the pervasiveness of its applications contributes to transforming many of the traditional features of the present-day **production, exchange** and **distribution systems.** This transformational potential is considered as the driving force allowing for the realization of a fundamental shift in the 'locus of competitive advantage', thus leading to a changing geography of competition (Goddard and Gillespie, 1987).

The economic significance of these technological developments to the competitive strategies of regions is nowadays greatly appreciated. NIT developments are taking the form of **new products, distribution systems, production processes, transaction innovation, management and decision making innovation, control innovation, etc.**

The transformation of information into a **commodity form** and its distribution via modern telecommunication networks have contributed considerably to the diffusion of information, reducing the barrier of distance and thus facilitating market expansion. Information availability is far less geographically constrained than before.

In the context of **production processes**, NIT has significantly contributed towards the integration of these processes, which have been automated and subjected to computer control. Computer integrated manufacturing (CIM) is nowadays a reality and has been considered by the various sectors as a key element which will enhance their future competitive position as to their ability to respond rapidly to changing market conditions and, most important, to exploit **economies of scope** - versus **economies of scale** - offered by the new technological developments, which will additionally allow the integration of all economic, organizational and technological functions within one organization, promoting in this sense the improvement of process predictability, timely delivery, quality and efficiency without losing flexibility and variety (Nijkamp, 1987; Goddard et al., 1985; Giaoutzi, 1987).

One of the most important applications of new technological developments is the **just-in-time production** and distribution systems (JIT) enabling a number of companies, which may be geographically dispersed, to establish a complex inter-organizational production and distribution system, thus creating chains of competitive advantage with considerable competitive strength, insofar as the aggregation of the competitive advantage of each member company is achieved.

In a managerial context, NIT increases the ability of multi-site organizations to control and integrate their activities over space through the opportunities offered by computer networks operating via modern telecommunication channels. This leads to a separation of the **location of activity** from the **location of control**, thus reinforcing the centralization of strategic functions, while depriving local areas of their initial relative autonomy. In this respect, NIT is **facilitating** the creation of such organizational structures characterized by strong power centralization patterns, and hence an even more complex spatial division of labour **is emerging**.

The availability of information and the interaction between decision-makers and information services available on the network market have considerably improved the **knowledge** of an organization or a region in the context of its environment, which - as a result of access to information networks - now refers to a significantly extended geographical space. A consequence of the above extension is that NIT - and more specifically its applications in the telecommunications field and access to information networks via telecommunication channels - will serve to facilitate the **internationalization of capital** and **goods markets** by making available distantly located opportunities (Goddard, Gillespie, 1987).

In the above context it is clear that NIT may play a steering role in re-shaping spatial relationships, given its potential towards altering the relative distances between places. This distance-shrinking effect is essentially considered on the basis of computer/telecommunications network innovations.

As a consequence of NIT making available distantly located opportunities and extending the geographical scope, new actors whose locational decisions are taken according to global rather than national objectives are going to emerge. The impact of NIT upon spatial structures and its potential to produce a new interregional geography of competition is expected to radically affect the **core-periphery relationships** in the political and social systems of national economies. So **core areas** are to a large extent maintaining their comparative advantage as to their access to networks. Firms in such a context, in their efforts to minimize risks, are oriented towards those locations which will provide them with a wide variety of polyvalent information, in other words towards the core regions. As a result core areas in a multinodal network context are becoming strategic nodes with increasing control power provided by NIT, through not only the quantity but also the quality of the information they convey.

Peripheral areas, on the other hand, are faced with new **prospects** associated with the possibility of having direct access to information (Giaoutzi and Nijkamp, 1988) as well as **threats** stemming from the uncertainty of the profound impacts upon our future ways of living, working, travelling, recreation and maintaining contacts, while their competitive position will be mainly determined by their ability to meet the locational requirements of the various agents at work.

The effects of NITs are usually multidimensional in nature and may be visualized as effects on employment, land use, industrial structure etc. at different geographical scales (Nijkamp and Salomon, 1988). In the light of the above discussion a strong issue at stake may emerge. will NIT increase the discrepancies between core and periphery or under certain conditions reverse the trends? The answer seems to rely more upon a smart regional development policy (e.g., emphasis on education and vocational training, incentive for Local Areal Networks (LAN) etc.) enabling a specific region to benefit from the use and allocation of the information sector, since technological development incorporates a set of antagonistic forces towards both concentration and deconcentration (Nijkamp, 1987).

4. New Perspectives for Core and Peripheral Areas

The role of New Information Technology and its dynamic potential in the process of economic restructuring and revitalization taking place at an international level has already been recognized. People, firms and countries as a whole are confronted nowadays with new challenges as to their locational and mobility patterns, organizational structures, technological functions, skill requirements, range of expertise, educational orientation, etc. (Giaoutzi, 1987).

Due to the profound restructuring observed in the world economies, an increasing competitive pressure is being exerted upon the various actors involved, leading them to a reorganization of their competitive strategies in order to take full advantage of the benefits to be gained from technological developments, as it has been realized that **international competition and**

efficiency are to a large extent related to the range of technological expertise.

In this context **access to and use of information** becomes a key force, as has been rather clearly reflected in the efforts of many countries when establishing either regional or national innovation policies in their territories, which also greatly supports the idea that the 'space of places' is gradually being superseded by a 'space of flows' (Castells, 1984). In this respect, the need for network advanced services, modern communication facilities and material and social infrastructure appear as essential factors for the emerging structural transformation in space.

Therefore, it becomes evident that the various actors are greatly inclined to locate their activities in those areas where all the above necessary infrastructure is available. This in turn implies that core areas, in the light of their ability to offer the above requirements, are to a large extent maintaining their comparative advantage as to best access to technological advances and networks to the detriment of the peripheral areas.

Core and **peripheral** areas in an information economy perspective are to a large extent going to compete on the grounds of the **quality of their regional information environment**. This quality is greatly determined by a number of factors such as the **stock of knowledge** of the region's labour force and enterprises based essentially on the existing education and training systems; the **quality** and **density** of the **information networks** establishing within the region, which will enable enterprises to take full advantage of the existing knowledge for their own benefit; and finally the **nature** and **variety of information and communication links** established between the region and the rest of the world, which will to a large extent determine the updating of the region's knowledge stock. In this respect, it becomes obvious that insofar as core areas are displaying a more structured information environment compared to peripheral areas, they will remain a more attractive location for the generation, adoption and diffusion of technological innovation as well as for the formation of new firms (incubators), which will continue reinforcing their central role while at the same time intensifying regional disparities.

Many research efforts have been devoted toward a better insight into the locational patterns of activities, the development of interregional trade patterns among regions and the corresponding regional division of labour in such a context.

The issue of comparative advantage, in this framework - associated with the potential of one region to specialize in the production and export of several commodities at a relatively more competitive cost vis-à-vis others - needs to be enriched by the new dynamics induced by the properties of a 'space of flows' which has been created by the recent logic of NIT.

The product cycle approach which disaggregates regional economic activities on the basis of their product life stage provides a promising explanatory ground in this respect. It also offers a sound exploratory framework which may contribute to a better understanding of the steering role that core areas are likely to play on the global scene in a long run perspective. According to the product life cycle theory, **core areas** - being characterized by a high level of R&D activities, better access to information and communication networks, high quality of scientific and technical labour force, academic and cultural activity, conditions promoting synergy and instability and an increasing capacity for geographical market extension - are more efficiently forming an environment which greatly supports the renewal and development of knowledge and competence, thus rendering these areas as centres of innovative information-oriented activities and attractive locations for both high-skilled scientific and technical labour force and high-tech activities. As a result of all the above, the new technology sector tends to agglomerate around the core areas, in order to take advantage of the various opportunities there.

On the other hand, **peripheral areas**, lagging in terms of access to technological developments and networks and infrastructure in general, are chosen by different actors as production sites for their routine activities, as far as these activities require a substantial input of low labour skill profile and have routinized and simplified their technological production requirements. In this context, the gap separating core and peripheral areas in a more disadvantageous position mainly associated with the lack of innovative and technological expertise necessary for an active rather than a reactive role vis-à-vis technological developments. What we actually observe is that a clear spatial distinction exists between routine production and non-routine R&D and decision-making activities, promoting discrepancies between core and periphery hosts.

The above polarization, coupled with the complementary role attributed to peripheral areas, reflects to a great extent the inability of peripheral areas - supply - to gain access to a number of requirements such as information networks, high skilled labour force, etc. needed by various actors in order to satisfy their locational requirements as imposed by the new technological regime. Such a situation obviously has undesirable effects on the economic base of peripheral areas in terms of technological expertise, sources of added value and employment opportunities, thus weakening even further their competition and in a way predetermining their future trajectories.

The spatial consequences of the above restructuring stemming from NIT developments are associated with an **international/interregional division of labour**, particularly in high-tech industries. These allow for a disjunction of different operations across space, given the fact that communication networks are able to restore the necessary links between different production units. What follows is a considerable acceleration of uneven development and disequilibrium of the spatial structure.

Significant changes are also going to occur within the large metropolitan areas, mainly associated with the growth of corporate services and high-tech manufacturing, the decline of traditional activities and the development of new economic sectors. The metropolises are acting as a real magnet at an international level, strongly attracting people, capital, minds, information, materials and energy and thus reinforcing their positions vis-à-vis regional structures (Andersson and Johansson, 1984).

In the light of the above discussion, it is evident that peripheral areas are facing considerable bottlenecks as to their development perspectives, which they will be able to overcome only as long as they manage to meet the locational requirements (e.g., educational facilities, knowledge networks and networks in general) enabling firms to profit from their environmental advantage.

On the other hand, core areas may lose the central role historically assigned to them, due to the economies of scale mainly dictated by the traditional technology requirements, to the extent that economies of scope will become the new emblem in the goals that corporate strategy has to set in an information age (Giaoutzi, 1987).

5. Location Aspects Reconsidered

In the light of the NIT developments and the potential offered by them in terms of competitive goals, it is rather clear nowadays that in order to develop, adapt and survive in a competitive environment, regional structures must innovate and have the grasp of up-to-date technology. In other words, a positive relationship appears to exist between technology - and especially its dynamic component, innovation - and the spatial organization of development. The innovative power of a region constitutes a prime factor for the development of regional potential and hence it becomes quite important to take a closer look at the locational aspects of technological change.

Usually firms seek a location in a certain place for a wide variety of complex reasons. The interplay between the needs and expectations of a firm on the one hand - the demand profile - and the potential and quality of a certain region on the other hand - the supply profile - determine jointly the locational patterns of new economic activities. The questions likely to arise in this respect may be formulated as follows: what are the major locational requirements enabling a region to attract high tech firms? Or stated otherwise, why are certain regions especially fertile in the production and diffusion of innovation?

In the last decades, in spite of the numerous policy efforts to disperse new technology and innovative economic activities, there still exists a persistent tendency towards agglomeration and concentration. Given the variety and diversity of activities and opportunities within agglomerations and given also our current understanding of non-routine economic activities, there appear to be sound reasons for this agglomeration economies as a necessary condition for regional high tech technology development. Nijkamp and Dieperink (1986) refer to agglomeration size and R&D infrastructure as the driving forces for innovative behaviour (the urban size hypothesis and the

R&D infrastructure hypothesis, respectively).

As far as high technology economic activities display several particular characteristics as distinct from traditional economic activities, it will be apparent that they will also follow distinct locational patterns. Currently, economic literature considers high tech industries as being footloose in terms of traditional locational factors such as transportation and markets (Camagni and Rabelloti, 1988).

These distinct locational patterns may be classified in three broad groups of explanatory factors, (see Nijkamp and Davelaar, 1989), viz.:

1. factors internal to the firm
2. factors external to the firm and
3. remaining factors

The main interest here is focused upon the second group, i.e., the factors external to firms. This group of factors consists of the necessary requirements to be offered by a region - the supply side - in order to attract high tech activities and thus reinforce its development potential and consequently its competitive position in the market.

The most straightforward factors in this respect is the **number and type of firms** within a region. The **regional sector** composition plays thus a decisive role which implies that in testing the fertility of various regions in terms of production and diffusion of innovation, we must be fully aware of the spatial mix of the various sectors. A great number of produced innovations in a region may to a large extent be attributed to a relatively high concentration of innovative sectors within a region.

The **total number of people** living in a region is thought of as the second important factor mentioned very often in the international literature. As Malecki (1979) claims there is an increasing relationship between R&D activity and city size, suggesting increasing economies of urban size for R&D. Also R&D personnel, namely the number of high skilled scientific and technical labour force, considered as a labour input in the process of innovation seems to have a positive correlation with the total number of people living in a region.

Information availability is regarded as a locational requirement of utmost importance in terms of the innovative power of a region. Information may become available through various channels of information flows. Discriminating according to the different 'senders' of information, information sources may be distinguished in: (a) information exchanged between firms in terms of interfirm contacts, (b) information provided by public research institutes, universities, institutes of technology in terms of technical and marketing consultancy and (c) information exchanged between firms and their customers in the context of the intensive spatial interaction patterns of persons. All the above information flows seem to enrich firms with a large amount of valuable information, which will exert a great impact on their innovative behaviour.

Environmental aspects, as a locational factor, are also playing a decisive role in terms of innovative behaviour of a region. The main reason for the increasing importance attached to environmental aspects, in terms of cultural and educational amenities and physical climate and environmental quality, is the locational preferences of R&D personnel and their tendency to concentrate in amenity-rich regions. In this context, as also the international literature suggests, environmental quality becomes one of the most important factors in the locational behaviour of highly educated scientific and technical labour force, influencing thus the location of new technology firms and consequently the innovative potential of a region (Aydalot, 1986; Stohr, 1987; Giaoutzi, 1986; Nijkamp, 1986; Camagni et al., 1988).

Agglomeration economies are thought to constitute another important key factor, as far as they play a multiple role in promoting technical progress and higher productivity (Malecki and Varaiya, 1985). Agglomeration economies are conveniently discriminated into **localization** economies and **urbanization** economies (Nijkamp and Davelaar, 1989).

Localization economies are referring to the advantages gained by the firms from the spatial concentration of similar firms (a single industry). These advantages are associated with lower operating costs, higher awareness of the competitors actions, higher necessity to innovate in order to attain a market share, etc. As a result, regions exhibiting a higher concentration of similar firms are displaying a considerable power in terms of innovative behaviour.

Urbanization economies on the other hand, are referring to the advantages to be gained by firms due to the spatial concentration of dissimilar firms (many diversified industries). These advantages are emerging by the diversity of buyers and suppliers close at hand, and also by the spatial clustering of R&D activities of different private firms and public R&D institutes and the positive effects resulting from such a co-operation.

The benefits gained from the agglomeration economies are subject to certain bottlenecks. Actually, there is a limit of spatial concentration of firms and people, beyond which diseconomies of scale are emerging due to the exhaustion of the capacity of the area at hand.

Another important factor influencing the innovative potential of a region is the **availability of venture capital**. This factor has been considered as an important explanatory variable for the existence of regional disparities in terms of innovative behaviour among regions.

Also, as the current regional literature suggests, much attention must be paid to the role of the **multi-locational firms** as to the fact that the spatial distribution of the various 'components' of these firms, namely the separation between the routine and non-routine and R&D activities, will considerably affect the spatial production of innovation.

Finally, **institutional arrangements** contain an important locational factor enhancing the innovative potential of a region in terms of various regulations which may stimulate or demotivate its innovative behaviour. As Perez suggests, periods of sustained economic growth exist when there is a good match between the new technological paradigm and the prevailing institutional climate (Gillespie et al., 1988).

As it is generally admitted, economic development of regions in the light of NIT developments is greatly dependent upon the **growth of new industries**, the carriers of the new technological paradigm, and the **transformation of production processes** in traditional industries. The impact of the above two processes upon the regional structures is likely to vary significantly between different regions due to the different industrial structures, the locational requirements of new industries, different rates of diffusion and differences in the adaptive capacity of socio-economic and political conditions.

Therefore, technological changes raise a number of important distributional issues related to the nature of the societies. A crucial question may arise in this respect: who gains and who loses in the information economy. The answer to this question is not quite straightforward as it depends very much on the role of and the control over technology (Goddard and Gillespie, 1988). In other words, it depends upon how technology is used, by whom and to what ends.

In an information economy perspective, knowledge and information become strategic resources which could potentially be used to attain **equity** and **efficiency** goals among regional structures. In practice, however, they turn out to be neither equitable nor efficient due to the way new technologies are applied.

On the one hand, information economy may become less equitable as a result of the strategic role played by **information** and **knowledge** as key resources. Access to information and knowledge networks is no more provided exclusively from the public, but becomes a marketable resource where information can be obtained only at a certain price. What will be the implications of this fact in terms of unequal access? In other words, which are due to the tarification policies encouraging even access to networks and which are the barriers imposed by diversified tarification systems upon development structures or spatial distribution of activities?

The information economy also appears to be addressing emphasis upon efficiency issues resulting from competition overlooking somehow the equity dimensions of a regional development context. The application of new information technologies favours particularly the large corporations, facilitating their geographical extension and control over global markets for their own benefits.

These issues should be faced with great concern from those involved in urban and regional policy, as it is evident that developments in information economy and use of new technologies without relevant policy framework are likely to exacerbate geographical divisions and to strengthen regional disparities. These policies should intend to ensure that benefits which can be potentially introduced by the use of information technology advances will be shared between the different cities and regions of our society.

6. Towards a Distributional Perspective

The above discussion has revealed a strong need for an operationalization of policies dealing with the core-periphery relations in an international level.

The 'space of flows' which has superseded the 'space of places' (Castells, 1984) in the context of NIT regime places information intensive activities and the use of NIT as the key element shaping 'economic power' relations over space. The above power relations are following hierarchical orders which do not conform with the tree-like ideas of Christaller type ranging from local to national upwards, but rather with supranational orders.

In this respect, competition of market niches is placed in another perspective since 'economic power' is referring to the ability to 'command' human and material resources not just in the locality but in the whole nation, and the complementary sphere of income distribution across space as well (Camagni et al., 1987). In Friedmann's effort to define economic power - as the ability of organizational and institutional actors, located in geographical space, to mobilize and allocate resources (manpower, capital and information) and internationally to structure the decision field of others (i.e., to constraint the decision of others by policies, rules and commands) - it has been revealed that power is not just decision-making power - the directional function - but also the relative position of ours vis-à-vis other actor's directional functioning (Camagni et al., 1987).

The new perspective of 'economic power' has either been in the form of 'potential' or 'actual' power - where the latter is more visible in its tangible effects (Dicken, 1976; Camagni et al., 1987) - identifying power both as a **stock** with uneven distribution and as a **flow** which is more or less reciprocal.

Spatial structures in such a reasoning are subjected to the irrevocable laws of gravity, the centripetal and centrifugal forces rendering inequalities as a structural component in the formation of space.

Figure 1 draws a path of relations in the context of the new economic

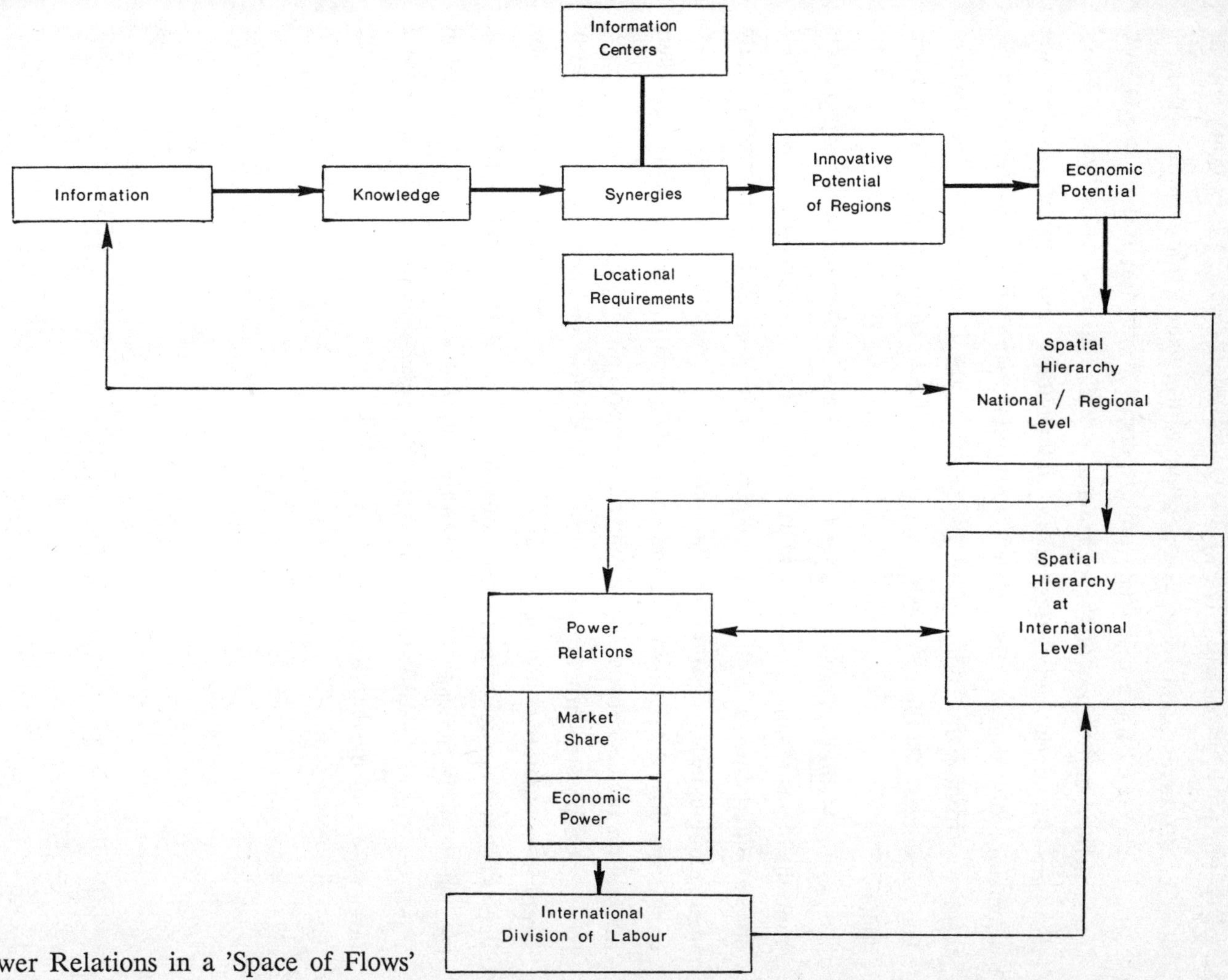

Figure 1. Power Relations in a 'Space of Flows'

power perspective imposed by the role of **knowledge** and **information** as strategic resources upon which competition and efficiency have to be based.

The need for a globalized policy perspective in this respect meets with the structural constraints of the peripheral scene. The cumulative advantage of the core in specialized infrastructure information and services evokes the dispersion of routine and lower rank activities while refrains strategic resources from being diffused in places with lower comparative advantages. The prospects of the periphery appear to have been at stake in such a context.

In Mydral's vision periphery will remain with unfavourable prospects until the development proceeds in certain advanced stages. But how can this be translated into a globalized context of relations? Will peripheries retain their complementary role in an economic power distribution ranking, competing for their survival? And if so, which are the prospects for their integration in the network economies of today?

Corporate structures in this new context are emerging with even stronger comparative advantages as to their access to advanced information networks, knowledge infrastructure, broader range of locational options as well as higher ability to capture the innovative potential of regions. How will this affect the forces at work in the market? The answer to this question lays to a large extent in the hands of policy makers!

References

Anderson, A. and B. Johansson, Knowledge Intensity and Product Cycles in Metropolitan Regions, Working Paper, IIASA.

Aydalot, P. (ed.) (1986), Milieux Innovateurs en Europe, GREMI, Paris.

Camagni, R. and R. Rabelotti (1988), Innovation and Territory: The Milan Information Technology, in: M. Giaoutzi and P. Nijkamp (eds.), Informatics and Regional Development, pp. 215-234, London, Gower.

Camagni, R. and T. Pompili (1987), Inter-regional and Inter-urban Economic Power Relations as Command of Local Resources: The Italian Case, Research Paper, Instituto di Economia Politica, Universita Bocconi, Milano.

Castells, M. (1984), Technological Change, Economic Restructuring and Spatial Division of Labour, Paper presented at the Symposium on Regional Development Processes/Policies and the Changing International Division of Labour, Vienna.

Dicken, P. (1976), The Multiplant Business Enterprise and Geographical Space: Some Issues in the Study of External Control and Regional Development, Regional Studies 10:401-412.

Friedmann, J. (1972), The Spatial Organization of Power in the Development of Urban Systems, Development and Change 4:12-50.

Giaoutzi, M. and P. Nijkamp (eds.) (1988), A Key Force for Spatial Dynamics, in: M. Giaoutzi and P. Nijkamp (eds.), Informatics and Regional Development, pp. 1-19, London, Gower.

Giaoutzi, M. (1987), Information Technology and the SMEs: A Spatial Perspective, Paper presented at the RSA Meeting, Athens.

Giaoutzi, M. (1986), Technological Change and Regional Dynamics, Paper presented at the GREMI meeting on Technological Change and Regional Development, Paris (mimeographed).

Gillespie, A., H. Williams and A. Thwaites (1988), Strategies to Encourage the Development of New Information Production in Europe's Less Favoured Regions, in: M. Giaoutzi and P. Nijkamp (eds.), Informatics and Regional Development, pp. 56-71, London, Gower.

Gillespie, A. and J. Goddard (1987), Information and Communications Technology and Regional Development, Paper presented at the OECD Seminar on Information and Telecommunications Technology for Regional Development, Athens.

Goddard, J. and A. Gillespie (1988), Advanced Telecommunications and

Regional Development, in: M. Giaoutzi and P. Nijkamp (eds.), Informatics and Regional Development, pp. 121-146, London, Gower.

Malecki, E. (1981), Technological Change and Regional Development, Papers of the RSA 47.

Nijkamp P. and E. Davelaar (1986), Spatial Dispersion of Technological Innovation: The Incubator Hypothesis, Research paper, Dept. of Economics, Free University of Amsterdam.

Nijkamp, P. and H. Dieperink (1986), Innovative Behaviour, Agglomeration Economies and R&D Infrastructure, Research Paper, Dept. of Economics, Free University of Amsterdam.

Nijkamp, P. and H. Dieperink (1986), Innovation and Regional Policy in a Spatial Context, Research paper, Dept. of Economics, Free University of Amsterdam.

Nijkamp, P. (1987), The Strategic Role of High Tech Firms in Peripheral and Border Areas, Research Paper, Dept. of Economics, Free University of Amsterdam.

Nijkamp P. and E. Malecki (1987), Technology and Regional Development: Some Thoughts on Policy, Research paper, Dept. of Economics, Free University of Amsterdam.

Nijkamp, P. and I. Salomon (1986), Telecommunications and the Tyranny of Space, in: I. Orishimo, G. Hewings and P. Nijkamp (eds.), Information Technology: Social and Spatial Perspectives, pp. 89-106, Springer-Verlag.

Oakey, R., High-Technology Industry and Agglomeration Economies, in: P. Hall and A. Markusen (eds.), Silikon Landscapes, pp. 94-117, Boston, Allen and Unwin.

Stohr, W. (1986), Territorial Innovation Complexes, in: P. Aydalot (ed.), Milieux Innovateurs en Europe, GREMI.

Swijngedouw, E. (1987), Planning!...What Planning?, in: J.J.M. Angenent and Bongenaar (eds.), Planning without a Passport, pp. 77-90, Utrecht, Elinkwijk.

Chapter 10

TECHNOLOGICAL CHANGE, UNCERTAINTY AND INNOVATION NETWORKS: TOWARDS A DYNAMIC THEORY OF ECONOMIC SPACE

R.P. Camagni

1. Introduction[1]

The theoretical and empirical literature on the relationships between space and technological change is literally immense, and scattered along different directions that may be listed tentatively in the following:

- the theory of innovation diffusion
- the spatial geography of R&D
- the spatial preconditions for (and obstacles to) innovation: presence of human capital, availability of producer services, 'urban' environment, industrial structure
- the characteristics of innovative environments: valleys, corridors, routes, parks; the 'Third Italy' phenomenon; the 'milieux innovateurs' of the new Gremi approach (see below)
- the regional differentials in productivity growth
- the effects of technological change on regional development
- the effects of technological change on urban development
- the spatial effects of specific technologies: industrial automation, information technologies, telecommunications,...

The aim of this paper is not really to make an overview of this literature; the task would be overwhelming in comparison to the scarce resources of a single researcher and the limited ambitions of a paper.

A different perspective seems however more fruitful in the present state-of-the-art of our discipline, as far as technological change problems are

[1] I am indebted for useful comments and suggestions on an earlier draft of the paper to many friends and colleagues; among them, I am particularly grateful to Olivier Crevoisier, David Keeble, Linda Kimball, Alfred Thwaites, and Peter Townroe.

concerned; namely:

a. to inspect in some depth the characteristics of the **new scientific paradigm** through which the field is approached by general, non-spatial economic theory, the **evolutionary paradigm**

b. to link it with the **new 'network' behaviour of firms** in their struggle for dynamic excellence, and

c. to highlight the **role of spatial variables** in the new interpretative context.

My approach starts therefore from the consideration of these three open issues, which in my view lie near the frontier of the present scientific debate and call for new substantial theoretical and empirical efforts. My general hypothesis is that, within the new 'evolutionary' paradigm, spatial variables are no longer relegated to a peripheral condition in the theoretical framework, no longer play the role of a simple extra-dimension of the problem, but represent central elements of the interpretative framework itself. This fact is particularly interesting in a context where the new firm behaviours of transnational cooperative agreements and network linkages at a first glance seem to annihilate space as a relevant economic operator.

In the theoretical framework that will be built throughout the paper, the local spatial context, or the local 'milieu', will emerge as a necessary and crucial element in the process of technology creation and as the 'operator' that allows the individual decision maker to cope with the problems of static and dynamic uncertainty which are intrinsic in innovative behaviours. The reflections presented here may therefore be intended as **prolegomena** to a new theorization of economic space in a dynamic context.

The characteristics of the emerging scientific paradigm in the study of technological change will be inspected first, keeping in mind both its actual limits of general consistency, coherence and completeness, and the important efforts made by the best representatives of the 'traditional' theory in enlarging its explicative power beyond abstract and often meaningless cases (section 2).

Then the new behaviours of firms will be analyzed when, in a dynamic and innovative context, they are facing inescapable problems of 'uncertainty', imperfect and costly information collection, limited forecasting capability and rationality (section 3). The central role of network relationships, developing both at a local-informal and at a formal-transterritorial level, will become clear in this context (sections 4 and 5).

Thirdly, the problems of innovation adoption and diffusion will be taken up in terms of evolving and competing technologies on the geographical space (section 6), and the most appropriate modelling approaches will be discussed.

2. The Emergence of the New 'Evolutionary' Paradigm in the Study of Technological Change

Traditional economic theory has directly transferred to a dynamic setting almost all the explicit or tacit assumptions that were employed to study optimizing firm behaviours in the static context of given and perfectly known technologies. Conditions of perfect knowledge are projected along the time horizon of the firm, hypothesizing perfect foresight on both technological advances and their economic outcomes, and a perfect and 'rational' utilization of the existing information. Standard-choice theory assumes therefore hyper-rational, never-failing agents which always select actions in order to maximize expected utility based upon observed and free information.

In this framework, technological change is not really explained, but only 'assumed' and instantaneously adopted by firms (if the new technology proves itself superior to all previous ones in all points of the factor-prices frontier, as often happens). In a world where technology is equated to perfectly free information (Arrow, 1962) and where actors' expectations are by definition 'rational', all that firms have to do is dip into the pool of technological know-how (become a sort of public good), optimizing an intertemporal objective function.

Needless to say, this vision trivializes both the concept of technology and the concept of 'time' in economics. In Prigogine's words, dynamics is reduced to 'a movement in a timeless time', with no role to irreversibility, 'memory', or history. Along the same lines, Frank Hahn, the champion of most advanced neoclassical theorizing, has questioned the rationality assumptions in presence of incomplete information and imperfect markets and suggested that, on the contrary, 'dynamics should be viewed as a learning process both about demand conditions and the strategies of near competitors. When an equilibrium is defined relatively to such (dynamic) processes, it seems that they are undetermined unless history - that is information - is explicitly modelled and known..... There is something essentially historical in a proper definition of equilibrium and of course in the dynamics itself' (quoted in Freeman, 1988).

The new 'evolutionary' paradigms in the scientific interpretation of technological change emerged because of similar dissatisfactions, and committed itself to the full consideration of such 'real life' elements as imperfect information, limited search capabilities, 'bounded' rationality, cumulative learning processes, static and dynamic uncertainty, even at the expense of a lower formalization and a limited prediction capability. Pioneered by the works of Nelson and Winter (Nelson and Winter, 1982), the new approach was quickly developed thereafter and recently received an important state-of-the-art presentation (Dosi et al., 1988).

As the new approach takes these elements of market imperfection and uncertainty directly into account and incorporates the very nature of the Schumpeterian creative innovation processes, its main concern appears almost at the opposite side with respect to that of the neoclassical one. In the latter case the evidence of a wide spectrum of differentiated, lagged and unexpected behaviours conflicts with the geometrical perfection of the theoretical model and its prediction of a unique optimal solution for all firms in each market. On the contrary, evolutionary approaches have to explain why and how the apparently anarchistic process of innovation creation and diffusion does not end up in a purely random aggregate phenomenon, but in a self-organized and ordered process showing regular patterns of change.

The rationale for this ordered pattern is found in the intrinsic **learning** nature of technological change, showing up on the double level of **microeconomic learning processes** (in both research units and firms) and **social-institutional learning processes.** These processes, embedded in the very nature of technological change, constrain its evolutionary path along ordered 'technological trajectories' and long-term, cyclical waves (Dosi, 1982; Dosi and Orsenigo, 1988; Peres, 1985; Freemand and Peres, 1986). At the microeconomic scale this effect stems from:

* the presence of specific technical properties, reducing the spectrum of possible behaviours
* the sharing of similar problem solving heuristics among firms
* the cumulative agreement in the society on the definition of relevant problems and targets
* the use of decision routines which limit the spectrum of possible actions
* the cumulative nature of 'incremental' innovations within each 'technological paradigm'.

At the macroeconomic scale, 'order' may come from socio-cultural and institutional resistances to change and from the stabilizing characteristics of the economic and political rules that define each 'régime de régulation' (Boyer, 1986).

According to the new approach, technological change may be interpreted and 'stylized' in the following way:

a) it is an irreversible, path-dependent and evolutionary process, stemming from the behaviour of economic agents which explore only a limited part of the set of theoretically possible actions, that part which is strictly linked to previous innovation adoptions and to already acquired know-how. This is far from the traditional view of technological change as a fast, flexible and optimal reaction to changing market conditions, choosing among a wide spectrum of perfectly known alternatives;

b) it lies therefore on a cumulative learning process, resulting in the 'creation' rather than the simple 'adoption' or imitation of already existing ideas (inventions or innovations);

c) it implies search and decision routines which limit the cost of information collection and the cost connected with the presence of uncertainty;

d) it implies the full commitment of all functions of the firm, and in particular a deep interlink among R&D, production, marketing and organization;

e) due to its dependence on internal learning processes (learning-by-doing, by-using, by-searching and, more indirectly, learning-to-learn) it cumulatively builds on tacit, firm-specific know-how and on 'intangible' assets: its transfer or imitation is therefore a highly difficult process;

f) its historical path may by no means be interpreted in terms of 'optimality'. From a macro point of view, in fact, its path-dependent nature and the non-linearities connected with the learning processes may act as dynamic 'entry-barriers' with respect to possible, theoretically more efficient, alternative technologies. Once a bifurcation point is overcome in the development path of a particular technology and a specific trajectory is chosen, cumulative processes reinforce and perpetuate that choice, highly reducing the spectrum of possible outcomes and alternatives (see, as an example, the 'genetic' limits of nuclear power technologies, which deeply influenced the subsequent trajectory). On the other hand, from a micro point of view, conditions of limited information and 'bounded rationality' limit (or change) the meaning of optimizing behaviours.

New reflections are still needed for the full development of the new scientific approach to technological change. In particular, the definition and the meaning of possible 'evolutionary equilibria' have not been stated in a sufficiently sound way. Analogies from other sciences suggest to employ in this respect the concept of an 'attractor', or a series of attractors, leading the evolution within each established technological paradigm towards some sort of stationary adjustment path (Dosi and Orsenigo, 1988). These attractors, however, are thought as partly endogenous, in that they, too, are path-dependent and behaviour-dependent: 'it is the very process of approaching any one attractor which may well change the value of the attractor itself' (ibidem). By this, the entire process being modelled may become excessively cumulative and 'hyper-selective', depending almost exclusively upon initial conditions and opening the door to a new sort of 'technological determinism', similar in principle, though in a different theoretical context, to the old determinism of the production function approach (Gaffard, 1986; Camagni, 1986a).

Emphasizing perhaps a teleological element of a more subjective nature, I might better utilize the concept of an attractor to indicate the final goal assigned to technology in a specific historical phase of the capitalist society, which informally leads the incremental development of technological innovations. Goals of this kind may be found in the full exploitation of economies of scale and division of labour in the 'fordist' society, and in the attainment of full managerial control over the production conditions in the information and flexible automation society (Camagni, 1986b) (see for an example in the latter case, the choice of programmable vs. 'play-back' factory automation technologies in the fifties at M.I.T., clearly inspired by the objective of limiting the labour-force responsibility, which highly constrained the technological trajectory thereafter).

3. Uncertainty and the Innovation Process

In spite of these and other difficulties, the new approach to technological change in industrial economics brings important insights into the genetic elements of technical advances, in a way that in my opinion allows to highlight the role of spatial variables in this context.

The key concept in this respect is the concept of **uncertainty** in its different forms. Uncertainty, and the correlated presence of imperfect 'information' prevents a pure price mechanism from allocating resources in an optimal way and driving economic activities to any kind of competitive equilibrium. In fact, as Arrow has shown (Arrow, 1974), uncertainty can be incorporated in a competitive equilibrium system only by assuming an equal (imperfect) access of all individuals to the same information, a condition which, in presence of highly differentiated firm sizes, market structures and spatial situations, is to be considered as highly unrealistic.

In their economic behaviour and decision making process, firms face five important kinds of uncertainty:

i) static uncertainty coming from an **'information gap'**, linked to the complexity, the width and the cost of the information collection activity; in the real world, the firm is usually left with a huge lack of relevant information on the occurrence of already known events;

ii) static uncertainty coming from an **'assessment gap'** linked to the difficulty of inspecting ex-ante the **qualitative**, mainly hidden, characteristics of inputs, components, production factors, technical equipment;
iii) static uncertainty coming from a **'competence gap'**, linked to the firm's limited ability of processing and understanding available information; the existence of technical problems whose solutions are obscure are an example of this wide category of situations;
iv) dynamic uncertainty coming from the so-called **'C-D gap'** (competence-decision gap); uncertainty regards, in this case, the impossibility of precisely assessing the outcomes of alternative actions, even in presence of full and free information on past events, due to the complexity of the decision problems themselves and inherently imperfect foresight. The probability of choosing a wrong or inferior technology is therefore large;
v) dynamic uncertainty coming from a **'control gap'**: the outcomes of present actions depend in fact on the dynamic interaction among independent decisions of many actors, on which the firm has by definition a minimum control.

All these forms of un-insurable uncertainty and, in particular, the dynamic ones call for mechanisms of reduction of the general cost they imply. The firm therefore has to develop new and specific functions, rules, routines and procedures which are not considered in the conventional neoclassical theory of decision-making under conditions of perfect information but emerge indirectly from the new evolutionary approach of technological change. These functions are designed to cope with each specific type of uncertainty, and may be listed as follows (Figure 1):
i') **search functions** and procedures regarding information collection, information organization, technological monitoring;
ii') **screening functions** of market signals and inspection of hidden characteristics with regard to inputs and equipment; **signalling functions** and quality certification with regard to outputs;
iii') **transcoding functions**, which translate external information into a language that the firm may understand. These functions are perhaps the most critical,

Sources of uncertainty	Type of uncertainty	Functions involved	Traditional instruments for coping with uncert.	Outcomes	New "0perators"
Information Gap (imperfect, costly information)	static	Search	Technology monitoring	Formation of beliefs on state-of -the-world	Local Environ- ment or "Milieu" / Firms' Networks and Network- - Firms
Assessment Gap (presence of hidden characteristics)	static	Screening/ Signalling	Quality contr / certification		
Competence Gap (imperfect inform. processing ability)	static	Transcoding	R&D	Know-how acquisition	
Decision Gap (C-D) (imperf. assessm. of decis. outcomes)	dynamic	Selection	Decision routines/ manag. style	Decision	
Control Gap (imperf. control on other actors' decis.)	dynamic	Control	Hierarchy	Reduction of complexity	

Figure 1. Uncertainty and firms behaviour: functions and operators

though widely overlooked by economic theory, in that they control the process of interfirm know-how transfer and information appropriation. Utilizing codified information, both freely available or costly, and merging it with tacit and informal information, transcoding activities convert a chaotic and unordered 'information' flow into a firm-specific 'knowledge' and possibly into potential business ideas at the disposal of the managerial decision-making. The main aim of the R&D efforts should be considered under this new perspective, rather than on the traditional and naive perspective of the 'invention' task (Foray and Mowery, 1988);

vi') **selection functions**, governing the proper decision making process through the adoption of decision routines and firm-specific management styles (Nelson and Winter, 1982);

vii') **control functions**, aiming at a drastic reduction of the complexity of the external environment, through an expansion of the power limits of the firm. Long since, Williamson has pointed out that the most likely firm strategy in presence of limited rationality, imperfect markets, dynamic uncertainty and risk of opportunistic behaviour is an expansion of the 'hierarchy', through acquisitions, mergers and any other form of equity participation in the direction of both customers/supplier firms (vertical integration) and competing firms (horizontal integration) (Williamson, 1985).

The first two types of functions address the creation of the firms' **'beliefs on the state-of-the-world'** under conditions of imperfect information, and are implicit in the behavioural models proposed by recent neoclassical approaches to decision-making under uncertainty such as search theory, market signalling theory and the economics of qualitative uncertainty (Hey, 1979; Spence, 1973; McKenna, 1986). The third kind of functions, the 'transcoding' ones, control the process of technology transfer to the firm and the development of its internal know-how: they are hidden, mainly tacit functions and processes, often overlooked by economic theory, as mentioned before.

The fourth and fifth kind of functions, selection and control, project the firm into a truly dynamic context, and aim at reducing complexity both in the decision procedures and in the external context itself.

The instruments utilized within all the previously mentioned functions in order to reduce uncertainty and complexity (information monitoring, quality control and certification, R&D, decision routines and equity control) may be labelled as traditional in that they stem from a standard interpretation of the firm as an **individual** agent, clearly separated with respect to all other agents, interacting with its external environment only through the canonic (but abstract) 'operators' of **markets** (and market transactions) and **organizations.**

But these operators have proved to be highly inefficient, particularly in a dynamic context, one which is relevant in the perspective of technological change. Therefore, some new, though equally imperfect, operators have to be found and added, both at the theoretical level and at the level of the real firm behaviours.

These **new 'operators'**, performing different but parallel tasks and, in particular, the task of reducing the degree of uncertainty in dynamic behaviours, may be found, in my view, in the **local environment** (the **'milieu'**) and in **cooperation networks** among firms. Both imply specific functions, procedures, costs and risks, as will be clarified in the next section, and are linked by the nature of their genetic principle: **synergy** and collective action, as opposed to (market) **competition** and (organizational) **power.**

4. The Firm and its Local 'Milieu'

In a world of free and perfect information, the boundaries between the market and the organization (firm) are clear and stable; in fact, in the case of zero information and transaction costs, these boundaries are defined only by the shape of the organizational cost curve (Figure 2). In order to communicate with the external environment, the firm utilizes internal interface functions such as marketing and procurement offices.

But the presence of inescapable static and dynamic uncertainty in the real world and in particular in the process of innovation and technical change implies the presence of:

* extra costs ('use costs of the market', in Williamson's terms) and therefore
* new functions to cope with these costs, as seen before, and therefore
* new 'operators' or institutions organizing these functions and shaping

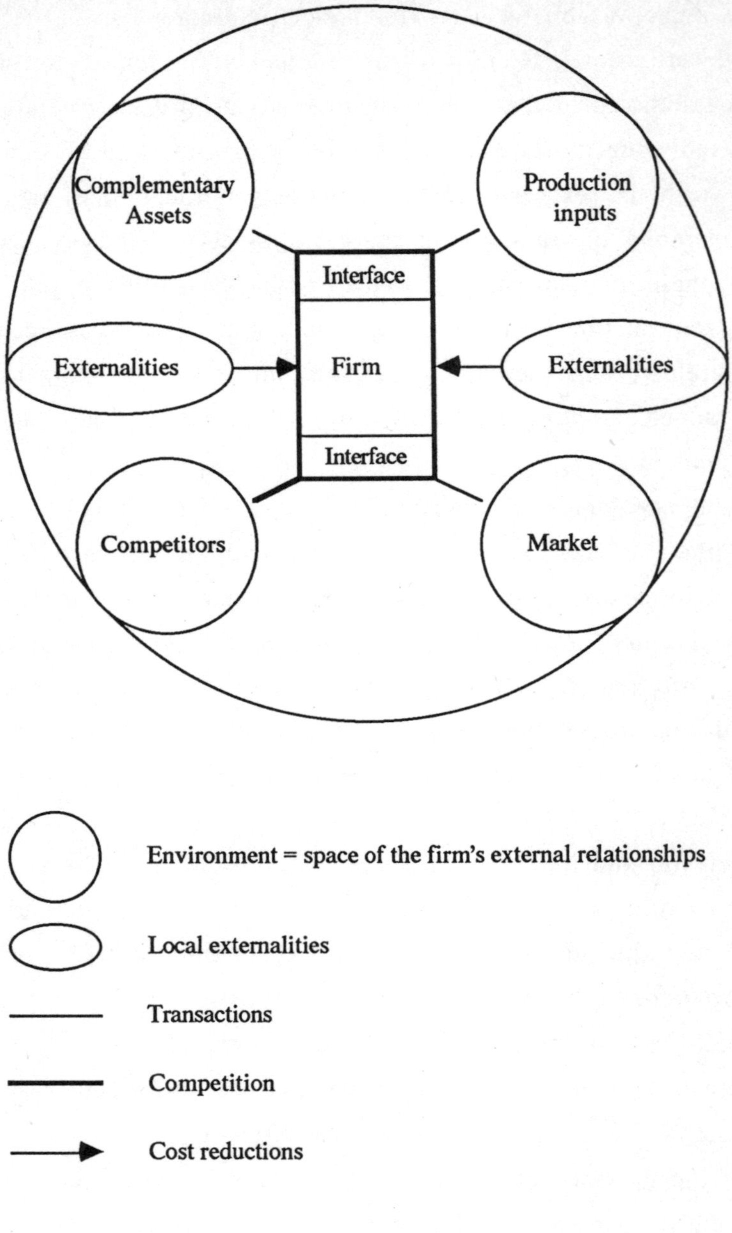

Figure 2. The firm and its environment in a condition of perfect information and a static setting.

factual behaviours, beyond perfect markets and hierarchies.

In my view, the **local environment** of the firm, or the local **'milieu'** as it is called by the GREMI Association approach (Aydalot, 1986; Aydalot and Keeble, 1988; Maillat and Perrin, 1990) may be considered as one, and perhaps one of the most important, of these uncertainty-reducing operators. In general terms, the local 'milieu' may be defined as a set of territorial relationships encompassing in a coherent way a production system, different economic and social actors, a specific culture and a representation system, and generating a dynamic collective learning process (Crevoisier et al., 1990).

In our specific theoretical context, the 'milieu' performs most of the functions mentioned in the previous section, in strict integration and 'synergy' with the firm, through a **collective and socialized** process **allowing cost reductions and enhancing the effectiveness** of the dynamic decision-making process of local firms (Figure 3). In fact, the local environment performs:

I) a **collective information-gathering and screening function**, through informal interchange of information between firms operating in the same markets and technologies, public or cooperative monitoring on factor markets and technical change, selection of information channels through repeated experience and 'memory' (**'search function'**);

II) a function of **'signalling'** in the direction of the market of local firms, in terms of product image and 'reputation', cooperative advertising, and supply of a sort of 'quality certification';

III) a **collective learning process**, mainly through skilled labour mobility within the local labour market, customer-supplier technical and organizational interchange, imitation processes and reverse engineering, exhibition of successful 'climatization' and application to local needs of general purpose technologies, informal 'cafeteria' effects, complementary information and specialized services provision (**'transcoding function'**);

IV) a **collective process of definition of managerial styles and decision routines**, through managerial labour mobility, imitative decisions, cooperative decision-making through local industrialists associations, complementary innovation processes (**'selection function'**);

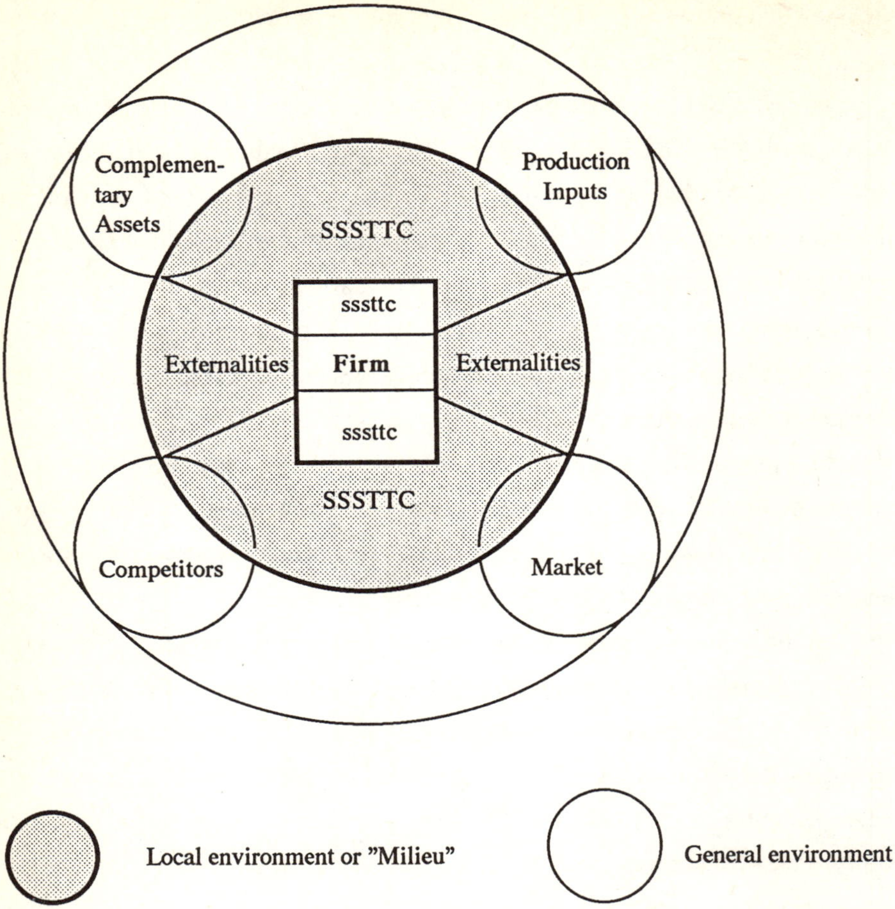

SSSTTC = uncertainty - reducing functions performed by the "Milieu":
search, selection, signalling, transformer, transcoding, control

sssttc = uncertainty - reducing functions performed by the firm (same)

Figure 3. The local and external environment of the firm and their functions in a dynamic setting

V) **an informal process of decision coordination**, through interpersonal linkages (families, clans, clubs, associations), easier and faster information circulation on innovative decision making, easier financial-industrial linkages, similar cultural background of decision-makers ('**control functions**').

Beyond these functions, linked to specific kinds of static and dynamic uncertainty, another important function is performed by the local 'milieu', contributing to enhancing local firms' effectiveness and innovation:

VI) a function of conversion of external energies to the needs of local firms, this function being particularly important in the labour market, human capital and educational sphere: in fact, not just information is decoded and collectively organized, but also potential energy, as represented by availability of generic production factors, is channelled and transformed in order to match with the qualitative claims of actual and potential demand of the local structure ('**transformer function**').

In the abstract neoclassical scheme, all these functions are performed automatically by the market. But, as we have mentioned earlier, many difficulties emerge for the firm, even in a static context, in the form of lack of transparency, presence of hidden qualities in the products (the market is full of 'lemons', whose characteristics are discovered only after purchase), opportunistic behaviours, imperfect knowledge of the codes and channels by which information may be gathered. In a dynamic context, which is the most important in the context of technology and innovation decisions, these difficulties are amplified, the signalling function of the market becomes weak, and the utilization of routines, the reference to widely accepted beliefs, the effort to control the decision-making process of the other actors become an inescapable must for the firm. The local 'milieu' may be considered under this respect as an extension and a specification of an 'organized market', where not just quantities and prices are fixed but also institutions, real actors, languages and codes interact with each other.

Consequently, the **definition that may be proposed of the 'milieu' or the local environment is that of a collective operator reducing the degree of static and dynamic uncertainty for the firms by tacitly or explicitly organizing the functional and informational interdependence of local actors and**

informally performing the SSSTTC functions (search, signalling, selection, transcoding, transformer and control) (Figure 4).

As far as the function of signalling is concerned and the parallel formation of accepted 'beliefs' in the case of quality uncertainty or dynamic uncertainty, it is important to remember that even in mainstream neoclassical models an explicit condition for the existence of an equilibrium solution is the good-matching of the two elements (signals and beliefs) (McKenna, 1986, chapter 8). The local milieu, through repeated experience and localized 'memory', performs exactly this function, attributing reliability to signals and spreading the acceptance of a common vision about the state-of-the-world.

From all the preceding arguments it becomes clear that 'proximity matters', and in fact, it does in a threefold way:

* because of the presence of local resources of human capital, that are quasi-immobile with respect to the external territory and highly mobile within the local territory; their presence accounts for much of the local collective learning process and, as far as it contributes to effectively enhancing the productivity of local firms and to creating a local external 'image', it cumulatively reinforces itself through polarization effects and attraction of external firms (the example of Silicon Valley is enlightening in this respect; see Gordon, 1989a);

* because of the presence of an intricate network of mainly informal contacts among local actors, building what Marshall called an 'industrial atmosphere' within industrial 'districts', made up of personal face-to-face encounters, casual information flows, customer-supplier cooperation and the like (see the contributions of Stöhr, Perrin, Quevit, Gordon and Dilts, Camagni in the Gremi publications);

* due to the presence of synergy effects stemming from a common cultural, psychological and often political background, sometimes enhanced by the effectiveness of some local 'collective agent'; the common cultural roots are highly important in that they contribute to the establishment of tacit codes of conduct, to the decoding of complex messages (Lundvall, 1988) and to the formation of common 'representations' and widely sheared 'beliefs' on products and technologies (Planque, 1983; Crevoisier et al., 1990).

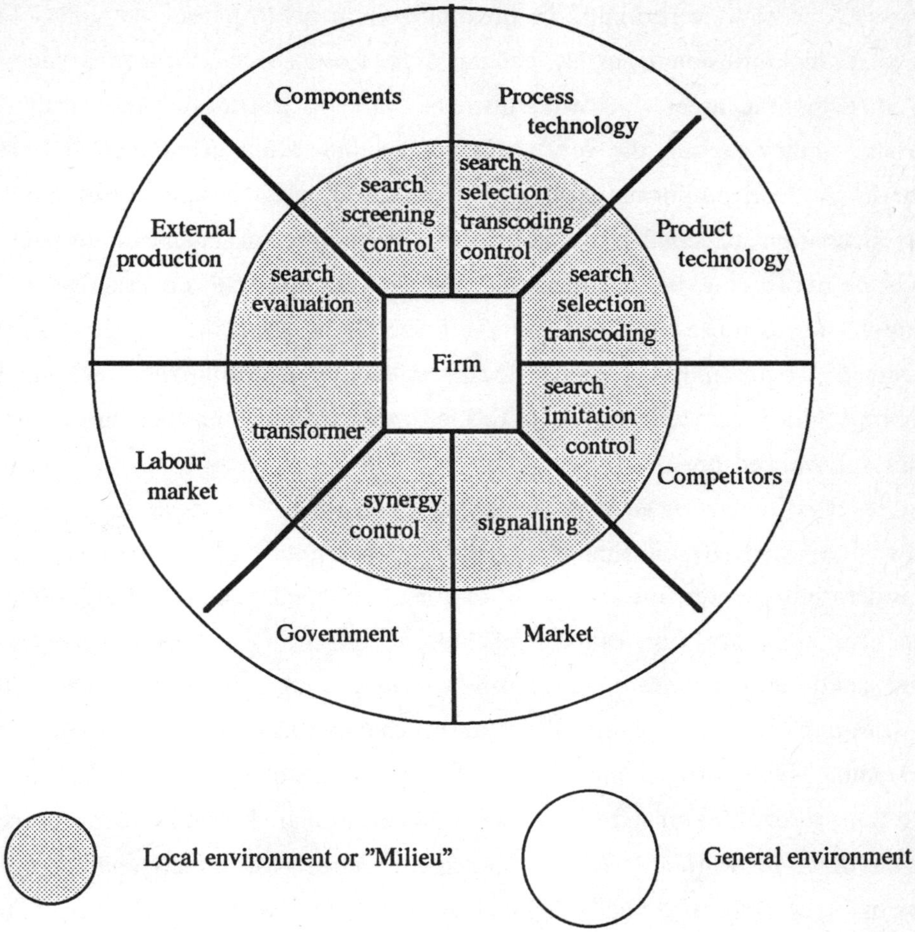

Figure 4. Main uncertainty - reducing functions performed by the 'milieu'

All these territorial and proximity elements explain why innovation creation and diffusion is highly enhanced in those special territories such as big metropolitan areas, industrial districts, 'valleys', 'corridors' and 'parks'. In particular, they explain the very nature of agglomeration economies and their role in the early, information-intensive phases of product life cycles, and in the 'incubation' of small firms which are particularly unarmed with respect to uncertainty. It may be affirmed that, if the existence of uncertainty in its multiple forms raises the minimum efficient firm size, the presence of an information-rich and synergetic local 'milieu' performing an uncertainty-reducing function, allows to keep this efficient size low enough to let small firms survive and prosper.

It is interesting to note that other branches of the social sciences have long since arrived to similar conclusions, even if not taking in direct consideration the territorial aspects of the theoretical problem. In particular, organization theory, organization psychology and strategic choice theory have highlighted the important relationships between the individual actor, the organization and its 'context' in a world characterized by uncertainty (Katz and Kahn, 1966; Friend and Jessop, 1969; Jahannisson, 1987). This approach has been utilized to analyze, inter alia, the locational decision-making process of the firm (Townroe, 1971). 'An adequate model for understanding policy making must start with the individual and its many types of **fallibility**, but it must also take into account the **collective** situation in which executives function' (Katz and Kahn, 1966, p. 274; our italics).

What is called the 'context of operations' allows the individual to overcome the unescapable presence of (static) uncertainty in the process of gathering and interpreting information, supplying him with both 'current views of how situations should be classified' and 'current objectives and appreciation of constraints' (Friend and Jessop, 1969). The consequent picture of a dynamic process of interaction and mutual modification of the 'context' and the 'organization' (a 'system learning process' in Townroe's words) is theoretically similar to the collective learning process taking place in our model within the local **milieu**.

5. Networking

5.1 Definition

The effectiveness of the local 'milieu' as an uncertainty-reducing operator has its limits, however.

Some of these limits are implicit in the nature of the relationships that constitute the milieu itself. These relationships are mainly informal and tacit relationships, operating better on information circulation and on imitating behaviours rather than on more direct linkages among economic actors. Therefore, the role of the milieu becomes weaker when control functions are directly concerned (as it may be seen in Figure 1).

Secondly, the behaviour of the milieu, even if it may be considered as the outcome of a collective learning process going far beyond the possibilities of individual firms, is subject to explicit risks of aggregate and generalized decline, especially in the case of very specialized and homogeneous local structures. The crisis of many specialized old industrial areas in the sixties and seventies, hit by sectoral crises (iron and steel, shipbuilding, textile and motor vehicle regions in the UK, Belgium, the USA, Germany) and the present crisis of a new success area such as Prato in Italy are examples of how local know-how and synergies may be unable to face big dynamic changes in markets or technologies. By the same token, diseconomies of scale and environmental problems may well overcome urbanization advantages in big metropolitan areas in particular historical circumstances.

Therefore, endogenous and exponentially growing locational costs, which may be considered as the opportunity cost of utilization of the milieu, and evident limits in the static or dynamic performance of the 'milieu' itself, push forward the creation of a new organizational and behavioural model, a new 'operator' enhancing the control capability of the firm upon its turbulent environment (Boissevain and Mitchell, 1973; Johannisson, 1987; Kamann and Nijkamp, 1988).

This new operator, superior in some respects to the local 'milieu' and the synergies it may develop, and intermediate between (market) competition and (organizational) power may be found in interfirm **cooperation**; its specific behavioural model is the 'network-firm'. With this new model - occurring through joint ventures, strategic alliances, consortia, technical cooperation, cross commercialization, licensing and franchising agreements - firms get easy access to important complementary assets, markets and technologies without incurring organizational or locational costs (which are typical of internal growth strategies), and free themselves from the limits of local (and internal) competence. In addition, through this strategy a wider control is acquired on both technological trajectories and competitors' conducts.

In our view, **a 'network' may be defined as a closed set of selected and explicit linkages with preferential partners in a firm's space of complementary assets and market relationships, having as a major goal the reduction of static and dynamic uncertainty.**

Network relations, of a mainly informal and tacit nature, also exist within the local environment, linking through **open** chains, firms and other local actors as we have seen before. Our proposal is, nevertheless, to use the term 'network' ('réseau') only in the case of explicit linkages among elected partners and to refer to the former as 'milieu' relationships (Figure 5).

At first sight, cooperation and networking on a trans-regional or trans-national basis represent a sound alternative to the exploitation of local synergies and seem to annihilate space in both its geographical and its relational dimension. In fact, 'certaines analyses se réfèrent à la notion de 'réseau pour caractériser une organisation de la circulation entre entreprises de plus en plus affranchie de la matérialité d'une configuration spatiale locale; l'organisation 'en réseau' marque le passage d'un espace de place à un espace de flux, ou encore celui d'un espace topographique à un espace topologique' (Plan Urbain, 1989). We will come back to this issue later on.

5.2 Relevance

The most important fact at this moment is to recognize the increasing or booming utilization of these new forms of external development by firms of various sizes, particularly in those areas of production characterized by fast innovation and technological change such as electronics equipment, telecommunications, semiconductors, software and factory automation devices; in a word, in the 'information technology' sectors. This new empirical evidence

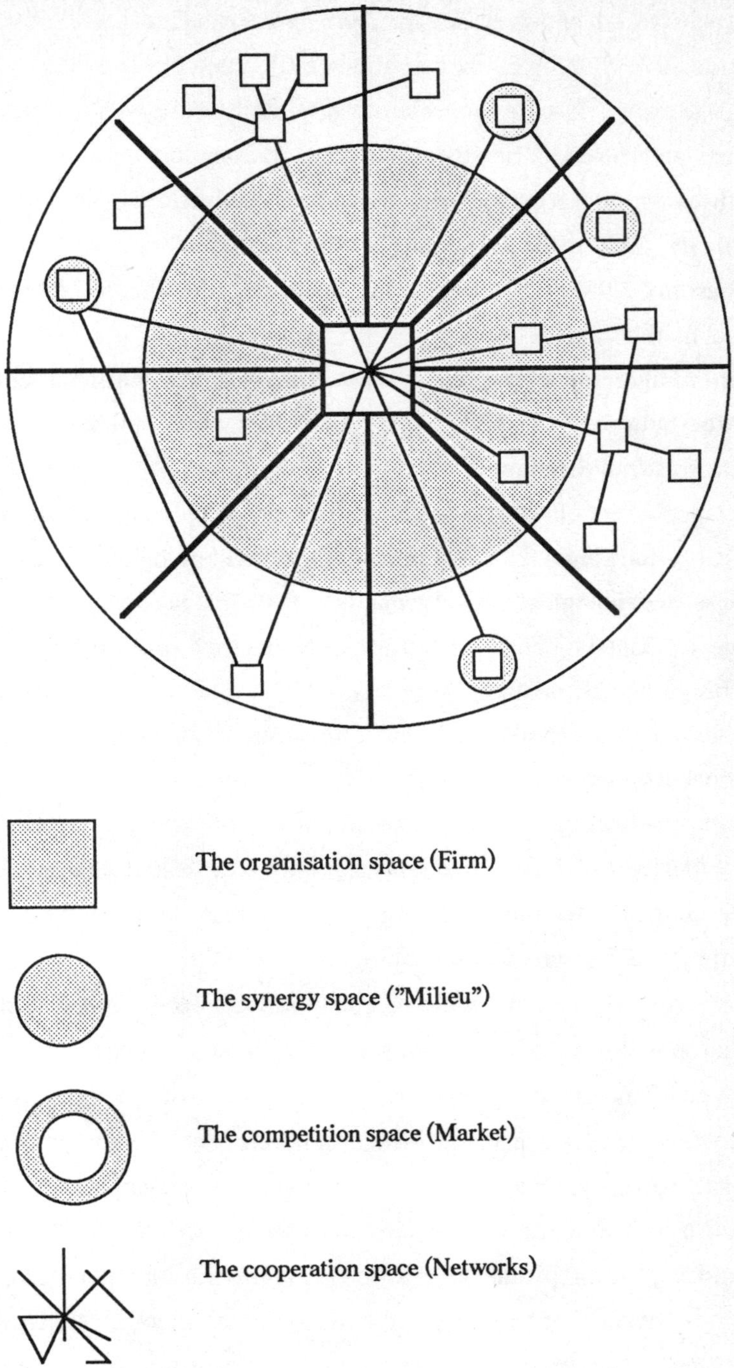

Figure 5. Networking and external environment of the firm.

has to be considered and included in all theoretical framework addressing the interpretation of technological and spatial development.

Up to now, due to the relative originality of novelty of the process, the relevant and already rich literature has mainly been addressed towards field inquiries or descriptive and taxonomic reflections (among the most recent: Foresti, 1986, OECD, 1986; Vickery, 1988; Chesnais, 1988; Camagni and Gambarotto, 1988). The main conclusion of the fertile debate may be summarized in this way:

* cooperation agreements represent **new** forms of international competition, **intermediate** between market resort and hierarchy, taking place within **oligopolistic** sectoral structures;

*they are specific of a context characterized by the emergence of a new technological **'paradigm'**, the information technologies one, featuring **pervasiveness, technological convergence** and fast innovation processes;

* they have as final objectives the traditional ones of **profitability** and **market power**, to be gained through:

- synergies and economies of scale in production, marketing and R&D
- scope economies and product differentiation
- cross fertilization and development of technological complementarities
- the increase of a fast-reaction capability to external shocks
- the control over those innovation assets that define future application patterns of information technologies
- the formation of new kinds of entry barriers (proprietary standards).

The possibility of incorporating the new 'network' behaviour in established economic models depends on the nature of the goals pursued by the firm. In this respect, we may distinguish three broad categories of goals:

A) The first category encompasses the goals of achieving **scale and scope economies**, through the merging of R&D facilities and resources, distribution channels and variety of products. This is the more traditional behaviour, easily interpreted in terms of standard microeconomic theory. Networking and strategic alliances in this case prove superior with respect to traditional behaviours such as equity participation and mergers in that they allow limited cooperation in well-defined fields or 'partial merging', leaving aside the possibility of competition in other fields (this element is important in the case of big conglomerates and multi-division firms).

B) A second and more important category of goals regards the stable utilization of complementary assets, the control of specific technologies and market channels, **avoiding the costs of search, screening and decoding** complex information and the uncertainty elements involved. Thus, we are back to the first three kinds of static uncertainty defined earlier, and we are therefore able to include networking and cooperation agreements within the same theory of evolutionary firm behaviour.

The new behavioural model overcomes the high use-cost of the market, or the high transaction costs which are caused by the presence of market imperfections and episodes of market failure; transaction costs may in fact be defined as 'the opportunity costs of any localized inefficiency in prices to deliver correct signals' (Antonelli, 1987a). In addition, the presence of important 'intangible assets' stemming from long internal learning processes prevents any localized technical progress from being easily transferred by means of simple market transactions (Camagni, 1989) and calls for closer cooperation between the donor and the accepting firm.

Cooperation imposes itself as the most efficient firm conduct with respect to market resort or internal development at particular levels of transaction costs and organizational costs (Figure 6). A third element to be considered in this picture is the appropriation regime of the technology concerned, in that a tighter appropriation condition, stemming from intrinsic complexity or from the presence of institutional barriers such as patents, may emphasize the case for direct cooperation.

C) The third category of goals are the most interesting ones: they regard the dynamic behaviour of the firm directly and confer the cooperation agreement a true nature of 'strategic' alliance. Here the objective is not just the control over a given technology or a given stock of complementary assets, but rather **the control over the development trajectory of these assets or technologies.** The agreement regards products which do not yet exist, and it seeks to control the processes that are considered as crucial for their conception and attainment.

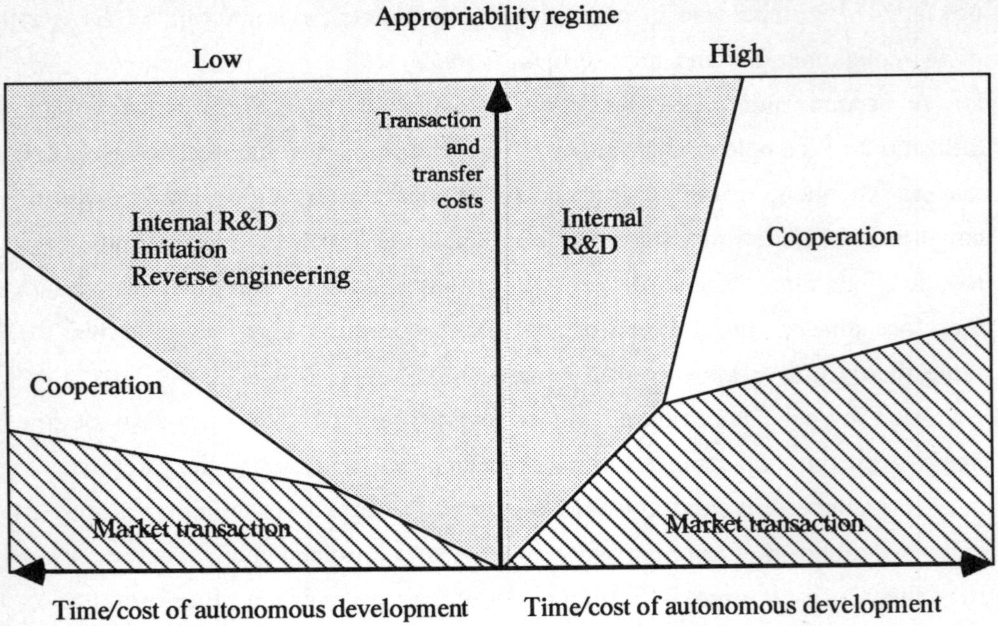

Source: our refinements on an idea by Teece and Pisano (1987)

Figure 6. Alternative firm strategies in technology acquisition

It is evident that in such a new context, which fits perfectly with the original and innovative characteristics of cooperation agreements and network behaviour, theories based on static efficiency and also transaction cost approaches seem completely useless. Future profits stem from a series of strategic decisions, oriented to fast-reaction and continuous innovation, to the early preemption of newly discovered market niches, to an aggressive marketing policy in order to discourage potential competitors. All these decisions are projected in a dynamic framework, incompatible with merely static approaches of allocative efficiency: markets (which do not yet exist) and (routine-oriented) organizations are intrinsically unable to produce the right signals on prices and the right standard for costs.

On the contrary, the theoretical areas to which it would be useful to address seem to be the analysis of strategic competition, negotiation theory and the theory of cooperative and non-cooperative games (Raiffa, 1982; Jacquemin, 1988). In fact, the concept emerging from this view is that of 'strategists firms that by force or by bluff try to control in a dynamic process their rivals and their environment to their own advantage. They calculate, anticipate, and invest in irreversible capital, thus segmenting markets, increasing their rivals' costs, tying up their suppliers and their clients, and manipulating information' (Jacquemin, 1987, pp. 123-124).

We are now back to the second type of uncertainty defined earlier, namely dynamic uncertainty, not just created but explicitly enhanced by the decentralized decisions of independent actors. Under these conditions, cooperation may stem tacitly from collusive behaviours (as in the prisoner's dilemma with repeated games: history and memory are once again important!) or derive from the explicit choice of the firm, trying to reduce the complexity of its decision parameters and to enhance the control on some of its supposedly strategic assets (and on their time trajectory).

5.3 Collective operators

A fourth behavioural space (and a fourth operator) is therefore at the firm's disposal, beyond the **organization space** (growth by internal development and acquisitions), **the competition space** (market transactions), and the **synergy space** (the local 'milieu'): we have called it the **cooperation space** and 'networking' represents its related behavioural model (Figure 5).

Two problems will be faced hereafter which may derive from the theoretical scheme hitherto presented: the relationships between the two new 'operators', viz. the 'milieu' and network cooperation, and the difference between our approach and the transaction cost approach applied to territorial analysis.

As it may be seen in Figure 5, 'milieu' relationships and network relationships appear as complementary and mutually reinforcing 'operators', the former linking the firm to its contiguous environment through mainly informal, tacit (and often even overlooked and apparently unappreciated) relationships, the latter linking it explicitly to selected partners in its operational environment.

Both operators bring an element of '**socialization**' into the picture of economic behaviour, stemming both from the collective learning process which happens at the local scale, and from the cooperation nature of network linkages: 'innovation does not emerge from the singular efforts of entrepreneurial firms or corporate research centers, for the contributions of individual actors are themselves produced within linkage networks that are collective in character and retain a critical territorial dimension' (Gordon, 1989a). Firms' networks work as a sort of 'collective property' of the complementary intangible assets shared by the network and as a 'collective participation' to the process of appropriation of quasi-rents and innovation profits stemming from the cooperative behaviour (Allen, 1983; Antonelli, 1987b).

The explicit nature of network and cooperation linkages may, at first sight, obscure the importance of local relationships and leave the researcher with the impression of a collapse of the concept of space, both in its geographical and relational meaning, into that of trans-territorial networking. On the contrary, the two concepts and related 'operators' are deeply interlinked and complementary.

On the one hand, the 'milieu' has to open up to external energy in order to avoid 'entropic death' and a decline in its own innovative capability;

firm networks seem the most important instruments (but hardly the only ones) to cope with the problem. On the other hand, when choosing a partner to link up with, not only does the firm choose a single partner, but also a 'collective' one (speaking allusively), at the same time linking itself with a 'local' culture and acquiring partial access to the synergies of its 'milieu'. A link-up with a firm located in Silicon Valley is more a link with the Valley itself than with a special firm, with which, if located elsewhere, no agreement would probably be made.

Sometimes, as an example, in the Third Italy or once again in Silicon Valley, the territorial specificities are so profound and crucial for the process of innovation and technical change that it may be rightly claimed that 'firms tend to be relatively contingent manifestations of technical projects developed in the region's professional culture' (Gordon, 1989a).

Will these new territorial and trans-territorial relationships be properly analyzed through the concepts of the Williamsonian Institutionalist School, and in particular, through the transaction-cost approach? As we have seen before, many of these concepts were in fact utilized in the preceding theoretical framework and recently some interesting works have regarded agglomeration economies showing up in industrial 'districts' and urban areas as the outcome of transaction cost-reducing processes (Lambooy, 1986; Scott and Angel, 1987, Cappellin, 1988).

The transaction-cost approach and the approach proposed here are, in fact, similar in some respects in that both point out the role of information and information gathering costs in determining allocative efficiency, and give spatial proximity (or networking) the role of reducing these latter costs. Nevertheless, in spite of the fact that information represents the bridge between static and dynamic behaviours, the transaction-cost approach remains basically a static one, addressing itself more to problems of allocative efficiency and design of organizational structures than to problems of dynamic efficiency and innovative behaviour. Through the related concepts, static but not dynamic uncertainty may be understood.

In fact, between the two general operators of market and hierarchy, there is the possibility of inserting a third one, cooperation, utilizing the same

Williamsonian general framework. However, in this case only the more traditional firm behaviours and goals may be grasped, those pertaining to complementary assets control. In contrast, the true dynamic objectives of 'strategic alliances' and their innovation enhancing role remain obscure in this context (Gordon, 1989b; Camagni, 1989). Furthermore, with respect to the general uncertainty reducing role of the 'milieu', a transaction cost-reduction hypothesis seems rather limiting and prevents us from truly incorporating spatial variables into an evolutionary theory of innovation and technical change.

6. From the Firm Space to the Technology Space

The problem now is how to model the previous relationships in explicit dynamic terms, passing from the space of single firms deciding upon innovation, to the space of competing and evolving technologies. In a word, passing from a (micro-economic) adoption perspective to a meso-economic perspective of technological diffusion.

Two main theoretical approaches may be followed, while still remaining within the context of ecological-evolutionary models:

A) an approach which, once again drawing from the biological analogy, we may call a **selection approach**, looking at the competition between mutually exclusive technologies and their '**substitution**' in space; and

B) an approach which we may call a **mutation approach**, looking mainly at problems of true **technological creation** within the context of development and diffusion.

The alternative stems from a mainly theoretical consideration of the nature of the technological evolutionary process: a process of competition between species or between already known technologies on the one hand, and a process of mutation and technology 'creation' along a trajectory of evolving (vintage) techniques as well as through changes of trajectories. From a modelling point of view this difference blurs a little because potential advancements in technology have to be, in one way or another, pre-defined into the space of the possible outcomes of the evolutionary process.

The first category encompasses mainly (but not exclusively) deterministic models considering and simulating a process of dynamic adjustment to an equilibrium condition where the most profitable technology completely replaces the competing ones. The second category encompasses mainly stochastic models of search and innovation creation, building mainly on a self-organization approach and on the consideration of collective learning and cooperation processes.

It is important to say that an explicit consideration of space and spatial relationships is still lacking in the relevant literature, as a sort of trade-off seems to exist between the inclusion of spatial variables and the explicit consideration of evolutionary processes. Simplifying assumptions are therefore generally adopted in spatial models which limit their heuristic capability.

6.1 Selection models

In Figure 7 a taxonomy of historical approaches to technological diffusion is presented according to the two dimensions previously proposed.

The main concern of non-orthodox models of technological selection is to explain the existence and causes of the lag structure of adoptions. At the firm level, the most comprehensive analysis of the process is made by Scherer (1980), introducing the variability among firms in the two variables determining the adoption (and the adoption time): **relative profitability** of the new technology (depending on factor price differences among firms, the depreciation share of the existing equipments, etc.) and **adjustment costs** from the old to the new technical structure (here, interfirm differences come from R&D commitment, internal labour relations in case of labour-saving technologies, managerial capability in handling the organizational aspects of change).

In determining the right timing for adoption, firms have to consider both the extra costs of an anticipated adoption and the profits which may come from market niches completion and early adopter strategy (Scherer, 1980, page 427; Cappellin, 1985).

Recently, an interesting insight into the problem of understanding the causes of the slow pace of technological adoptions of advanced and profitable

Approaches \ Dimensions	**Firm (adoption)**	**Technology (diffusion)**	**Space (spatial diffusion)**
Selection models **(technology substitution)**	Adoption of a new technology a) presence of adjustment costs (Scherer, 1980) b) presence of imperfect decisions and expectations (Heiner, 1988 a,b)	Competition between known technologies a) profitability (Gibbon - Metcalfe, 1989; Iwai, 1984) b) adopters heterogeneity and perfect information (David, 75; Davies, 79) c) information exchange between technologies (Sonis, 1986) d) dynamic learning (Metcalfe, 1981; Camagni, 1985)	a) The firm level (Cappellin, 1985) b) The single technology level (Hägerstrand, 1967; Camagni, 1985; Capello, 1988)
Mutation models **(technology creation)**	Evolutionary models a) search and decision - making models (Nelson - Winter, 82)	Self-organisation models a) Stochastic models of technology evolution (Montano- Ebeling, 80; Silverberg et al, 1988)	Urban self-organisation models a) urban innovation (Camagni,Diappi, Leonardi, 1986: Soudy 1) a) urban synergies (Camagni, Diappi, 89: Soudy 3)

Figure 7. Some ecological - evolutionary models of technological development

technologies came from Heiner's works (1988a and b). Even in a world of perfect information and no adjustment costs, if there exists what we previously called C-D uncertainty (a gap between competence and decision) and consequently the possibility of decision errors, the 'optimal' strategy for the firm will be one of imperfect, delayed and sluggish adjustment to changing external technological possibilities.

At the 'meso' level of the diffusion of competing technologies and their mutual substitution, the main approaches through which the process has been modelled may be sketched as follows:

* deterministic adjustment models where **relative profitability** is the central element for convergence towards the best technology: this process may stem from the simple reinvestment in the same technology of the higher profits it allows (Metcalf and Gibbon, 1989) or come from investment shifts from technology i to j, according to higher relative profitability of technology j and to its market share (acting as an information element) (Iwai, 1984);

* equilibrium adjustment models with **adopters heterogeneity** and instant-perfect information: reaction characteristics among the potential adopters may come from size and efficiency diversities (David, 1975) or be stochastically distributed as in probit models (Davies, 1979);

* dynamic models of **information interaction** between technologies; in the most advanced model of this family (Sonis, 1986) the time variation of the relative market shares of different competing technologies is linked to the interaction of an antisymmetrical 'competition' matrix A presenting the values of an iterated game among technologies (the effects of the adoption of i technology on j with an 'information' matrix M (presenting the information which originate from the adoption of i and go to j). The substitution curves generated by this interaction are proved to be generalized logistics, and from their empirical estimation in simplified cases, an estimation of the two matrices is made possible. Particularly interesting is the fact that the superior technology (gaining in the competitive game) defines the asymptotic shares of all technologies through the information it releases on them (Colla and Leonardi, 1984). Differently from 'mutation' models, matrices A and M, and consequently also the characteristics of the competing technologies in time, are

assumed as constant;

* dynamic models where endogenously determined variations in the **price of the new technology** and **learning processes** of their adopters determine a shift in the size of its adoption potential (Metcalfe, 1981) or where this potential more simply evolves through a logistic expansion of the number of sectors or firm size classes interested by the technology (Camagni, 1985).

All these models bear only an implicit spatial dimension. When this dimension is made explicit, as in the well-known Hägerstrand model (1967), the simplified assumption of equal contact probability of all actors, underlying all epidemic diffusion models, is made. This assumption highly reduces the coherence of this approach: a spatial context is in fact characterized by the opposite condition of differentiated profitability, adjustment costs and willingness to accept risks. A possible escape from this problem could be that of fitting single models in different but homogeneous regional spaces, and to interpret lag and diffusion speed parameters in a second step on a cross-regional base (Camagni, 1985; Capello, 1988).

6.2 Mutation models

More recently the element of technology evolution and technology 'creation' along the diffusion path has been taken into full account through stochastic models incorporating Schumpeterian innovation. The technological frontier evolves in time in a way that is only imperfectly anticipated by firms.

Still from Figure 7 we see that in this case the natural quotation as far as the behaviour of the single firm is concerned, is the work by Nelson and Winter (1982, chapter 9) on **evolutionary search behaviours**. Here the firms, dissatisfied with actual profit levels, may enter a process either of imitation of existing external technologies or a process of search on a metric space of potential technologies (ordered in terms of distance with respect to the present know-how of the firm).

Dynamic self-organization models have enlarged the view from the firm space to the technology space. A 'master equation' approach, defining the change in the probability of finding a specific distribution of technologies at a given time, may give us the evolution of the entire spectrum of actual and potential technologies, starting from the individual transition probabilities of technology imitation, improvement, creation (Jimenez Montano and Ebeling, 1980; Silverberg, 1988). Perhaps the most advanced attempt in this direction is the Silverberg, Dosi, Orsenigo self-organization model (1989) in which transition probabilities are modelled carefully in terms of both the choice of a particular vintage within a specific technological trajectory and the change of trajectory happening at certain points in time. These changes are a function of the evolution of the internal know-how and of an external 'public' skill made available to all actors.

The inclusion of spatial and collective learning processes allowing an easier innovation decision could easily be introduced in the model, allowing the transition probabilities to be influenced by the decisions taken in previous times by surrounding firms.

In a different context, one of interurban competition for innovating functions, a similar kind of self-organization model was presented some years ago (Camagni et al., 1986) where a master equation controls for the transition probabilities from a set of urban functions to an upgraded one, and urban size represents the control function in the same way as firm productivity does in the preceding models.

A further refinement of this model in the same logic was presented recently, allowing spatial synergies to explicitly affect the innovation probabilities of each area (Camagni and Diappi, 1990, in this volume). Synergies may come from vertical, 'filière-type' integration or from horizontal interaction among similar productions, two elements that represent an early quantitative treatment of the spatial relationships which produce an innovative 'milieu'.

7. Conclusions

'Technological progress is in the first instance the reduction in uncertainty. The product of a research and development effort is an observation on the world which reduces its possible range of variation' (Arrow, 1969). Nobody could have stated in a more concise and effective way the central role that uncertainty plays in any theory of technical change.

The main point made in this paper is that the inescapable presence of static and dynamic uncertainty in any dynamic model of economic behavior calls for the development of specific 'operators' that, well beyond pure market and hierarchy, may limit its paralysing impact on firms dynamic behaviour, complement the imperfect signalling function of the price system, organize a viable learning process for both individual firms and society and enhance their creativity potentials.

Two important 'operators' of this kind are found, on the one hand, in the local 'milieu' or the 'synergy space' that it potentially represents, and on the other hand, in the 'cooperation space' and the possibility of trans-territorial network linkages between firms. Both these operators act as uncertainty-reducing devices, performing the functions of information searching and screening, signalling, transcoding of complex messages, selecting appropriate decision routines and controlling other actors' economic conduct, in a collective and socialized way.

The territoriality of the first operator is apparent, as proximity plays a necessary (but not sufficient!) role in the creation of local synergies. The second operator, that apparently departs from a strict territoriality, in fact also bears a territorial nature: in fact, the assets firms bring into the network agreement are often the outcome of a complex local culture and of localized social learning processes.

Therefore, on the one hand, territorial relationships and the local 'milieu' emerge as necessary and crucial elements in the innovation process; on the other hand, a proper vision of the evolutionary process of technology creation has led us towards the construction of a new, intrinsically dynamic interpretation of economic space. From a modelling point of view, dynamic self-organization models seem to be the most apt in incorporating the new approach and its related concepts into an operational and coherent mathematical framework.

References

Allen, R.C. (1983), Collective Invention, Journal of Economic Behaviour and Organisation, 1:1-24.

Antonelli, C. (1987a), New Information Technology and Industrial Organisation: Experiences and Trends in Italy, OECD, Directorate for Science, Technology and Industry, mimeo.

Antonelli, C. (1987b), Dall'economia industriale all'organizzazione Industriale', Economia Politica, 2:277-320.

Arcangeli, F., P. David and G. Dosi (eds.) (1989), The Diffusion of New Technologies, vol. 1, Oxford, Oxford University Press (forthcoming).

Arrow, K. (1962), Economic Welfare and the Allocation of Resources for Invention, in: The Rate and Direction of Inventive Activity, NBER, Princeton University Press.

Arrow, K. (1969), Classificatory Notes on the Production and Transmission of Technological Knowledge, American Economic Review, 1:29-35.

Arrow, K. (1974), The Limits of Organisation, New York, W.W. Norton and Co.

Aydalot, Ph. (ed.) (1986), Milieux Innovateurs en Europe, pp. 361, Paris, Gremi.

Aydalot, Ph. and D. Keeble (eds.) (1988), High Technology Industry and Innovative Environments: the European Experience, pp. 241, London, Routledge.

Boissevain, J. and J. Mitchell (eds.) (1973), Network Analysis: Studies in Human Interaction, The Hague, Mouton.

Boyer, R. (1986), La Théorie de la Régulation: une Analyse Critique, Paris, La Découverte.

Camagni, R. (1985), Spatial Diffusion of Pervasive Process Innovation, Papers and Proceedings of the Regional Science Association, 58:83-95.

Camagni, R. (1986a), Dinamica Industriale, Tempo, Memoria, Economia e Politica Industriale 51:159-166.

Camagni, R. (1986b), The Flexible Automation Trajectory: the Italian Case, paper presented at the International Conference on Innovation Diffusion, Venice; forthcoming in F. Arcangeli, P. David and G. Dosi (eds.)(1989).

Camagni, R. (1989), Alleanze Strategiche: Nuove Comportamenti e Nuovi

Approcci Teorici, Economia e Politica Industriale 59:255-265.

Camagni, R., L. Diappi and G. Leonardi (1986), Urban Growth and Decline in a Hierarchical System: a Supply Oriented Dynamic Approach, Regional Science and Urban Economics 1:145-160.

Camagni, R. and L. Diappi (1990) Soudy 3: a Supply Oriented Urban Dynamics Model with Innovation and Synergy Effects, in this volume.

Camagni, R. and Gambarotto, F. (1988), Gli Accordi di Cooperazione Come Nuove Forme di Sviluppo Esterno delle Imprese, Economia e Politica Industriale 58:93-138.

Capello, R. (1988), La Diffusione Spaziale dell'Innovazione: il Caso del Servizio Telefonico, Economia e Politica Industriale 58:141-175.

Cappellin, R. (1985), Regional Factors in Productivity Growth, in: R. Camagni and R. Cappellin (eds.), Sectoral Productivity and Regional Policy, Document, Commission of the European Communities, Bruxelles, pp. 152.

Cappellin, R. (1988), Transaction Costs and Urban Agglomeration, Revue d'Economie Régionale et Urbaine 2:261-278.

Chesnais, F. (1988), Technical Cooperation Agreements between Firms, STI Review, OECD 4:51-120.

Colla, L. and G. Leonardi (1984), Modelli di Diffusione dell' Innovazione, in: R. Camagni et al. (eds.), Cambiamento Tecnologico e Diffusione Territoriale, pp. 59-107, Milano, F. Angeli.

Crevoisier, O., D. Maillat and J-Y Vasserot (1990), L'apport du Milieu dans le Processus d'Innovation: le Cas de l'Arc Jurassien, in D. Maillat and J-Cl Perrin.

David, P. (1975), The Mechanisation of Reaping in the Ante-bellum Midwest, in: Technical Innovation and Economic Growth, Cambridge, Cambridge University Press.

Davies, S. (1979), The Diffusion of Process Innovations, Cambridge, Cambridge University Press.

Dosi, G. (1982), Technological Paradigms and Technological Trajectories, Research Policy 3:147-162.

Dosi, G. et al. (1988), Technical Change and Economic Theory, pp. 646, London, Pinter.

Dosi, G. and L. Orsenigo (1988), Coordination and Transformation: an Overview of Structures, Behaviours and Change in Evolutionary Environments, in: Dosi et al., pp. 13-37.

Foray, D. and D.C. Mowery (1988), Industrial Research and Technological Change: Towards a New Economic Representation, paper presented at the European Summer Institute of the RSA, Arco, pp. 27.

Foresti, J. (1986), Une Toile d'Araignée se Referme Autour de l'Industrie Européenne: Les Accords de Coopération Industrielle, Commission des Communautées Européennes, mimeo.

Freeman, C. (1988), Introduction to Part 1, in Dosi et al., pp. 1-8.

Freeman, C. and C. Peres (1986), Innovation, Diffusion and New Techno-Economic Paradigms, paper presented at the International Conference on Innovation Diffusion, Venice; forthcoming in F. Arcangeli, P. David and G. Dosi (eds.).

Gaffard, J.-L (1986), Restructuration de l'Espace Economique et Trajectories Technologiques, in: Ph. Aydalot (ed.) (1986), pp. 17-28.

Gibbons, M. and J.S. Metcalfe (1989), Technological Variety and the Process of Competition, in: Arcangeli et al.

Gordon, R. (1989a), Alleanze Strategiche e Socializzazione del Capitale, Economia e Politica Industriale 59:267-270.

Gordon, R. (1989b), Entrepreneurs, Firms and the Social Foundation of Innovation, Sociologie du Travail 1.

Grandori, A. (1989), Le Reti Interorganizzative: Progettazione e Negoziazione, Economia e Management, March-April.

Hägerstrand, T. (1967), Innovation Diffusion as a Spatial Process, Chicago, University of Chicago Press.

Heiner, R.A. (1988a), Imperfect Decisions and Organisations: Towards a Theory of Internal Structure, Journal of Economic Behaviour and Organisation, Febr.

Heiner, R.A. (1988b), Imperfect Decisions and Routinized Production: Implications for Evolutionary Modeling and Inertial Technical Change, in: Dosi et al. (1988), pp. 147-169.

Hey, J.D. (1979), Uncertainty in Microeconomics, Oxford, M. Robinson.

Iwai, K. (1984), Schumpeterian Dynamics: an Evolutionary Model of Innovation and Imitation, Journal of Economic Behaviour and Organisation 5:159-190.

Jacquemin, A. (1987), The New Industrial Organisation: Market Forces and Strategic Behaviour, pp. 217, Oxford, Claredon Press.

Jimenez Montano, M.A. and Ebeling (1980), A Stochastic Evolutionary Model of Technological Change, Collective Phenomena 3:107-114.

Johannisson, B. (1987), Organising: the Network Metaphor, International Studies of Management and Organisation 1.

Kamann, D. and P. Nijkamp (1988), Technogenesis: Incubation and Diffusion, paper presented at the European Summer Institute of the RSA, Arco, pp. 40.

Lambooy, J. (1986), Information and Internationalisation: Dynamics of the Relations of Small and Medium Sized Enterprises in a Network Environment, Revue d'Economie Régionale et Urbaine 5:719-731.

Lundvall, B.A. (1988), Innovation as an Interactive Process: from User-Producer Interaction to the National System of Innovation, in: Dosi et al., pp. 349-469.

Maillat, D. and J.-Cl. Perrin (1990), Entreprises Innovatrices et Réseaux Locaux, GREMI, Paris, ERESA-Economica (forthcoming).

McKenna, C.J. (1986), The Economics of Uncertainty, pp. 129, Brighton, Wheatsheaf Books.

Metcalfe, J.S. (1981), Impulse and Diffusion in the Study of Technical Change, Futures, Oct., pp. 347-359.

Nelson, R. and S. Winter (1982), An Evolutionary Theory of Economic Change, Cambridge Mass., Harvard University Press.

OECD (1986), Technical Cooperation Agreements between Firms: some Initial Data and Analysis, Paris, May, mimeo, pp. 49.

Peres, C. (1985), Microelectronics, Long Waves and World Structural Change, World Development 3:441-463.

Planque, B. (1983), Innovation et Développement Régional, Paris, Economica.

Plan Urbain (1989), Les Entreprises et Leur Environnements: Flux et Réseaux, Consultation de Recherche, Paris.

Raiffa, H. (1982), The Art and Science of Negotiation, Cambridge, Harvard University Press.

Scherer, F.M. (1980), Industrial Market Structure and Economic Performace, Chicago, Rand McNally.

Scott, A.J. and D.P. Angel (1987), The US Semiconductor Industry: a Locational Analysis, Environment and Planning A, 19:875-912.

Silverberg, G. (1988), Modelling Economic Dynamics and Technical Change: Mathematical Approaches to Self-Organization and Evolution, in: Dosi et al., pp. 531-559.

Silverberg, G., G. Dosi and L. Orsenigo (1988), Innovation, Diversity and Diffusion: a Self-Organization Model, The Economic Journal 98, Dec. pp. 1032-1054.

Sonis, M. (1986), A Unified Theory of Innovation Diffusion, Dynamic Choice of Alternatives, Ecological Dynamics and Urban/Regional Growth and Decline, Ricerche Economiche 4:696-723.

Spence, M. (1973), Job Market Signalling, Quarterly Journal of Economics, Aug., pp. 355-374.

Teece, D. and G. Pisano (1987), Collaborative Arrangements and Technology Strategy, paper presented at the Conference on New Technologies and New Intermediaries, Stanford University, June, pp. 33.

Vickery, G. (1988), A Survey of International Technology Licensing, STI Review 4:9-50.

Williamson, O. (1985), The Economic Institutions of Capitalism, pp. 450, New York, The Free Press

Chapter 11

TECHNICAL CHANGE AND REGIONAL DEVELOPMENT: SOME FURTHER DEVELOPMENTS WITH THE SEQUENTIAL INTERINDUSTRY MODEL

E. Romanoff
S.H. Levine

1. Introduction

An understanding of effects of simple processes of technical change on processes of economic development is offered, using cross-industry information flows for technical diffusion within the dynamics of interindustry production. Greater understanding and policy guidance are suggested through integrated modeling of these processes. Schumpeter's view of technical change as an evolutionary process leading to economic development was examined in Nelson and Winter (1982), Batten, Casti and Johanssen (1986) and Dosi et al., (1988), among others. Another explanation of regional economic development relies on urbanization and other external economies stemming from locational advantage. Progress in examining regional technical change was presented in Camagni (1989) and progress in examining regional development was presented in Suarez-Villa (1989), both at this conference.

Processes of technical change, gains in externalities and regional development, all appear related in the broad context of interindustry analysis. The departures from equilibrium which these processes represent are not amenable to satisfactory examination by the use of the input-output model. Technical change may also violate the model's homogeneity assumption, and externalities which do not flow along interindustry linkages cannot be addressed.

For the task at hand the dynamics of all three processes and departure from equilibrium during transitions are recognized. These require a more detailed view of industry production than is possible by the formal input-output model, and require accepting non-homogeneous changes in the production function during transitions and recognizing cross-industry flows. Modeling this view is made possible by the use of an enhanced time-varying Sequential Interindustry Model, SIM, developed for the analysis of transient behavior. In SIM, account is made of the time required for each industry to produce its product beforehand and supply its own final demand and that of the directly demanding other industries, for the latter to use as input to their own production (Romanoff and Levine 1977, 1981, 1986 and 1989; Romanoff 1984a; and Levine and Romanoff 1989). Because of the complexity of economic behavior only selected variables and other features are modeled in each SIM presentation, and their behavior is not necessarily uniform in each case. SIM may thus be represented by a family of submodels, each designed to explore specific aspects of this rather complex system. SIM contributes to the recently revived interest in modeling the dynamics of interindustry production, e.g., Mule (1983), Ten Raa (1986) and Cole (1988).

At the aggregate level technical change pertains both to product change and to process change; only the latter is emphasized in this paper. Yet the two are not always independent of each other; the association between them is noted further in the text.

In section 2 a discussion of technical change in the context of our model, SIM, is presented, focusing on simple replacement of one production technique with another in one of the production processes of an industry. In section 3 a mathematical overview of the model is outlined. In section 4 the implication of this technical change on regional development is presented, emphasizing the role of cross-industry information flows of technical diffusion in the analytical approach. Section 5 contains concluding remarks.

2. Technical Changes

2.1 Technical change in the Leontief model

Structural Change and Technical Change. Economic change, as noted by Leontief (1953, 17), can be explained either as a structural change or as a dynamic process. Essential to such explanations with the static model is a satisfactory definition of industries that give rise to consistent patterns of interdependence upon which observed structural relations are based. It has long been recognized, however, that product mix, process-mix, and product multidimensionality combine to introduce indeterminance in industry input-output coefficients. Suggestions for overcoming these difficulties were put forward quite early in Holzman (1953). Since then, much research has been undertaken on the product-mix problem, and improvements in industrial classification and model disaggregation have been made. The process-mix and product multidimensionality difficulties received much less attention because of their complexity, leading in turn to studies of process analysis pioneered in Manne and Markowitz (1963). These and subsequent studies explored complex interindustry relationships using linear programming methods, thereby shifting the emphasis away from the input-output model. They focused on an industry or a group of industries because, even at the detailed 4-digit SIC level, the definition usually encompasses more than one production process.

These difficulties appear compounded when one focuses on a subset of all possible coefficient changes, namely on technical change. Rose (1984), in a review of the input-output literature on this subject, defined technological change as 'a change in physical input combination in production', (p. 305) and noted the limitations of the various approaches, with the use of process analysis and the production function appearing the more advantageous.

Industry Production Processes. The limitations of using even the detailed industry definition must be underscored. Some industries are defined by their key process or major input used, as among the primary metal industries, while others are defined in terms of their products. It appears that examinations of technical change using the static model were more successful for the former, and less so for the latter. This is because information on the production processes is compressed within the industries of the input-output model, even more so for those defined by product, so that change in different production processes are extracted in the aggregate. Further, for some industries technology appears inherently unstable with time. Responsive industries, responding to orders and to specifications of buyers, may alter input combinations in specific production processes more readily than anticipatory industries, anticipating orders and producing to their own specifications (Romanoff, 1987).

2.2 Production process and their production techniques

To enable modeling transient processes associated with technical change on must go beyond the list of input requirements to the description of production 'recipes', thus enabling examination of industry production processes and their production techniques. One ought to address the question of 'why' 'what' 'how' and 'when' of an industry production. In the currently enhanced SIM, the 'why' is addressed by the producing industry designation, j, whose purpose is producing a product. The 'what' is addressed by the inclusion of the different production processes in a time-phased production function, indicating what is to be done in the industry production schedule, such as mixing (of few inputs), kneading (dough) and baking in bread-making, measured backwards in time from completion (Nikaido, 1962). See Figure 1. Baking bread, crackers and pizza as well as baking bricks are identical processes, as are petroleum cracking and boiling of maple sap.

The 'how' is addressed by each of the production techniques used in processing specific inputs at a given time for a production process. In the production process of baking bread a batch-feed production technique may be used or, alternatively, a line-feed technique with conveyors running through the oven may be used. From the economic standpoint a technique is usually identified by the combination of few inputs used, a dynamic recipe for specific input applications (boil eggs at 350o F for 3 min.), the required specific equipment (kind of oven in bread-baking), a complement of occupational skills, with other factors contributing (e.g., scale of operation). The pasteurization technique is used in the production process for preserving food products, such as beer, cheese, milk and wine; techniques are not unique to a specific product or industry.

The 'when' is addressed by an array of time indices, indicating the

Time-phased flows

i	h_{ij}	SIM			Static I-O
1	3	$x_{1j}(-2)$			$= X_{1j}$
2	2		$x_{2j}(-1)$		$= X_{2j}$
3	2		$x_{3j}(-1)$ +	$x_{3j}(\emptyset)$	$= X_{3j}$
4	1			$x_{4j}(\emptyset)$	$= X_{4j}$
5	3	$x_{5j}(-2)$ +	$x_{5j}(-1)$ +	$x_{5j}(\emptyset)$	$= X_{5j}$
V A	3	$x_{vj}(-2)$ +	$x_{vj}(-1)$ +	$x_{vj}(\emptyset)$	$= X_{vj}$

Production Schedule

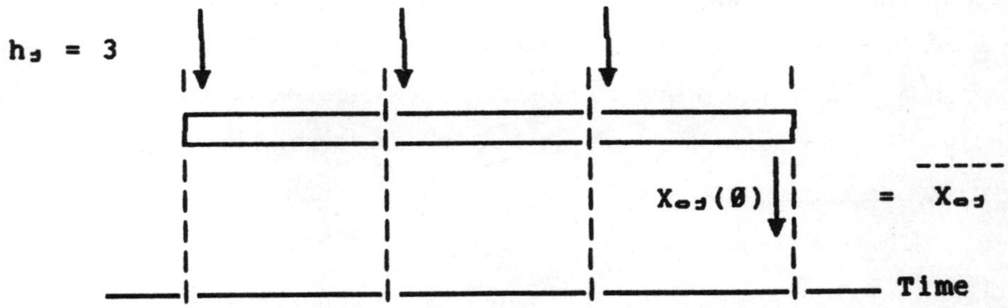

Figure 1. Time-Phased Flows and a Production Schedule

timing of input application referenced to the time of product completion, as well as other events relevant to the industry production schedule in the time-varying presentation of SIM, noted further below.

The association between product change and process change discussed in the more aggregate literature seems to be related to the hierarchy of changes in the manner of production. At the elementary level, the change takes place by replacing one production technique with another in a given production process of an industry; this is the change mostly addressed in this paper. At a higher level a production process may be changed, added, or eliminated for an industry, in the last case obviating subsequent changes in production techniques. For example, the substitution of aluminum for steel in motor housing (e.g., lawn mowers) may eliminate the need for the coating process.[1] That is, a change in technique for one production process may affect the need for a subsequent process and may lead to product variation. At the still higher level a new product may need a different set of processes and an industry may be added or eliminated in an economic system.[2] In this hierarchy, the higher the level of change in the manner of production the more closely it is associated with a change in an industry product.

2.3 Changes in production techniques.

To examine industries undergoing technical change and explain the change, it appears advisable to specify the associated production technique for each of the production processes of an industry. Rules by which techniques may be improved or replaced for a production process could then be specified, and processes of technical change could be examined. However, the link between production processes and production techniques is quite complex, involving a host of engineering and economic considerations. To model this linkage two enhancements are necessary. First, coupling engineering methods linking production processes and the model is required. Information on schedules of complex construction projects was thus coupled with SIM (summarized in Romanoff and Levine, 1980) and further enhanced by accepting engineering data from the critical path method and other network-based techniques used by complex multi-process industries (Levine and Romanoff, 1989).

Second, the technique used for each production process is identified. In this way two production processes used concurrently, and even of differing duration, may be identified by their techniques and a change in one of the processes may be specified in terms of an alternative technique. During a replacement episode the mix of old and new techniques may be specified. Similarly, an input used by two techniques in two distinct processes at the same time may be identified by the percentage required by each. The mathematical formulation builds on Burmeister and Dobell (1970, 228-266) for the Leontief production. The time-phased interindustry flow, $x_{ij}(\cdot)$, for a production process represents a component of a production technique, μ, applied at t for input i, μ_{ti}, where μ = 1...m. The vector $\underline{\mu}_t$ (with components μ_{ti}) represents a technique for using a specific combination of inputs for a particular production process. The industry j set of vectors $\underline{\mu}_t$, of techniques used in the production schedule of a product completed at σ is specified by matrix $M_{\sigma j}$. In turn, the collection of matrices $M_{\sigma j}$ (for all j) represents the system's record of technology, M_{σ}, used by all industries in the production of products completed at σ, and is only part of the list of available technologies, or the economy's 'book of blue prints'.

By decomposing the column vector of a Leontief production function into a time-phased technology matrix, with columns associated with production processes, each using a specific production technique at a given time, significant flexibility is achieved in modeling technology and its changes. This approach fuses economic and engineering information, while still residing within neoclassical economic doctrine. It affords examination of several processes of technical change. During a transition in techniques the resulting Leontief production function is non-homogeneous and may be influenced by scale effects for a subset of the inputs associated with the changing techniques. By the end of the transition, once all substitutions have been made, in conformity with Samuelson (1951) a new homogeneous production function emerges and lasts until the next disturbance. The use of $\underline{\mu}_t$ affords examination of technical change internal to an industry and of linkages through information flows across industries which can or do use identical production techniques. By requiring identical occupational skills the $\underline{\mu}_t$ introduce a measure of externality for cross-industry technical diffusion presented further in the text.

3. The Model

3.1 A modeling issue

An insight into modeling the dynamics of interindustry production is provided to those mostly familiar with the static model, in terms of the issue of the information available to industries.

The Basic Model. The time-invariant model (presented in Romanoff and Levine, 1986) determines the sequence of total output vectors, $\underline{X}(t)$, required to supply final demand in terms of a sequence of final output vectors, $\underline{Y}(t)$. By accounting for the time required for production, a temporal relationship between these two sequences (or chronologies) is established. Production time is accounted for in the time-phased technical matrix $A(t-\sigma)$, which specifies the time interval, t, at which inputs are required for output completion at interval σ. The general formulation of this model, with no inventories of any kind, and ignoring for a moment transportation delays, is:

$$(1a) \quad \underline{X}(t) = \Sigma\, A(t-\sigma)\, \underline{X}(\sigma) + \underline{Y}(t).$$

Use of Z transform converts this equation to:

$$(1b) \quad \underline{X}(z) = A(z)\, \underline{X}(z) + \underline{Y}(z).$$

If transportation delays are explicitly considered, (1b) becomes

$$(2) \quad \underline{X}(z) = [\Phi(z) \cdot A(z)]\, \underline{X}(z) + \underline{Y}(z),$$

where $\cdot$ is an element by element product, $\Phi(z)$ is the transportation delays matrix in shipping goods from industry i to industry j. The solution of (2) leads to:

(3a) $\underline{X}(z) = [I - \Phi(z) \cdot A(z)]^{-1} \underline{Y}(z)$

(3b) $= B(z) \underline{Y}(z).$

The two sequences $\underline{X}(t)$ and $\underline{Y}(t)$ are thus related by:

(4) $\underline{X}(t) = \Sigma_\alpha B(t-\alpha) \underline{Y}(\alpha),$

where α is the time interval at which final demand occurs.

In this mechanical model total output is computed on the assumption of perfect information that requires having:

1. complete knowledge of future final demands, $\underline{Y}(t)$;
2. complete knowledge of the technology of production, $A(t-\alpha)$, including magnitude and timing;
3. complete knowledge of transportation delays, $\Phi(z)$.

Clearly non of these assumptions is realistic. A more realistic but still highly simplifying assumption holds that total product or output, $\underline{X}(t)$, is estimated from imperfect information requiring, for example, knowledge of only a part of $A(t-\sigma)$, such as industry i having information about the 1th row, and perhaps of future values for key components of $\underline{X}(t)$, $\underline{Y}(t)$ and $\Phi(z)$. The difference between the estimated total output to be produced and actual shipments, $\underline{X}(t)$ - $\underline{S}(t)$, is compensated for by a complex mechanism of inventory management, excluded from present discussion. The shipments can be thought of as a feedback mechanism leading to corrections of the estimated output, thereby reconciling supply and demand. This simple description bypasses many important subleties of the model.

Time-Varying Model. The time-varying time-phased technical matrix is written $A(t,t-\sigma)$ to express its explicit dependence on time, t. Since technology changes with time, this leads to more complex relationships between the sequences of total and final product, $\underline{X}(t)$ and $\underline{Y}(t)$. The determination of $B(t,z)$, where $A(t,z)$ is the Z transform of $A(t,t-\sigma)$, is quite complex. Symbolically the relationship may be written:

(5) $B(t,z) = f\ [\Phi(z), A(t,z), IM],$

where IM represents alternative possibilities of inventory management, and B(t,z) is a generalization of the Leontief inverse. In general,

$$(6) \qquad \underline{X}(t) = \Sigma_{\alpha} B(t,t-\alpha)\ \underline{Y}(\alpha),$$

relates the sequences $\underline{x}(t)$ and $\underline{y}(t)$.

The inclusion of time-varying technology requires an additional assumption to the perfect information implied by the model in (4), namely:

4. Complete knowledge of future technologies.

However, errors of imperfect knowledge of future technologies contribute to the difference between output and shipments and further require the use of inventories and perhaps other adjustments. Since this system departs from equilibrium during transitions in techniques, closed form solutions no longer apply. One must resort to numerical simulations to estimate industry production paths.

3.2 Key features of the model

To provide an insight into properties of the model (Romanoff and Levine, 1989) some of its key features are presented, focusing on anticipatory production.

Model Indices. In addition to the normal indices one encounters in input-output models, in the time-varying SIM specific indices are used. These are given in parentheses and pertain to production techniques in use and to temporal events or periods (to simplify, indices for production processes are omitted). The temporal indices are:

t - time interval of input application.

σ - time interval of output or product completion.

h_{ij} - application period of an input from industry i used by industry j, referenced from the initial application interval t, to product completion σ.

h_j - production period of industry j, equivalent to the longest application period, h_{ij}.

t-σ - input duration, indicating the period from an input application interval t to the time of output completion, σ, measured backwards in time from σ, (which is equal to or less then h_{ij} in absolute duration).

ϕ_{ij} - transportation delay associated with the shipment from industry i to industry j, representing the time goods are in transit; a component of the transportation delay matrix, $\mathbf{\Phi}$.

Anticipatory Production Model. If producers' prices are used, the price of the input from industry i, p_i, is determined not at the time of input application by industry j, but earlier involving the transportation delay in shipping from i to j, ϕ_{ij}. Assuming also for simplicity sake just-in-time production (no input inventory), we have: $p_1(t-\phi_{ij})$. The input quantity $q_{ij}(\mu_{ti};t,\sigma)$, records the quantity of input i used by j, processed by technique μ at t in order to complete the product at σ. The value of an input from i applied by j is:

(7) $$x_{ij}(\mu_{ti};t,t-\phi_{ij},\sigma) = p_i(t-\phi_{ij})\ q_{ij}(\mu_{ti};t,\sigma).$$

The output of industry j, x_{oj}, completed at σ is priced at the time of product completion for anticipatory producers; the time indices of output completion and output pricing are the same, σ:

(8) $$x_{oj}(\sigma,\sigma) = p_j(\sigma)\ q_{oj}(\sigma).$$

The time-phased technical coefficients represent the division of (7) and (8) and the resulting time indices are given in pairs, for the inputs from i and for the output of j, for a total of four:

date of input application	date of input pricing	date of output completion	date of output pricing
$(t,$	$t-\phi_{ij},$	$\sigma,$	$\sigma)$, i.e.,

(9) $${}_{ij}a_{ti}(\mu\ ;t,t-\phi_{ij},\sigma,\sigma) = \frac{x_{ij}(\mu_{t1};t,t-\phi_{ij},\sigma)}{x_{oj}(\sigma,\sigma)}$$

In equilibrium, the coefficients represent the relative costing out of i for component techniques of a production schedule, so that for each i they sum over time to the coefficient of the Leontief model.

Consider some of the implications of the assumptions embedded in these coefficients. If inputs are valued in producers' prices then they are dependent on ϕ_{ij}, and hence on the temporal efficiency of the transportation system, which is external to industry production economies. In addition to location effects, the explicit modeling of these delays suggests that infrastructure improvements, of lack thereof resulting in congestion, may have repercussions on industry productive efficiency, by altering the coefficients and even causing production delays. While this externality is accepted in transportation and economic development quarters, it has not been formally integrated into tradition input-output modeling. In regional models it bears particularly on interregional trade.

Consider prices. Since pricing inputs precedes the pricing of output, the coefficients are conditional on $p_j(\sigma)$. They may be viewed as ex ante priced coefficients.

Total intermediate output produced by industry i becomes:

$$(10) \qquad w_i(t,t) = \Sigma_\sigma \Sigma_j \Sigma_\mu a_{ij}(\mu_{ti};t,t-\phi_{ij},\sigma,\sigma)\ x_{oj}(\sigma,\sigma),$$

which is a generalized convolution describing the dynamics of intermediate production.

Total output of industry i consists of three components: intermediate output in (10), output to inventory, $u_1(t,t)$, and final output, $y_i(t,t)$. The accounting identity is then:

$$(11) \qquad x_i(t,t) = \Sigma_\sigma \Sigma_j \Sigma_\mu a_{ij}(\mu_{ti};t,t-\phi_{ij},\sigma,\sigma)\ x_{oj}(\sigma,\sigma) + u_i(t,t) + y_i(t,t).$$

In matrix notation (11) becomes:

$$(12) \qquad \underline{X}(t,t) = \Sigma_{\sigma} A(M_{\sigma};t,t-\phi_{(\cdot)},\sigma,\sigma)\ \underline{X}(\sigma,\sigma) + \underline{U}(t,t) + \underline{Y}(t,t),$$

where M_{σ} is the system's record of technology, and $\phi_{(\cdot)}$ represents the appropriate elements, ϕ_{ij}, of the transportation delay matrix, Φ. Unlike $\underline{Y}(t,t)$, output to inventory $\underline{U}(t,t)$ is generally not independently specified, but is a function of the different inventory policies of anticipatory producers.

The solution for $\underline{X}(t,t)$, given $\underline{Y}(t,t)$, depends on inventory policies. Functionally, this can be expressed as:

$$(13) \qquad \underline{X}(t,t) = \underline{\beta}\ [A(\cdot),\ \underline{U}(t,t),\ \underline{Y}(t,t)\ \ldots]$$

where $\underline{\beta}[\cdot]$ is a vector function. This function, or calculation rule, represents a general extension of the Leontief inverse and requires specific algorithms, depending on the specification of inventory and other considerations, e.g., capacity.

The model includes responsive producers (who do not carry product inventory, and who receive orders before production for ordered product starts) and is slightly more complex in its time indices. The fuller model includes inventory management, and would include linkages between total and final product and still other features.

3.3 Two views

The dynamics of production may be interpreted in two ways, each view with its own conceptual and operational implications. Consider (6). **In the temporal view**, production is associated with the timing of input applications and with the duration of each of the industry's direct production schedules, each planned for a different date of product completion. In such chronology of production all industries are examined with reference to a **specifically dated final output**. The process of technical change pertains to changes in magnitude, in timing of input applications and in the duration of direct production schedules of the A(t,z) matrix embedded in B(t,z). In this view a transient process of technical change may start from a steady state using one production technique, and reach the other steady state using the new technique; during the transition the B matrix of the old technique is gradually transformed into the matrix of the new technique prevailing at the other steady state. During the transition the B matrix represents features of both techniques and is unique to each time interval.

This temporal view must be distinguished from the cross-sectional or 'snapshot' view obtained at a specific time interval t. **In the cross-sectional view** one observes at a specific time, t, different 'slices' of progressively 'forward moving' technically changing A and thus also B matrices, where each of the 'slices' describes production with reference to differently dated outputs and **a time series of final outputs**. During a transition in production technique the hybrid generalized inverse thus perceived is quite complex, reflecting the mix of techniques that contributes directly and indirectly to a time series of final outputs.

Prices may also be interpreted differently in the two views. In the temporal view production is described with reference to a set of prices dated at any t, for a specific final output $\underline{y}(\alpha)$. In the cross-sectional view, production is described with reference to a specific set of prices prevailing at the interval of observation for different phases of product completion produced to meet a time series of final outputs.

Theoretically, the temporal and the cross-sectional views of production are associated with two distinct schools of economic thought. Within the interindustrial framework the production schedules contain features that have their origin with Böhm-Bawerk and the Austrian school, who viewed production as one continuous temporal process bringing the product nearer to the final consumer. Cross-sectional production represents the interpretation by Walras and the Laussanne school, who viewed production as a series of cross-sections through time (Kuenne, 1963, 237-264). An illustration of these two views using single-interval production schedules in SIM was presented in Romanoff (1984a) and an examination of some of the underlying Austrian concepts in Durand and Romanoff (1986).[3] While this Trans-Alpine Controversy has subsided by now, one of its ramifications may have been to deter development of dynamic interindustry modeling.

SIM may be viewed as an attempt to integrate some aspects of these two schools of thought within a dynamic framework stemming from the Leontief static input-output model. While these two symmetrical interpretations are inconsequential for a system in steady state with not technical change, for then they represent identities, the distinction becomes critical for a system departing from the steady state and undergoing technical change, for then the system departs from equilibrium and the results for the components of the two representations diverge during the transition.

4. Technical Change and Regional Development

4.1 Internal and external effects

Consider the ramifications of technical change on development. Focusing on the transition during which an industry replaces one production technique with another, one may address the changes taking place in the innovative industry and the changes in other industries. The effects on the innovative industry were modeled in Romanoff and Levine (1989) using the logistic function. The examination revealed the effect of the industry production period, h_j, on the function, as well as some of the variations from normal production that may occur during the transition episode stemming from considerations internal to industry operations.

Several externalities affect the other industries. Some may be examined along interindustry linkages; in using SIM the adjustment processes of supplying industries may be modeled, indicating changes in production paths for both industries adversely affected as the demand for their product decreases, and for industries benefitting as the demand for their product increases; these are consistent with the adjustment in Robinson and Markandya (1972). Adjustments in suppliers' production levels may also be examined using a static input-output model, comparing the 'before' and 'after' conditions, implicitly assuming that a complete transition takes place during the intervening period; information is lost in this comparison. In SIM technical change and birth and demise of industries may also be addressed, as may still other adjustments; pricing still to be included. All these represent an aspect of the development process.

Another externality pertains to those industries which are not directly linked interindustrially. These industries were examined in Lichtenberg (1960), including the apparel industries. In the input-output model the apparel industries are recipients of major inputs from the textile industries, but trade little if any among themselves. They appear linked mainly by a variety of cross-industry information flows, representing other parameters in a dimension beyond that of the input-output model. The challenge in examining effects of information flows within the input-output model was recently noted in Carter (1989). The cross-industry flows usher modeling of technical diffusion processes across industries.

4.2 Cross-industry technical diffusion

Cross-industry technical diffusion may take place among industries having an identical production process, Θ, and which are possible adopters of a new technique, $\underline{\mu}^*$, successfully used by an innovative industry. Success may be demonstrated, say, by accounting for one-half of innovative industry saturation, measured on its logistic curve. The parameters for the diffusion algorithm may be modified and calibrated; our present focus is on the analytical approach to modeling the underlying process.

Several approaches were used in modeling diffusion; they were reviewed in Davies (1979) who introduced his own enhanced model. These models, aimed at examination of firms, are based on the assumption that firms are either homogeneous within an industry, however broadly defined, or can be sorted into a few such groups using key attributes, e.g., firm size. However, if one is to hold to an underlying assumption of the Leontief model, industries are distinct from each other in input combinations and in output. In SIM the several production parameters as well as the various linkages may also impose a unique behavior on each industry. That is, in examining technical diffusion among industries they cannot be considered a homogeneous set, as are human populations in contagion models using logistic or related functions, and which may be applied to intra-industry analysis. The departure from accepted modeling practice affords inclusion of some recently suggested ideas; these appear amenable to SIM modeling.

The diffusion process may be viewed in two parts. First, feasible adopting industries are identified. They use older substitute techniques at any t during concurrent h_j production periods terminating at σ, since cross-industry linkages need not be among processes identically positioned in production schedules. See Figure 2. Identification is amenable to computer-based search routines. The potential market for $\underline{\mu}^*$ would still include industries which, for a variety of reasons, either will not adopt the new technique or must incur adoption delays. An alternative to pasteurization is dehydration; while dried milk and dried cheese are being produced, the beer and wine producers may be considered non-adopters.

The second part involves examination of the set of feasible adopters for the exclusion of remaining non-adopters, and for the others an examination of the sequential decision to adopt, introduced in Arthur (1989) and modified for industries' heterogeneity. The modification takes into account the external and internal influences on industry decision to adopt. The external influences stem from the regional labor pool of the required occupations for $\underline{\mu}^*$. In migrating across industries, initially from the innovative industry, workers in such a pool bring with them the technical know-how for the new technique which enhances the formal knowledge available. As the Massachusetts data suggest, this migration generates gains from the externality of the labor pool, thereby reducing initial adoption costs. Modeling this process builds on concepts of network externalities introduced in Katz and Shapiro (1985) and applies to economies with free labor mobility. Internal influences on feasible adopters stem from both engineering compatibility considerations associated with coupling a new technique to a production schedule, and from economic considerations associated with the anticipated profitability of the change. SIM's internally generated data can be used in the examinations of the changing industry characteristics for a two-stage model. In the first stage cross-industry diffusion is modeled using the external and

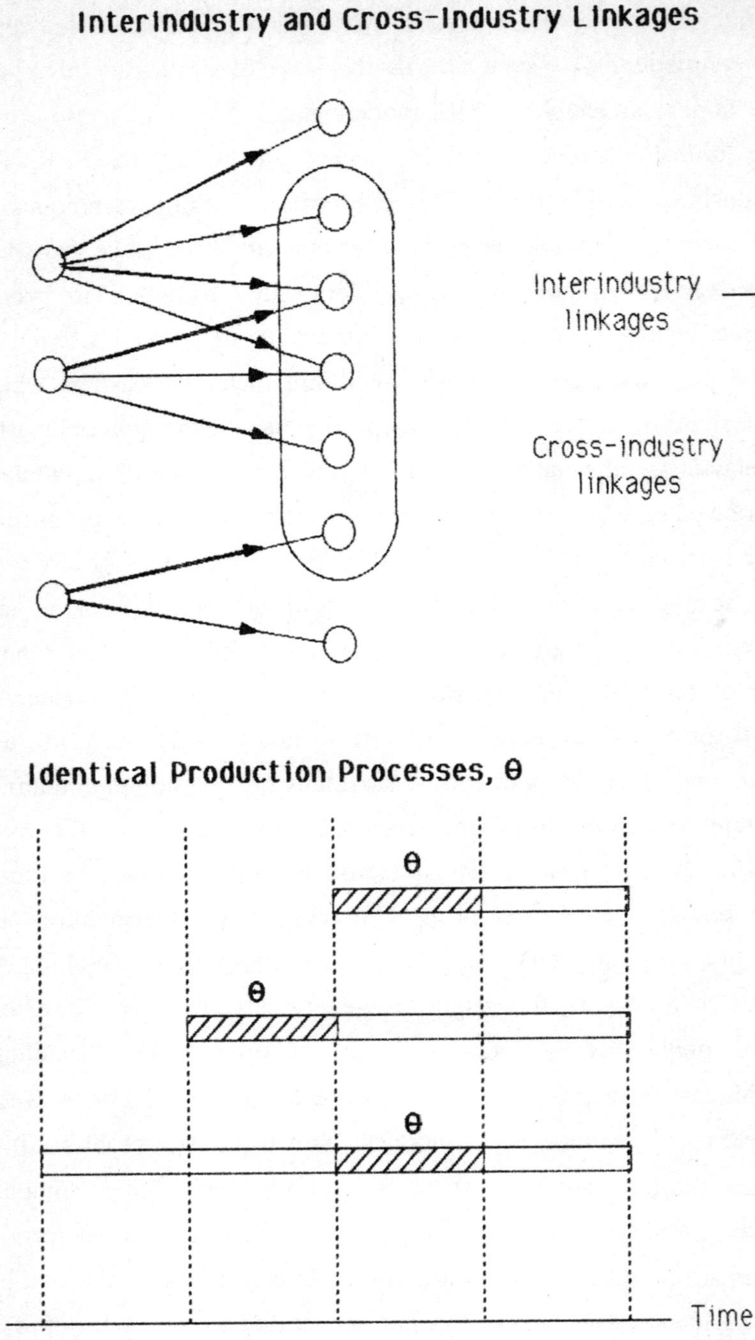

Figure 2. Cross-Industry Diffusion for Identical Production Processes

internal influences to indicate the different delays in the timing of adoption among the heterogeneous industries. In the second stage the spread of the technique within each adopting industry with homogeneous accounting units is modeled, using logistic functions modified by the industry's own production period.

4.3 Diffusion and development

Modeling purposes impose alternative approaches. If the interest in modeling diffusion is not in guiding technological forecasting but in a SIM module with feedbacks guiding the development process, then the module described would pertain to only an initial phase of a combined diffusion-development process. To address the modeling of a combined diffusion-development process a much broader perspective than hereto undertaken is explored.

Consider just one aspect of what appears useful in modeling, namely new techniques. The emphasis thus far was on a single replacing technique, the most elementary technical change. Building on Boyer (1988) and others, a broader view requires the extension of the planning horizon in considering external and internal effects for industry behavior as may also be modified by social institutions (government) to avoid technological determinism, and to account for economic fluctuations which may stem from other than technical change. That is, in a longer horizon one ought account for a broader range of linkages where the processes themselves may change with time. This task is consistent with the time-varying SIM.

Consider two modeling features: changing properties of a technique and alternative diffusion paths. From the time of adoption, techniques appear to be modified as experience leads to increasing returns through learning-by-doing (Davies, 1979). This means that $\underline{\mu}_t$ differs from $\underline{\mu}_{t+k}$, where k is a period enabling significant distinction. But as Arthur (1989) noted, not all techniques enjoy increasing returns with adoption; as inputs are bid up in price diminishing returns may accompany adoption with late adopters among others suffering the consequences. These mechanisms ought to be included in the model, but the assumption of incomplete information in SIM requires some departure form the method advanced by Arthur. Perhaps the little understood adoption choice mechanism ought to be expanded to include regional variations in cultural acceptance of and the conditions for change, suggested by anthropologists. All these suggest that adoption paths may differ for an identical industry located in different regions. Regions lagging in the sequential choice in adoption (technically, economically or culturally) may not benefit as expected from the change.

For a replacing production technique withdrawn from the accumulated inventory in the Book of Blue-Prints, it is possible that one may be more adoptable than another, and that there may be several modes by which a technique spreads among the industries along a network. Since some paths are more feasible than others and industries may have several possible ways of adjusting production to changing demand, with some more feasible than others during different economic conditions, the number of feasible alternative paths of economic development during the long diffusion-development process may be quite large; they usually differ among regions. Moreover, two or more new competing techniques may also appear, such as competing techniques for the process of managing computer operating systems. Also government intervention may modify diffusion along a path, with self-reinforcing feedbacks perhaps magnifying the effects of small shifts (Arthur, 1989). Furthermore, innovation begets innovation, leading to multiplicity of diffusion processes. All these alternative and changing processes converge toward fewer, stable and better understood patterns by the end of the long transition. To model these changes SIM ought to be further expanded to include, for example, the dynamic linkages between output and final demand; an initial examination of this topic was presented in Romanoff (1984b). In the evolutionary development of SIM the prospect of modeling an environmental component would enable inclusion of a still broader set of externalities to indicate the benefits and costs of the diffusion-development processes.

A partial assessment has been presented of what is to be modeled before addressing the 'how' of modeling, suggesting that the diffusion-development module leads to quite a complex model in its own right. Still, the effort appears worthwhile since during such long transitions the different adjustment paths would indicate significant dislocations that may occur. These are but consequences of alternative development processes with the model serving as guide to greater understanding and guidance to policy-making.

5. Concluding Remarks

For regional development agencies searching for immediate economic development policy guidance in this paper, two principles are offered.

1. Expand scope of operations to consider an industry technology and changing production techniques for the implications on many industries, however linked, in terms of changing requirements for capital, occupational labor, other effects, and tax-paying ability.
2. Avoid the naive practice of imitating a success story from other regions. Without careful local examination and proper timing the success story in one region may spell embarrassment in another (citations are intentionally omitted).

In the time-varying Sequential Interindustry Model (SIM) outlined in this paper a detailed view of industry production was presented to examine a simple technical change where one production technique replaces another in one of the production processes of an industry. This enabled addressing processes of technical change for effects on processes of regional development. The introduction of techniques ushers understanding of cross-industry information flows among those which either use or can use a specific production technique, while adding another dimension to the input-output model. This leads to modeling cross-industry diffusion of simple technical change. However, the modeling of diffusion as a SIM module with feedbacks to economic development requires a longer horizon and more inclusive modeling than before for a combined diffusion-development process. A partial assessment of such requirements for the long transition in technology suggests a quite complex model in its own right. A challenging research agenda is thus suggested. Coupled with a further enhanced SIM, such a model would provide greater understanding of the possible alternative development paths and guidance to policy-making.

Finally, records of technology are of interest to economists, planners, regional scientists, and others concerned with economic development. Viewing from the windows of this conference center in Jerusalem, the historic results of uncounted generations producing sequences of total output to satisfy sequences of final demand, perhaps one may be permitted the observation that, if the periods between dates of product completion, σ, are lengthened to span centuries then past records of technology are of interest to anthropologists and archaeologists and historians. Since the pace of technical change has accelerated with time, changes in such historic records of technology to the present, despite many disruptions, would be evident over ever shortening (or logarithmic-like) periods. Since not all of the old techniques have been discarded while new ones are continually being added, the list of techniques in use over such periods would be ever lengthening. Combining, the resulting function would portray a key aspect of the changing historic record of regional economic development.

Notes

1. To reduce air emissions the U.S. Environmental Protection Agency has encouraged the use of non or less polluting techniques for the coating process.

2. An examination of the impact dynamics of the addition of an industry to a regional economy was presented in Romanoff (1984a). Examinations of the impact dynamics of industry closing were presented in Romanoff and Levine (1977) and in Cole (1988), using a somewhat different approach.

3. Recently a Marxian interpretation of the various forms of capital represented in the dynamics of interindustry production was suggested (Ten Raa, Chakraborty and Das, 1989).

References

Arthur, W.B. (1989), Competing Technologies, Increasing Returns, and Lock-In by Historic Events, The Economic Journal 99:116-131.

Batten, D., J. Casti and B. Johansson (eds.) (1986), Economic Evolution and Structural Adjustment, New York, Springer-Verlag.

Boyer, G. (1988), Technical Change and the Theory of 'Regulation', in: G. Dosi et al. (eds.), Technical Change and Economic Theory, pp. 67-93, London, Pinter.

Burmeister, E. and A.R. Dobell (1970), Mathematical Theories of Economic Growth, New York, Macmillan.

Camagni, R. (1989), Technological Change and Regional/Urban Dynamics, Third World Congress, Regional Science Association, Jerusalem, Israel.

Carter, A.P. (1989), Input-Output Reicpes in an Information Economy, Economic Systems Research 1:27-43.

Cole, S. (1988), The Delayed Impact of Plant Closures in a Reformulated Leontief Model, Papers of the Regional Science Association 65:135-149.

Davies, S. (1979), The Diffusion of Process Innovations, Cambridge, Cambridge University Press.

Dosi, G. et al. (eds.) (1988), Technical Change and Economic Theory, London, Pinter.

Durand, D. and E. Romanoff (1986), Capital Intensity, Time, Social Accounts, & Input-Output Analysis: An Austrian-Tinged View, Lexington, Regional Science Research Center. Discussion Paper 86-1.

Holzman, M. (1953), Problems of Classification and Aggregation, in: W. Leontief et al. (eds.), Studies in the Structure of the American Economy, pp. 326-356, New York, Oxford University Press.

Katz, M.L. and C. Shapiro (1985), Network Externalities, Competition, and Compatibility, American Economic Review 75,3:424-440.

Kuenne, R.E. (1963), The Theory of General Economic Equilibrium, Princeton, Princeton University Press.

Leontief, W. (1953), Dynamic Analysis, in: W. Leontief et al. (eds.), Studies in the Structure of the American Economy, pp. 53-90, New York, Oxford University Press.

Levine, S.H. and E. Romanoff (1989), Economic Impact Dynamics of Complex Engineering Project Scheduling, IEEE Transactions on Systems, Man, and Cybernetics 19,2:232-240.

Lichtenberg, R.M. (1960), One-Tenth of a Nation, Cambridge, Harvard University Press.

Manne, A.S. and H.M. Markovitz (1963), Studies in Process Analysis, New York, John Wiley.

Mules, T.J. (1983), Some Simulations with a Sequential Input-Output Model, Papers of the Regional Science Association 51:197-204.

Nelson, R. and S. Winter (1982), An Evolutionary Theory of Economic Change, Cambridge, Harvard University Press/Belnap.

Nikaido, H. (1962), Some Dynamic Phenomena in the Leontief Model of Reversely Lagged Type, Review of Economic Studies 29:313-323.

Robinson, S. and A. Markandya (1972), Complexity and Adjustment in Input-Output Systems, Oxford Bulletin of Economics and Statistics 35:119-134.

Romanoff, E. (1984a), Interindustry Analysis for Regional Growth and Development: The Dynamics of Manpower Issues, Socio-Economic Planning Sciences 18,5:343-363.

Romanoff, E. (1984b), The Dynamics of Income and Product, Cambridge, Regional Science Research Center. Discussion Paper 84-1.

Romanoff, E. (1987), Structure of Final Producers Under Changing Final Demand: Apparel, Computers and other 'High-Tech', Lexington, Regional Science Research Center. Discussion Paper 87-2.

Romanoff, E. and S.H. Levine (1977), Interregional Sequential Interindustry Modeling: A Preliminary Analysis of Regional Growth and Decline in a Two-Region Case, Northeast Regional Science Review 7:86-101.

Romanoff, E. and S.H. Levine (1980), Impact Dynamics of Large Construction Projects: Simulations by the Sequential Interindustry Model, Cambridge, Regional Science Research Center. Discussion Paper 80-2.

Romanoff, E. and S.H. Levine (1981), Anticipatory and Responsive Sequential Interindustry Models, IEEE Transactions on System, Man, and Cybernetics 11,3:181-186.

Romanoff, E. and S.H. Levine (1986), Capacity Limitations, Inventory and Time-Phased Production in the Sequential Interindustry Model, Papers of the Regional Science Association 59:73-91.

Romanoff, E. and S.H. Levine (1989), Technical Change in Production Processes of the Sequential Interindustry Model, Lexington, Regional Science Research Center. To be published in Metroeconomica.

Rose, A. (1984), Technological Change and Input-Output Analysis: An Appraisal, Socio-Economic Planning Sciences, 18,5:305-318.

Samuelson, P.A. (1951), Abstract of a Theorem Concerning Substitutability in Open Leontiefs Models, in: T.C. Koopmans (ed.), Activity Analysis of Production and Allocation pp. 142-146, New York, John Wiley.

Suarez-Villa, L. (1989), The Evolution of Regional Economies: Sectoral Restructuring and Regional Development over the Long Term, Third World Congress, Regional Science Association, Jerusalem, Israel.

Ten Raa, T. (1986), Dynamic Input-Output Analysis with Distributed Activities, Review of Economics and Statistics 68:300-310.

Ten Raa, T., D. Chakraborty and T. Das (1989), Working Capital in an Input-Output Model, Economic Systems Research 1:53-68.

Part C. ECONOMIC RESTRUCTURING AND SPATIAL DYNAMICS

Chapter 12

THE EVOLUTION OF REGIONAL ECONOMIES: SECTORAL RESTRUCTURING AND REGIONAL DEVELOPMENT OVER THE LONG TERM

L. Suarez-Villa

1. Introduction

Long-term socioeconomic change is at the core of the most significant contemporary concerns in regional development. The emergence of new industries and the decline of others, the diffusion of new technologies and organizational forms, and the changing spatial distribution of the population are all, for example, very much related to the dynamics of long term change. Despite its relevance for most contemporary concerns in our field, however, the study of long term spatial economic change remains both an underresearched and daunting endeavor. The uncertainty and complexity of the factors affecting long term change, and the fact that a comprehensive understanding of such processes is usually beyond the grasp of any one academic discipline, are major obstacles to the study of this topic.

A multidisciplinary approach to the study of long term change requires a solid grounding in the historical processes affecting innovation at both the macro and micro levels. The technological and organizational innovations that are at the root of long term change have historically been realized and diffused through the development of entrepreneurship. At the micro-level, the diffusion of new techniques and decision processes associated with entrepreneurship have been instrumental in the restructuring of organizations and their productive activities. These have, in turn, determined the historical emergence of new sectors, the restructuring of others, and the spatial redistribution of employment and population.

Tracing the historical process through which entrepreneurship emerges, and the macro-level sectoral and regional changes that it promotes, will be the prime objective of this paper. For this purpose, a typology of various historical entrepreneurial roles will be developed. These will be related to the broader sectoral and spatial macroeconomy, by providing an overview of several historical eras and major changes in the dynamics of the various economic sectors and their spatial structures. A selective empirical illustration will then follow, analyzing U.S. regional macroeconomic change over a 94-year period (1889-1983).

2. Innovative Entrepreneurship: Historical Roles

Innovation is now acknowledged by many to be a central force behind the process of economic change. Trends in productivity growth, employment, competition, and international trade, among others, are thought to be substantially influenced by the application of inventions and their diffusion (Hägerstrand 1967; Boserup 1981; Molle 1983; Rosenberg 1986; Scherer 1986; Thomas 1986; Batten et al. 1987; Kleinknecht 1987; Nijkamp 1987; Solomon 1987; Sweeney 1987; Suarez-Villa 1987, 1988; Weiss 1988). But the recognition of the importance of innovation has long predated any attempts to understand the nature of the process through which it occurs, or the socioeconomic agents bringing it about.

A study of the process of innovation must start with, and take into account, the changing historical roles of the entrepreneur. For these roles have determined not only the character of the innovations that are produced and applied, but their long term impact on the economy and its space. Tracing the emergence of these roles in historical time is not an easy task, however, since recorded scholarly interest does not often coincide with the functional emergence of each role. In some cases, the lags were substantial enough to leave significant gaps in our understanding of the details surrounding the emergence of a role, and its initial impact on the sectoral and space economy which it affected.

In most instances, such lags are due to be the absence of a formalized academic field in which a role could be researched. Thus, for example, the absence of the field of managerial science prevented a better understanding of the nature of industrial entrepreneurship when mass production techniques were being introduced early in this century. The problem was then further compounded in the spatial literature, since significant interest in the effects of any innovative role often did not appear until after the topic had been seminally dealt with in the mainstream social science fields, especially in economics.

The early role of **capital investment and accumulation**, so essential to the Industrial Revolution, drew the greatest attention in the classical economic literature, through the work of Cantillon in the middle of the eighteenth century and, later, Adam Smith's and Say's contributions (see Suarez-Villa 1989, ch. 2, for an extensive review of the literature). The latter's explicit attention to risk-bearing would be instrumental later on, in expanding the classical definition to consider other aspects of the entrepreneurial function. The innovative possibilities of this and other roles would then not be extensively considered until Schumpeter's work appeared early in this century. During the eighteenth and much of the nineteenth centuries, however, the attention given to this role monopolized the literature, as becomes obvious, for example, from Marx's own critique and virtually exclusive concern with the process of capital accumulation.

Initially, the application of this role in industry occurred through the most rudimentary workshops. Typical of the earliest stages of the Industrial Revolution, these could be characterized as single product-single function operations in Alfred Chandler's typology of organizations (see Chandler 1962, 1977). It should then not be surprising that the most innovative impacts of this role occurred in manufacturing, without which the very emergence of the Industrial Revolution would not have occurred (see, e.g., Ashton 1948; Cochran 1977; Berg 1985). Later on, the establishment of financial markets would become the most significant institutional innovation resulting from this role, as the availability of capital resources for investment in organizations and ventures became systematized.

Historically very significant in the development of industrialization, **intermarket linkage** became essential in the establishment and coordination of input sources to capital goods industries. Its innovative qualities would involve such activities as the discovery and development of new and substitute input sources, ensuring the reliability of input quality and supply through new transactive arrangements, and the structuring of vertical integration in innovative ways to take advantage of scale and scope economies (see Suarez-Villa 1989, ch. 2).

The substantial lag found between the emergence of this role in practice and its recognition in the scholarly literature, can be attributed to the slow development of economic theorizing on organizations throughout the nineteenth century. It was virtually not until Schumpeter's acknowledgement of the 'opening up of new markets and sources of inputs' as an innovative entrepeneurial endeavor that this role would acquire some currency in the economic literature (Schumpeter 1934). Later, the most significant interest in the functional character of this role would occur through the economic development literature of the nineteen fifties and sixties. Albert Hirschman's (1958) conceptualization of forward and backward linkages in the process of development envisioned entrepreneurs 'filling in' the missing components of productive operations, from extraction to final demand.

More recently, Harvey Leibenstein's (1978) definition of 'gap filling' entrepreneurship in his X-efficiency conceptualization also addresses some of the functions of this role. Thus, the ability to maintain input supply specifications and efficiency through the marhsalling of organizational talent and leadership, which is always an imperfectly marketed entrepreneural attribute, can become part of the definition of this role. Where input markets would not even exist, as in the case of entirely new industries and productive activities, the innovative capabilities of this role would be in greatest demand, as previous experience and routinized knowledge might not be of much help in structuring all the necessary operations.

There is little doubt that the emergence of the intermarket linkage role allowed the establishment of the factory system in the United States by the middle of the nineteenth century, by managing a large and stable supply of inputs to the new capital gods industries. This role was crucial, for example, in the establishment of large scale operations in the manufacture of steel and heavy transportation equipment, such as locomotives and ocean-going vessels. Without it, it would have been impossible for manufacturing to advance beyond the rudimentary workshops of the seventeenth and early eighteenth centuries. At the same time, the onset of vertical integration within individual plants brought about the establishment of what Chandler categorizes as multifunction-single product organizations in the industrial sector.

Crucial in the effort to attain ever greater productivity has been **productive coordination**, a role that encompasses middle-level and line managerial activities. Among the innovative aspects of this role have been the creation and adaptation of inventions in production, and the devising and implementing of new organizational forms to raise productive efficiency. Finding new ways to motivate the labor force to produce with better quality and efficiency, and developing transactive arrangements between management and labor, can also be considered to be part of this role's functional spectrum.

One of the earliest explicit references to this role is in Say's (1803) distinction between strategic risk-bearing and managerial decision-making. Later, Alfred Marshall would consider managerial entrepreneurship to be the fourth factor of production, along with land, labor and capital. In our own century, Schumpeter drew attention to the innovative character of this role, as being one which would devise and implement 'new organizational forms' (Schumpeter 1934). Despite these references, however, this role has endured substantial neglect in the economics literature (see, e.g., Leibenstein 1987).

The functional emergence of productve coordination became essential for the implementation of mass production techniques early in our own century, and the development of process integration in multifunction-multiproduct organizations that consolidated various productive operations within a single plant. A significant time lag would then be found between the institutionalization of this role in practice and the development of scholarly research on its functioning and impacts. It was not until managerial science became established as a field of academic inquiry that significant scholarly research on this role would emerge, through the work of such scholars as Alfred Chandler. His early typology of managerial levels, developed in the early 1960s, which distinguished between line, middle and top level management, would help its functional definition (Chandler 1962).

Chandler saw the first two levels (line and middle level management) as being primarily tactical rather than strategic, though nevertheless possessing much innovative capacity. In the 1970s, Leibenstein (1978) then placed the coordinative role at the core of the X-efficiency paradigm, elaborating on its ability to make innovative use of imperfectly marketed factors of production, where contracts or performance can never be completely specified. Such quantitatively elusive characteristics as leadership and motivational ability were then factored into the X-efficiency concept. These contributions would help in linking up with the Schumpeterian perspective, and would go some ways toward correcting the mistaken view, prevalent in much of mainstream economics, of management as being 'routine' and devoid of any innovative capability.

Essential for the emergence of the large mid-twentieth century business organization, the **strategic planning** role was also instrumental in promoting what has come to be known as the 'activist' corporation of our own times. Operating in multiregional and multinational markets, these large, multifunctional and multiproduct organizations influence not only the strategies of their competitors, but have also had substantial impacts on the policies of governments, financial institutions, and international organizations. The functional focus of this role therefore involves top-level authority over corporate decision-making and planning. It is different from the intermarket linkage and productive coordination roles in that most of the decision-making and implementation apparata it controls tend to deal with product markets and linkages, marketing strategies, and the resolution of any pressures that affect these activities, such as regulation. This role is therefore essential in the firm's successful adaptation to market demand pressures (Mintzberg 1979; McDermott and Taylor 1982; Hoenack 1983; Donaldson and Lorsch 1986).

The innovative character of this role usually revolves around the design and implementation of new marketing and planning strategies to increase market share, profits, or to reduce corporate risk. Given the complexities of modern corporate decision-making, it may not be under the control of one, but of several individuals. The need for a greater division of labor in decision-making can result in substantial delegation of top-level authority, depending on corporate size, organizational structure, the spatial dispersion of its units and their functional differentiation, and the availability of decision-making skills. All these factors will affect the degree of organizational entropy found in the corporate structure, since excessive delegation of authority can create substantial problems in pursuing coherent and effective marketing strategies, or in reducing institutional pressures on corporate activities.

Historically, the definition of this role is indebted to Say's early and rudimentary nineteenth century consideration of management as an entrepreneurial task. Its innovative aspects were later emphasized by Schumpeter when he acknowledged the 'opening up of new markets' as an important outcome of innovative entrepreneurial action. Alfred Chandler, in particular, ascribed a definite spatial causality to the emergence of this role, by demonstrating that the implementation of modern corporate administrative structure in the U.S. began with the geographical dispersion of corporate units, which made it necessary to further differentiate top-level corporate planning and strategy-making from the other managerial tasks (see Suarez-Villa 1989, ch. 2).

The development of the multifunctional-multiproduct organization in the second half of our own century has contributed to the extension of this role, as the complexity of organizational structures increased. This has occurred, for example, through the horizontal integration of the disparate activities of various firms. In other cases, the expansion of the scope of many industrial enterprises, even when not tending toward horizontal integration, has also added greater importance to this role (see, e.g., Teece 1980).

In the spatial literature, some connection between this role and regional economic change can be found through the literature on market area changes dating back to von Thünen's seminal contribution. In our own century, Lösch assumed the entrepreneurial capability associated with this role, along with natural resource endowments and population densities, to be a major determinant of the market areas of firms, especially in new ventures and industries. At a more aggregate level, Lösch further assumed the market areas of cities to result from their advantages in attracting and supporting innovative entrepreneurs within their boundaries. Significantly, locational decision-making in the modern corporation remains a top-level activity that is intimately tied with marketing strategies, and the logistics of production.

Essential to the introduction of innovations, **invention** has supported the historical emergence of every other entrepreneurial role. It was crucial, for example, in producing the technological advances which made the Industrial Revolution a reality, through the creation and development of the steam engine (see, e.g., Usher 1954). It would later be essential to the development of heavy industries through the advances developed in mining and smelting, and then again to the development of mass production through the organizational techniques initially proposed by Frederick Taylor before the turn of the century. Few would doubt the importance of this role in the development of high technology industries in our own modern times, through such advances as the transistor and the semiconductor (see, e.g., Jewkes et al. 1961; Mansfield 1968; Davies 1979; Goodman et al. 1987).

Although much emphasis is placed on the scientific and technological impacts of this role, it is important to note that its much less well documented organizational effects have been quite substantial. Organizational inventions are more difficult to appropriate and safeguard than scientific and technological inventions, however. They often result from the efforts of numerous actors, rather than from the ingenuity of a single individual, and they occur not within the confines of laboratory settings, but through the shaping of human activities and behavior over extended periods. The specialized knowledge required to diffuse organizational inventions has often been less restrictive than that of their scientific or technological counterparts, diffusing more swiftly and making them harder to trace back to their original source.

It should be worthwhile to differentiate between invention and innovation at this point. Innovation, in this study, has been used to refer to the application of scientific, technological or organizational inventions. As such, the innovative character of each entrepreneurial role has been based on the preexistence of inventions, regardless of their appropriability or the temporal dimension elapsing between the creation of any invention and its eventual application in any entrepreneurial venture.

Schumpeter, in particular, viewed inventive entrepreneurship as a generator of knowledge that would eventually be applied by innovative entrepreneurs. The application of such knowledge would, in the Schumpeterian perspective, occur in two basic ways. One would be the applications that occur within an existing or already structured process and which, while improving the process or its outcomes, do not change the structure of the process itself. In the second instance, the inventive application would change the very structure of the process to which it was applied, resulting in radically new and better ways of obtaining similar or better outcomes.

The diversification and expansion of the inventive role have been quite remarkable over the second half of the twentieth century. Scientific and technological invention are now far from being individual endeavors, being undertaken through corporate R&D, non-profit and governmental research laboratories, and academic institutions. Institutional support, in the form of patent and property rights mechanisms, have been extended and perfected. These are key incentives, as the qualitative and quantitative development of this role depend greatly on the prospects of appropriating the returns on any eventual applications.

Still, the uncertainty and risk involved in most any inventive venture are usually greater than those found in any of the other entrepreneurial roles. Unintended and unforeseen outcomes have often become revolutionary inventions, casting many doubts on the possibility of planning any major discoveries, or the process through which they may occur (see, e.g., Kuhn 1970; Nelson and Winter 1982). Their application is, on the other hand, much more predictable, being either embodied in a product or incorporated in a process that usually involves 'new ways of doing things' to some degree. The former usually facilitate imitation, thereby reducing the prospects of appropriating any returns. The latter are usually more complex, requiring special skills to understand and implement, thereby retarding their diffusion and reducing their potential benefits until the learning process becomes a routine activity.

3. Sectoral-Regional Change and the Historical Process

The historical emergence of each entrepreneurial role is closely related to the process of sectoral restructuring, occurring as innovations are translated into new activities that grow rapidly, while others become less favored or decline altogether. This supply-side linkage between entrepreneurship and macroeconomic change has been much neglected in the mainstream economics literature, which customarily assumes entrepreneurial capabilities to be automatic and ubiquitous. Changes in consumer and producers' preferences are then also assumed to be automatically translated into new goods and services. This leaves little space to consider how entrepreneurs respond to the signals that stimulate the decision-making heuristics and risk-taking behavior involved in any economic action (see, e.g., Baumol 1968; Simon 1979; Cyert and Simon 1983). Little thought is then given to the social costs of shortfalls in entrepreneurial capabilities in the process of development.

While this study has neither the objective nor the space to go into the microdynamics of entrepreneurial decision-making and action, nor the social costs of any supply shortfalls, it is nevertheless important to conceptualize the linkage between the development of the entrepreneurial function in all its various roles, and the macroeconomic changes that lead to major shifts in the spatial distribution of economic activities and population (see Suarez-Villa 1989, ch. 3). For this purpose, a conventional sectoral taxonomy was thought to be most appropriate, not only because it is widely understood and applied, but because it also facilitates the collection of historical data.

The fact that demand for new goods and services has historically been income-elastic is a powerful incentive for entrepreneurship in any role. It is, however, the demand for **certain** new goods and services that has historically promoted the emergence of a specific role. Without the perceived industrialization and the functions associated with the intermarket linkage role could not have emerged. It is then the attempt to fulfill such demand that leads to the establishment of 'new frontier' activities within a given sector. Such activities are then quick to seize the growth leadership of their respective sectors and, in many cases, their dynamism and impact on related activities can propel their sector to grow faster than any of the other sectors. It is thus that innovations are translated into economic change, determining the fortunes of national and regional macroeconomies (Olson 1982; Maddison 1982; Baumol 1985; Andersson 1986; Chinitz 1986; van den Berg et al. 1987; Freeman 1987; Engerman and Gallman 1987).

The 'new frontier' sectors may thus experience the fastest growth in, for example, employment, gross output, and contribution to national income. The socioeconomic effects of such dynamics would reflect the impact of an emerging entrepreneurial role, in mobilizing to satisfy the preferences for new goods and services that determine the new frontier sector's growth. In our own times, the entrepreneurial roles most suitable to such a sector would very likely be strategic planning, investment, and intermarket linkage. Invention would also be very important, though perhaps temporally disconnected, as we draw from the stock of available knowledge. Here it is significant to note that lead times between an invention and its application as a practical innovation can be as long as 100 years, as Mensch's (1979, pp. 124-129) work has shown.

For the less favored sectors, long term productivity increases become imperative, to overcome the effects of a lower demand elasticity and, possibly, greater domestic or international product market competition. The long term restructuring of the less favored sectors involves searching for more efficient technology and organizational forms to reduce production and transaction costs. Such efforts may involve relocating industrial production to lower cost regions or nations, which can result in substantially reduced production costs without technological adjustment. In the agricultural sector, a comparable outcome may involve a shift toward higher value crops, while imports substitute those of lower value.

Reallocation of labor toward the new frontier sector becomes inevitable, resulting in a reduction of employment growth in the less favored sectors, or even in absolute employment losses. The spatial impact of the reallocation process will depend on how different the spatial orientations of the new frontier sector and that which it replaces from the growth leadership are. Thus, for example, the transition from an agrarian to an industrializing society involves much rural-urban migration. Then, the transition from a manufacturing to a services-driven economy will most likely result in faster growth for certain urban places, if the fastest growing service activities are more spatially dispersed (see, e.g., Simon 1947; Jones 1984).

The restructuring of the less favored sectors will place greater demand on some entrepreneurial roles over others. In our own times, for example, productive coordination has been essential. Organizational changes aimed at increasing productivity which involve worker motivation, better labor-management relations, or the structuring of new tasks are central to this role. The intermarket linkage role can then be essential for cost reduction through input substitution, by finding lower-cost or more efficient alternatives. Strategic planning can also be important to the larger enterprises of a less favored sector, by reducing uncertainty through diversification, or inducing horizontal integration with activities in the new frontier sector.

Spatially, the locales and regions that concentrate the entrepreneurial roles and skills most important to a new frontier sector will have the greatest comparative advantages. These will then likely experience the fastest growth. Over the long term, they are bound to attract a great deal of capital resources, labor and population from less favored regions and locales, and will experience substantial qualitative improvements in their social and institutional infrastructures. If these regions were less developed at the start of the new sectoral dynamic, and the previous era's advanced regions do not share well in the distribution of new frontier sector activities, a long term reduction of interregional inequalities over the new era will very likely occur. It is thus that Williamson's (1965, 1986) well-known interregional dynamic can unfold over the course of the new era.

The long term effects of restructuring in the less favored sectors are bound to promote the outmigration of labor and population, barring major institutional rigidities. Only a massive and sustained effort to increase productivity in regions concentrating the less favored sectors will prevent substantial economic decline. In this sense, productivity improvement becomes merely an adjustment mechanism for coping with long term uncertainty, if not adversity. Thus, regions that achieve substantial productivity gains in the less favored sectors will not be the most successful performers in the new sectoral division of labor. Contrary to much conventional wisdom, achieving productivity gains, even if substantial, will only prevent further decline. It is, rather, the regions capable of growing rapidly through the new frontier sector activities that, regardless of their productivity performance, will be most successful over the long term.

A general outline of the historical association between the emergence of each entrepreneurial role and the sectoral and regional dynamics discussed above can illustrate the relationship between the micro and macro components of this study. In Table 1, each sectoral and regional era is closely related to the emergence of a role, with brief indications of the demographic and intersectoral impacts. Manufacturing becomes a central force in the long term process of development, catalyzing inventions and innovations that are incorporated in its own activities as well as those of the other sectors (see Suarez-Villa 1989, ch. 3, for a more complete discussion of this aspect). The analytic details of this framework are not worked out in this study, but it should nevertheless be interesting to examine its general assumptions. Hopefully, this exercise may become a first step toward the eventual modeling of the long term relationships outlined in Table 1.

The **Industrial Revolution** provides a starting point for this framework, where the investment and inventive roles become crucial to the emergence of manufacturing. In the absence of institutionalized financial markets, the importance of the investment role is paramount, where the marshalling of large amounts of investment capital becomes a creative and largely ad hoc endeavor. All accounts of this era also point to the enormous importance of key inventions, such as the steam engine and the new iron and textile manufacturing techniques. These promoted a restructuring of the linkages between the manufacturing and the agricultural sectors, as mined commodities, such as coal, replaced wood and charcoal as the prime industrial fuels.

The rudimentary workshops through which these roles made their impact can best be characterized as single function organizations, utilizing Chandler's typology. Most of the accessory activities later incorporated in industrial plants, such as distribution, marketing and services, were subcontracted to independent entrepreneurs. Technological limitations then pretty much constricted the workshops of the time to single-product

Table 1. Long Term and Regional and Sectoral Impacts of Innovative Entrepreneurship

	Industrial Revolution (U.S. 1780s-1840s)	**Capital Goods Industrialization** (1840s-1890s)	**Mass Production Industrialization** (1890s-1950s)	**Services-Oriented Industrial Change** (1950s-)
Entrepreneurial Roles	capital investment (invention)	intermarket linkage (invention)	productive coordination (invention)	strategic planning (invention)
Enterprise Form	single function/single product	multifunction/single product (factory system)	multifunction/multiproduct (process integration)	multifunction/multiproduct (conglomerates)
Sectoral Dynamics				
New Frontier sectors	agriculture/ manufacturing	manufacturing	manufacturing/services	services
Intersectoral Emphasis				
Outputs to manufacturing	a: farm products (e.g. fibers) s: finance	a: mined raw materials (e.g. ore) s: energy	a: processed raw materials (e.g. fossil fuels) s: marketing	a: farm products and processed raw materials s: information
Inputs from manufacturing	a: durable goods (e.g. farm implements) s: durable goods (e.g.apparel)	a: durable goods (e.g. mining equipment) s: durable goods (e.g. railroads)	a: durable/nondurable (e.g. farm equipment/ chemicals) s: durable/nondurable (e.g. autos/chemicals)	a: nondurable goods (e.g. biotech products) s: durable/nondurable (e.g. computers/medical supplies)
Regional Impacts				
Economic	H: manufacturing h: agriculture	H: manufacturing/services h: agriculture	H: services/manufacturing h: services/manufacturing	H: services h: services/manufacturing
Demographic	H: urban concentration* h: rural dispersion	H: urban concentration (rural-urban/inter-regional migration) h: rural dispersion/urban concentration* (interregional/rural-urban migration)	H: metropolitan concentration h: urban concentration* (rural-urban migration)	H: metropolitan deconcentration h: metropolitan/urban concentration* (inter-regional migration)

*population growth leadership

a: agriculture and mining, s: services, H: heartland regions (Northeast, Midwest), h: hinterland regions manufacturing, with very limited possibilities for differentiation. Chandler (1977), in particular, notes that the expansion of workshops in the newer activities usually occurred through internal employment growth rather than through the 'putting out' system. This is obviously consistent with the notion that a new frontier activity would show rapid employment growth, even at the expense of productivity, which would most likely occur through subcontracting.

In the U.S. heartland, comprising the industrializing Northeastern states, the growth leadership would pass from the primary sector to manufacturing. At the same time, the backward linkages between these two sectors were bound to become stronger, as an increasing number of manufactured goods began to be used in farming and mining, helping to raise productivity in these activities. The spatial impact of this sectoral dynamic would involve a greater concentration of population in cities with significant industrial agglomerations, through rural-urban displacement and immigration from abroad. It should not surprise us that the cities which most successfully industrialized during this era were good ports with substantial ocean-going traffic, that were also important transshipment centers (see Suarez-Villa 1989, ch. 3).

In the U.S. hinterland, which at this time comprised most of the states and the frontier territories, the primary sector and its related commercial activities continued to lead the regional economies. Thus, the pressure to increase productivity primarily became a heartland problem. It is reasonable to assume that the incorporation of frontier and unsettled lands into agricultural production would postpone this problem in many hinterland areas for many decades. In the hinterland regions, population would thus remain rurally dispersed, with any significant urban settlement occurring because of its nodal functions and commercial activities linked to agriculture and mining.

The innovative strides which characterized the era of **Capital Goods Industrialization** promoted major changes in distribution, through the establishment of the railroads as a new transport mode. Iron and coal became the predominant raw materials of this era, being utilized as inputs to virtually every manufacturing activity. They became essential for the emergence of steel production as the most important capital goods industry of the time, with applications that include a wide range of products, from locomotives and ships to bridges and household furniture and equipment (see, e.g., Ratner et al. 1979, p. 186).

The larger size of most capital goods industries promoted the develoment of the factory system, as opposed to the workshops of the previous era. Greater vertical integration and a larger scale thus required greater and more diverse input quantities from more distant sources. The innovative practices required to coordinate and manage the more complex input structure would have been impossible to implement without the intermarket linkage role. Spatially, the greater the distance between input source and factory, and the larger the value of these inputs as a proportion of total input value, the more important their management and coordination became. In this respect, for example, Chandler (1977) notes that, for the first time, many U.S. factories in the 1850s and 60s began to establish input accounting and purchasing functions as separate administrative units.

The intermarket linkage role thus became essential to the 'filling in' of the missing linkages in the forward and backward linkage chain. This role was also instrumental in the development of the multifunctional industrial enterprise, which required the integration of many diverse productive operations within a single plant. The enormous demands this placed on input coordination initially resulted in major bottlenecks, when inadequate supplies of many raw materials, such as coal, became a major obstacle to industrial expansion in the 1830s and 40s. Inadequate transport and interfirm coordination were thus preventing the integration of productive operations in the larger enterprises. It is well documented that these problems led to the purchase of raw materials from more reliable European sources, despite the lower cost and plentiful domestic sources of similar inputs (see, e.g., Chandler 1977, pp. 75-76).

In the U.S. heartland, comprising the Northeastern industrial states, manufacturing became the most important sector, with substantial deepening and intensification of the linkages with extractive activities located in this region and in the hinterland. In contrast, agriculture and extraction were still very predominant in the hinterland regions. Spatially, this continued the trends of the previous era, with two important qualifications. One is that a greater spatial dispersion of industrial employment occurred within the heartland industrial states and in adjacent states both within this region and in the Midwest, as resource-based industries were established. These industries would typically favor locations that were close to the raw material source. At the same time, the pace of urbanization accelerated in the heartland's major cities, along with the expansion of commerce and various services, some of which supported the new factory system (see Suarez-Villa 1989, ch. 3).

Major inventions and discoveries of the late 19th century, such as the diesel engine, electricity, the automobile and the development of fuels technology, and the emergence of the coordinative role, were essential to the onset of **Mass Production Industrialization**. This would bring many new and older goods within reach of the average individual consumer. Serving multiregional markets became typical for industries implementing mass production techniques, as their output volumes would surpass local and regional demand for their products. Advancing urbanization in both the heartland and hinterland regions, through rural-urban and interregional displacement and immigration from abroad, would increase the size of potential consumer markets substantially. Thus, the long term rise in the income elasticity of demand of many manufactured goods would ensure the leadership position of the manufacturing sector through at least the middle of this stage.

The coordinative entrepreneurial role was essential to the development of this era, as plant size expanded. The increasing separation between production and management in the growth of the multifunctional enterprise, and the vertically integrated production of various products in a single plant, made the functions of this role essential to most of the new industries. Various institutional innovations were also extremely important in promoting and safeguarding the inventive role. The development of the U.S. patent system was an important one, as the number of patents virtually doubled between 1880 and 1910. This era also witnessed the establishment of the first organized research units in some manufacturing enterprises, as the need to innovate increasingly became a routine component of production. The development of nationwide financial markets was also important in increasing the availability of capital to many ventures (see Suarez-Villa 1989, ch. 3).

The long term decline of the agricultural sector's importance became obvious, as this sector's employment amounted to less than 50 percent of the total U.S. labor force by the start of this era, despite the frontier expansion of previous decades. The need to increase productivity meant that much actual and potential employment was thus reallocated from this sector toward manufacturing and services, in both the heartland and hinterland regions. As a result, urbanization increased substantially through rural-urban migration. The shift of the sectoral leadership from manufacturing toward the services sector then became increasingly apparent toward the latter part of this era, as the demand for urban-based services multiplied.

The emergence of a manufacturing belt in the Midwest would come to rival that of the Northeastern seaboard. In these heartland regions manufacturing would retain substantial importance through most of this era, supported in part by the increased demand generated by the two world wars during the 1910s and 40s. Many new and older hinterland resource-based industries, such as petroleum refining and lumber, became more strongly linked with the heartland's industrial base. This would also promote the growth of related services in both heartland and hinterland. At the same time, the metropolitanization of the heartland contined throughout this era, as many of its industrial cities expanded with better employment opportunities in some manufacturing activities and in services.

The internationalization of corporate strategy and the horizontal integration of industrial and service enterprises are among the most important characteristics of our current era of **Services-Oriented Industrial Change.** The passing of the sectoral leadership to services can be seen as a major motivation for many established industrial firms to integrate horizontally with service activities. Reducing uncertainty by investing and acquiring firms in the more dynamic service activities thus became part of industrial corporate planning. This also helped extend corporate decision-making and strategy to the U.S. hinterland regions, where services experienced their fastest growth. Substantial strides in marketing and product planning also formalized many top-level corporate functions as separate units within the larger enterprises. Even the newer and smaller advanced technology firms began to devote substantial attention to their linkages with service activities. Product planning in both large and small industrial firms thus encompasses many considerations that are far removed from the internal complexities of production.

The new efforts at marketing and product planning came to adopt the product cycle dynamics as articles of faith in predicting the long term performances of corporate fortunes. It is indeed hard to find any other concept so pervasively applied in the marketing literature over the past four decades. At the same time, internal corporate structuring became more closely linked with product market changes and expectations than ever before. The reduction of internal inefficiencies and entropy has, for example, become more closely linked with product quality control and the efforts to preserve or increase market share. For this and the other top corporate functions, the emergence of the strategic planning role, and the knowledge base it generated, became essential. The inventive role has also gained substantial importance, as corporate R&D became more strategic in expanding market share and long term product and process viability.

Spatially, the greatest contrasts of this era occur between the dynamic areas of the hinterland and the mostly stagnant or declining heartland. In the Sunbelt regions, the dynamism of urban services and some manufacturing activities promoted greater urban and metropolitan growth, through interregional migration from many heartland states. In the traditional agricultural states of the hinterland, the long term need to increase productivity is leading to the emergence of large agribusiness holdings and the decline of family farming. Production thus becomes more technology-intensive, through the greater application of manufacturing inputs, such as biotechnology, chemicals, computers, and more sophisticated machinery. For both heartland and hinterland manufacturing, the ever-present need to increase productivity has affected capital goods industries most, as productive capacity is lost to more competitive nations. At the same time, the rise of advanced technology industries in some of the hinterland's Sunbelt regions may signal the long term reversal of regional roles in manufacturing.

4. Sectoral-Regional Change in Perspective: The Long Term Trends

An overview of the long term trends found over the course of the third and fourth eras (Mass Production Industrialization, Services-Oriented Industrial Change) can provide some perspective on the sectoral and regional changes discussed in the previous section. Because of the paucity and incompatibility of much data for the first and second eras, their trends will not be considered in this analysis. Starting with the third era also provides greater consistency, since the state and regional boundaries had already become crystallized, after frontier occupation came to an end. The distributive sector is considered separately, to provide greater disaggregation and a clearer perspective. Additionally, the industrial and services sectors are further subdivided, to assess the effects of construction and governmental activities on their performance.

Employment is the prime variable utilized in this analysis, for two reasons. One is that it is a better indicator of the socioeconomic impact of long term change, especially at the regional level. A second consideration is its better availability throughout the 94-year period being considered. Nevertheless, this analysis will not attempt to test the conceptual framework developed in the previous section in an exhaustive way. Instead, it will provide some indication of the most important trends reflecting the sectoral and regional changes discussed before. A more comprehensive treatment of these trends and historical periods may be found in Suarez-Villa (1989, chs. 4, 5).

The trends shown in Figure 1(a) provide an indication of the changing income elasticities, as reflected in each sector's share of national income. The primary sector provides the greatest contrast, with a sustained decline through most of the third, and throughout the fourth era. The rapid rise of the services sector's share does not start until after the late 1940s, with rates unmatched throughout the 94-year period. The total and non-governmental services' shares actually surpass that of manufacturing by the late 1950s and 70s, respectively, providing further confirmation of that sector's passing from the growth leadership. Manufacturing's share increases modestly from the late 1900s until the onset of the Great Depression, and then briefly but rapidly through the recovery and on to the end of World War II. As expected,

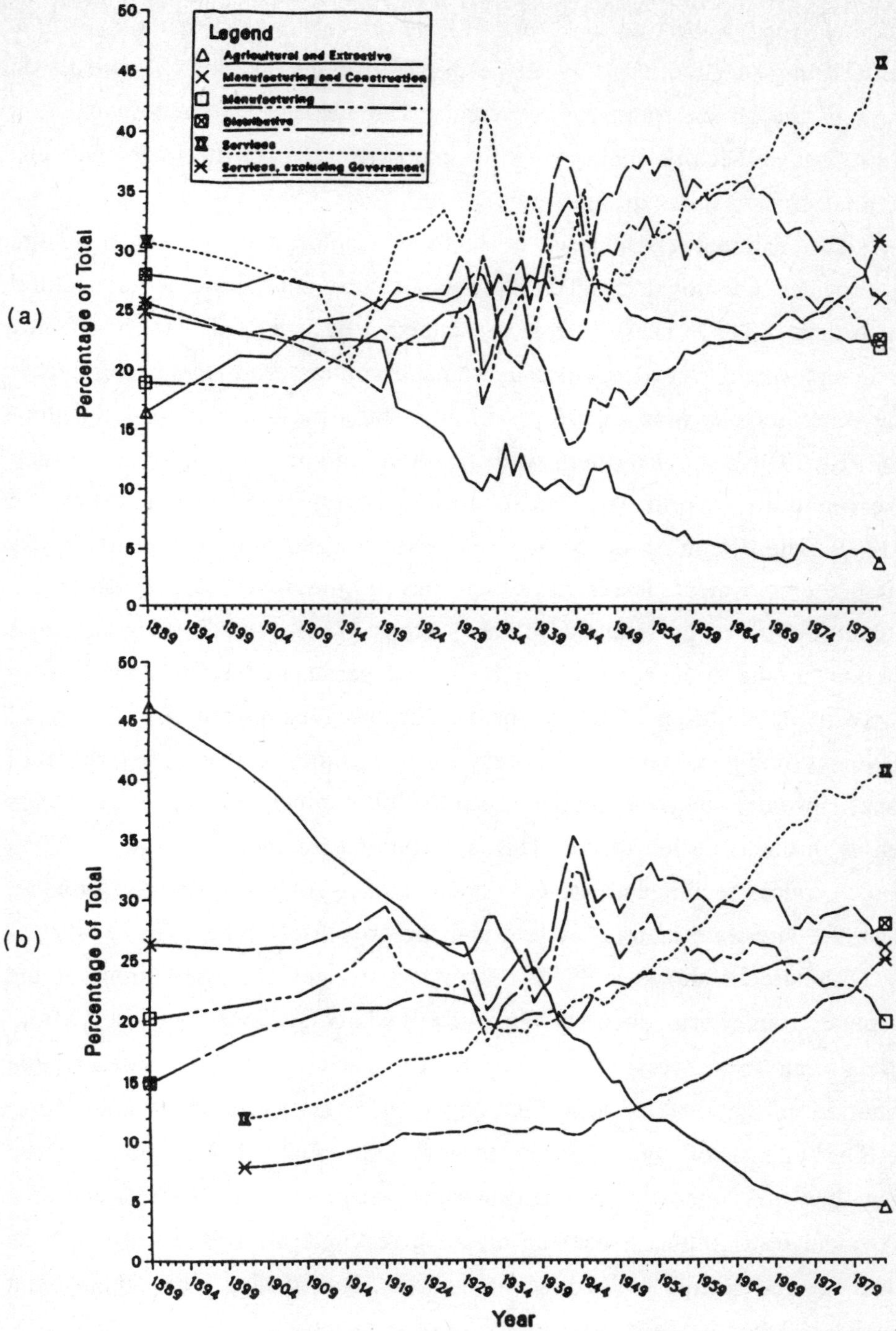

Figure 1. Distribution of National (a) Income and (b) Employment, by Sector.

Data sources: U.S. population and economic censuses, various years.

stagnation and some decline in its share sets in after the 1940s. Manufacturing's decline also affects the distributive sector after the mid-1940s, because of the strong transport and commercial linkages between them. Still, the distributive sector's linkages with many service activities prevent any substantial erosion in its income share.

The national distribution of sectoral employment, shown in Figure 1(b), provides additional insights on the socioeconomic impacts of sectoral restructuring. The acute and sustained decline of the primary sector'semployment share provides a dramatic contrast with the performances of the other sectors. Manufacturing's decline starts by the middle of the third era (1910s-1930s), as expected, with respites through the post-Depression recovery and the World War, while decline continues throughout after the mid-1940s. The dynamism of the services sector is again reflected in its rapidly increasing employment shares after the start of the fourth era (post-1940s). The contribution of government to this sector's growing employment share is significant, as the application of the Keynesian paradigm increases the role of government at all levels. This can be readily observed in the rising share of total versus non-governmental services. Despite some decline in its share of national income, the distributive sector's employment share nevertheless increases through the fourth era. This is accounted for by its linkages with the dynamic services sector, and the development of major infrastructural projects, such as the interstate highway system and the growth of commercial aviation.

The trends shown in Figure 1 support the general assumptions of the conceptual framework discussed in the previous section. The shift from manufacturing to services is reflected by these sectors' performances in the distribution of national income and employment over the third and fourth eras. The impact of the changing income elasticities of demand of these sectors becomes obvious in their income share trends. These, and the employment distribution, reflect the on-going restructuring and the attempts by the less favored sectors to increase productivity, by reducing their employment shares while labor is reallocated to the most dynamic sector.

The rankings of input magnitudes from the 1919, 1939 and 1977 U.S. input-output matrices, shown in Table 2, provide additional insights on long

Table 2. Input Rankings: Interindustry and Intersectoral, 1919, 1939, 1977

	Interindustry			Transportation+			Services+ excluding Government	
	1919	1939	1977	1919	1939	1977	1939	1977
Iron and Steel Products	1	3	4	1 (0.26)	4 (0.02)	7 (0.007)		
Food Products	2	1	1					
Textiles	3	7	8					
Apparel	4	4	6					
Chemicals	5	6	5				5 (0.002)	8 (0.009)
Motor Vehicles	6	5	3					
Petroleum Refining	9	2	2	4 (0.01)	1 (0.05)	1 (0.16)	3 (0.004)	3 (0.02)
Processed Coal and Coke Products				2 (0.07)	3 (0.03)	*		
Lumber and Wood Products				3 (0.02)	*	*		
Printing and Publishing				5 (0.005)	n.a.	10 (0.005)	1 (0.07)	1 (0.02)
Non-Metallic Minerals Processing				6 (0.002)	10 (0.002)	++		
Transportation Equipment				* (Autos)	2 (0.03)	2 (0.04)	++	2 (0.02)
Rubber/Plastics				*	5 (0.004)	3 (0.02)	6 (0.002)	6 (0.01)
Electrical Equipment				n.a.	n.a.	6 (0.007)		
Fabricated Metals				n.a.	n.a.	4 (0.009)	n.a.	5 (0.01)
Machinery				n.a.	n.a.	5 (0.007)	4 (0.003)	14 (0.005)
Furniture							2 (0.006)	++
Pharmaceuticals							n.a.	4 (0.01)

*no appreciable inputs. +inputs from manufacturing. ++very low ranking; exact rank impossible to determine.
n.a.: not available.
In parentheses: Inputs per dollar of total purchases (1977: per dollar of total intermediate inputs).

Data sources: Leontief, W. W. (1941), *The Structure of the American Economy, 1919-1939*, New York, Oxford University Press, for 1919 and 1939 data ("Net Total Outlays"). U.S. Department of Commerce (1984), "The Input-Output Structure of the U.S. Economy, 1977," *Survey of Current Business* 64:42-84, for 1977 data ("Total Intermediate Inupts"; restaurants and food services excluded from Services estimates).

term sectoral restructuring. The six highest ranking inputs in 1919 and 1977 are traced and contrasted with each other. Among the interindustry and transportation inputs, the decline of the steel industry is a harbinger of the condition faced by many U.S. heavy industries in our own time. The substitution of metals by lighter weight plastics and synthetic rubber products in many goods, compounded by increasing international competition after the 1950s, are major contributors. Textiles and apparel also decline, as these industries moved toward lower cost nations with the advance of their product cycle dynamic. The rise of refined petroleum and motor vehicle inputs reflect these industries' much interrelated importance, and their strong linkages with the most favored service activities.

The rise of the automobile industry throughout the third and part of the fourth era is most obvious in the transportation input rankings. Oil thus replaces coke, coal, and lumber and wood inputs as the most important fuel. Fewer changes are apparent among the non-governmental services' inputs, where this sector's enormous reliance on printed matter for most transactions allow printing and publishing to retain the leading rank. The importance of transport goods is also obvious, as the rise of services relies substantially on distribution and communications. The rise of pharmaceuticals by 1977 then reflects the importance of health services, and the population's growing concern with most aspects of health care over the life cycle.

Among the regional farm employment trends, in Figure 2(a), the Southeast experiences the greatest decline. Long the U.S.'s most backward region, changes in land tenure patterns and the technification of farming promoted substantial rural-urban migration to its urban places, and to various Northeastern and Midwestern cities over the course of the third and early fourth eras. Declines in absolute farm employment are also obvious in all the other regions since the middle of the third era (1920s), but are much less pronounced in the Western (Pacific, Mountain) regions. In these, the delayed effect of late frontier exploration and settlement resulted in lower rural population per land area and, consequently, in a much lower potential rural-urban migrant pool.

Trend contrasts are less apparent in the farming employment share

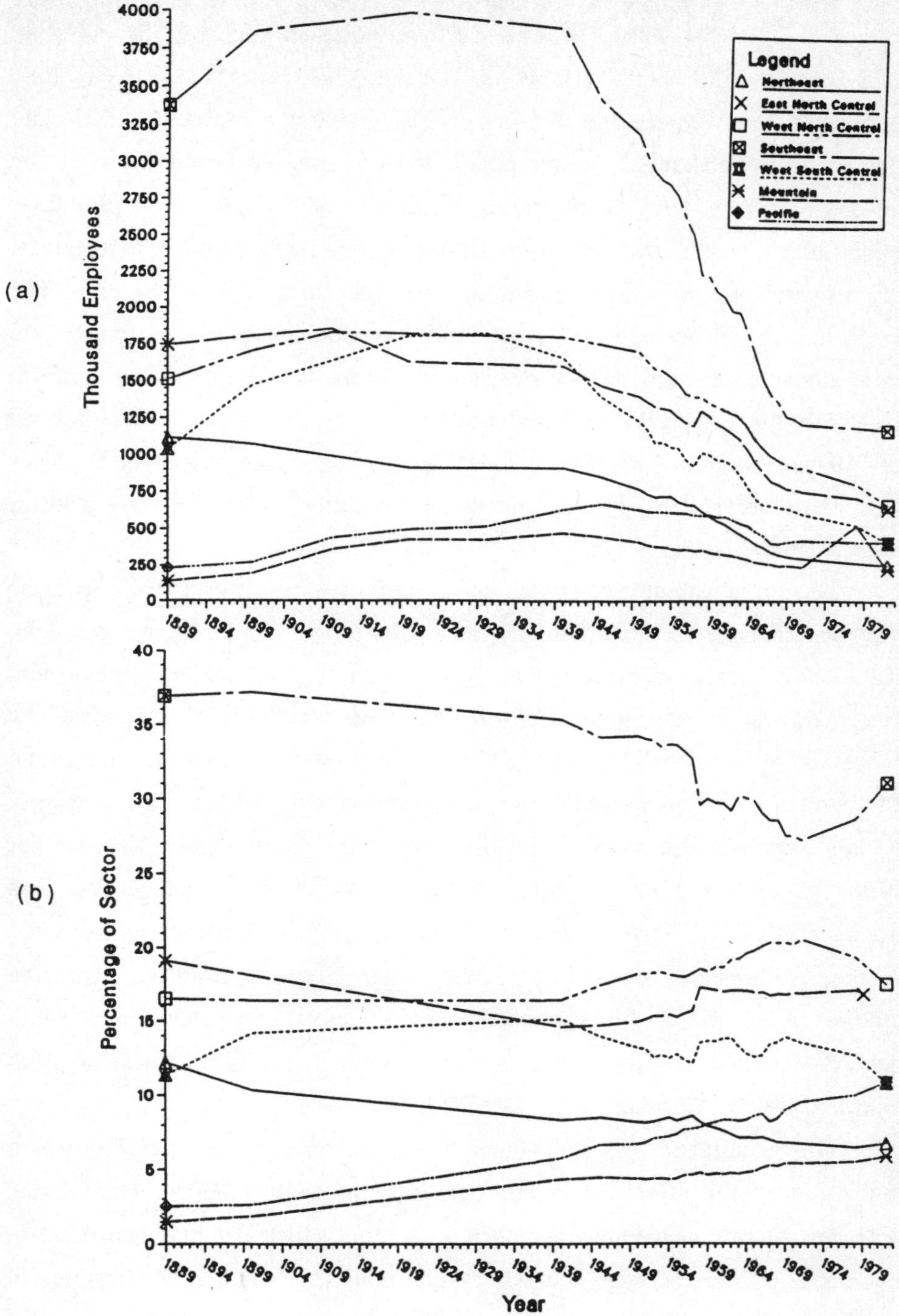

Figure 2. Farming: Regional (a) Employment and (b) Employment Distribution.

Data sources: U.S. agricultural and economic censuses, various years.

estimates of Figure 2(b). Despite its pronounced decline in absolute employment, the Southeast retains the largest share over the 94-year time period, while the Western states' modestly rising shares again reflect their late settlement, with substantially lower population per suitable farmland area. The fairly constant long term share trends of all the other regions, despite their declining absolute employment, again reflect agriculture's position as the least favored sector in the national economy over the third and fourth eras. The attempt to raise productivity through more technology-intensive farming, and labor reallocation toward the more dynamic sectors, is therefore fairly uniform across both the heartland and hinterland regions. For the least favored sector, fewer contrasts in the regional division of labor, vis-à-vis theother sectors, therefore reflect greater innovative limitations in overcoming a lower income elasticity for its goods.

The employment trends in Figure 3(a) show long term manufacturing employment decline to be primarily a regional rather than a national problem. Absolute employment decline in the Northeast throughout the fourth era, and in the Midwest (East North Central) since the mid-1970s, are offset by increases in all the other regions. The more modest performances are found in the central U.S. regions with lower urbanization rates (Plains, Southwestern, Mountain regions). The employment distribution trends, in Figure 3(b), further confirm the emerging new regional division of labor in manufacturing. The sharp decline in the Northeastern share throughout the third and fourth eras, and the Midwestern decline throughout the fourth, contrast with the Southeastern, Pacific and Southwestern performances. The rising trends are largely determined by the more dynamic Sunbelt states in these regions (California, Texas, Florida, North Carolina).

The manufacturing employment trends reflect its shifting position as a less favored sector after the middle of the third era (1920s). This occurs despite the rise of new industries, such as motor vehicles in the heartland in the 1910s and 20s, and high technology electronics in some Sunbelt states in the 1950s and 60s. As expected, its displacement from the leading sectoral role starts to become obvious after the services sector begins its long term takeoff, by the middle of the third era. Employment decline in the industrial

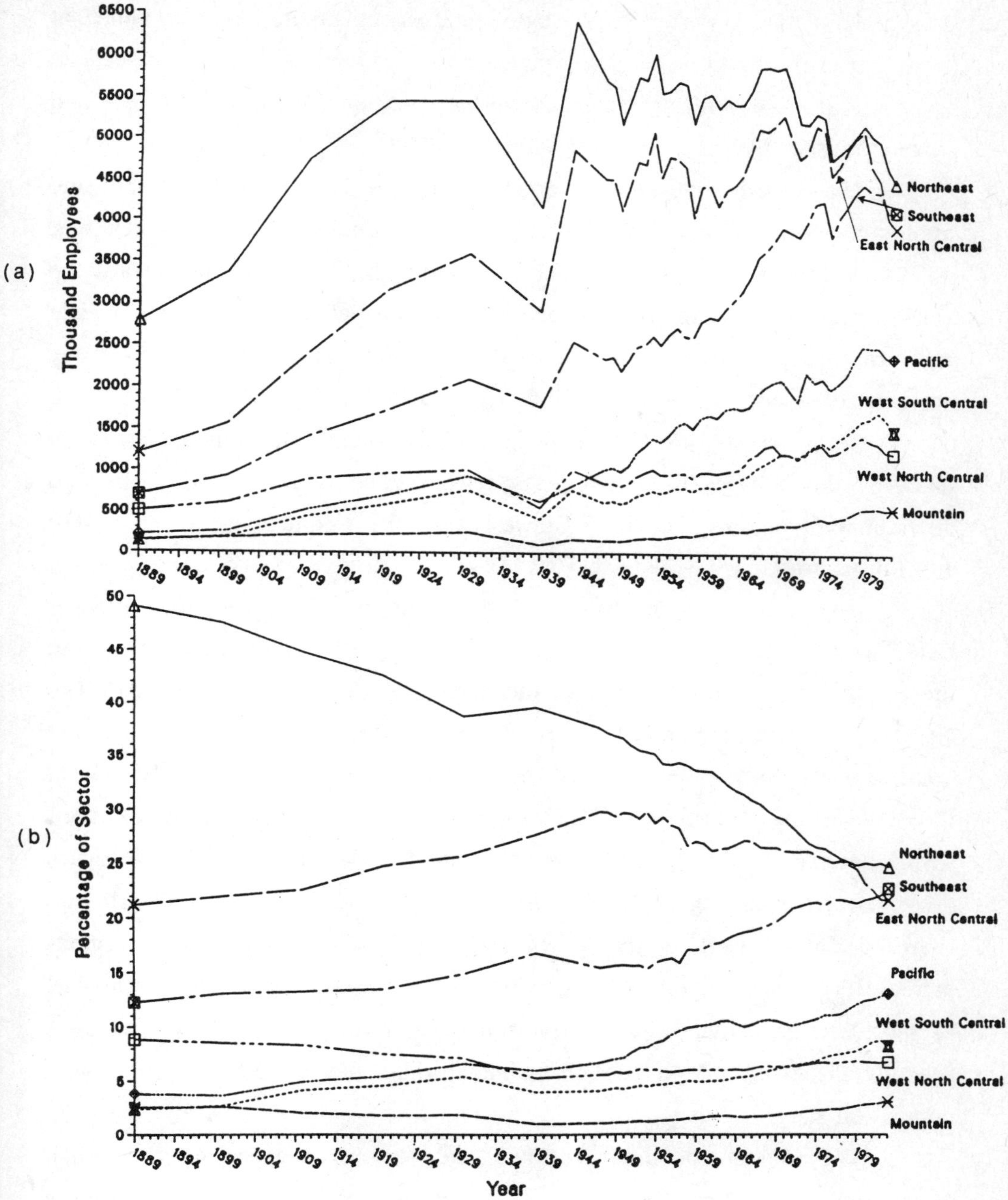

Figure 3. Manufacturing: Regional (a) Employment and (b) Employment Distribution.

Data sources: U.S. manufacturing and economic censuses, various years.

heartland then results from technological and organizational changes aimed at increasing productivity, and from plant closings, downsizing, and relocations brought about by increasing international competition, and the lower cost and growing market advantages of the Sunbelt regions.

The distributive sector's employment trends, comprising retail and wholesale commercial activities, transportation, utilities, and communications, reflect the shift of population growth away from the heartland and toward the Sunbelt regions. Most of its activities are well correlated with population size and are strongly linked with services. In Figure 4(a) and (b), the rising importance of the Southeast through the fourth era becomes obvious, as it passes the Midwest and Northeast in both absolute employment and total share by the late 1960s and 70s, respectively. The share trends of the two other Sunbelt regions (Pacific, Southwest) also rise significantly over the third and fourth eras.

All regions gain in absolute employment in non-governmental services reflecting this sector's most favored status and growth leadership throughout the fourth era, and briefly through the middle of the third [Figure 5(a)]. The Northeast retains the largest absolute employment throughout, while the Southeast surpasses the Midwest after the late 1960s. Only the Great Depression and World War II disrupt the rise of this sector, as their socioeconomic impact is felt across all regions. The share trends of Figure 5(b) reflect a changing regional division of labor, however, as the heartland's shares decline throughout the fourth era, while those of the Sunbelt regions increase. The shift of population growth toward the latter is a major influence, as most service activities, being traditional central place functions, are more spatially dispersed than those of manufacturing, and correlate strongly with population size.

The impacts of these regional and sectoral changes on the urban population distribution are reflected in the heartland-hinterland contrasts shown in Figure 6(a). The Northeastern share, in particular, peaks by the middle of the third era (late 1920s), following closely the shift of the sectoral leadership from manufacturing to services. As the most favored sector, the rise of services is also closely linked with the shifting distribution of urban

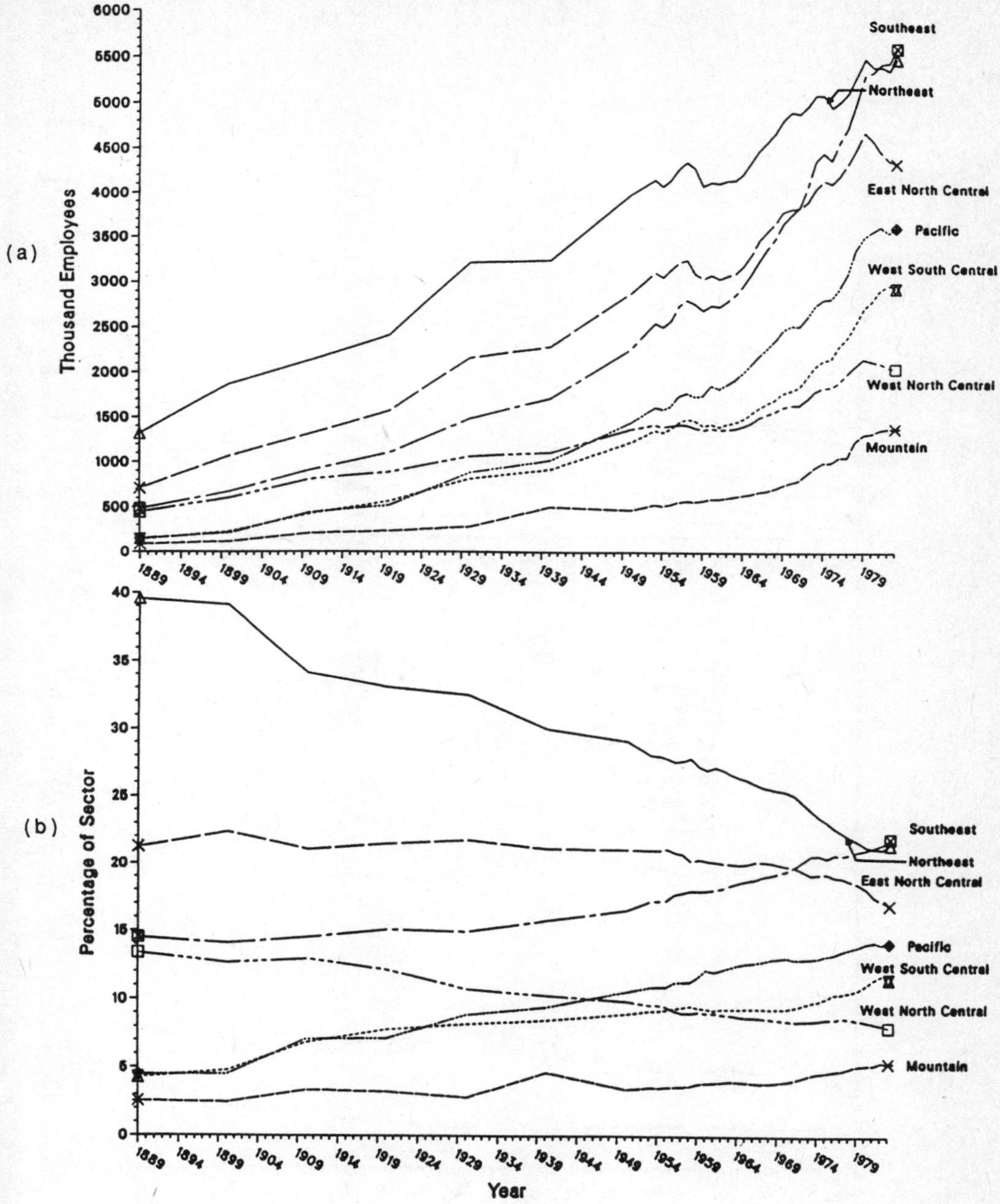

Figure 4. Distributive Sector: Regional (a) Employment and (b) Employment Distribution.

Data sources: U.S. economic censuses, various years.

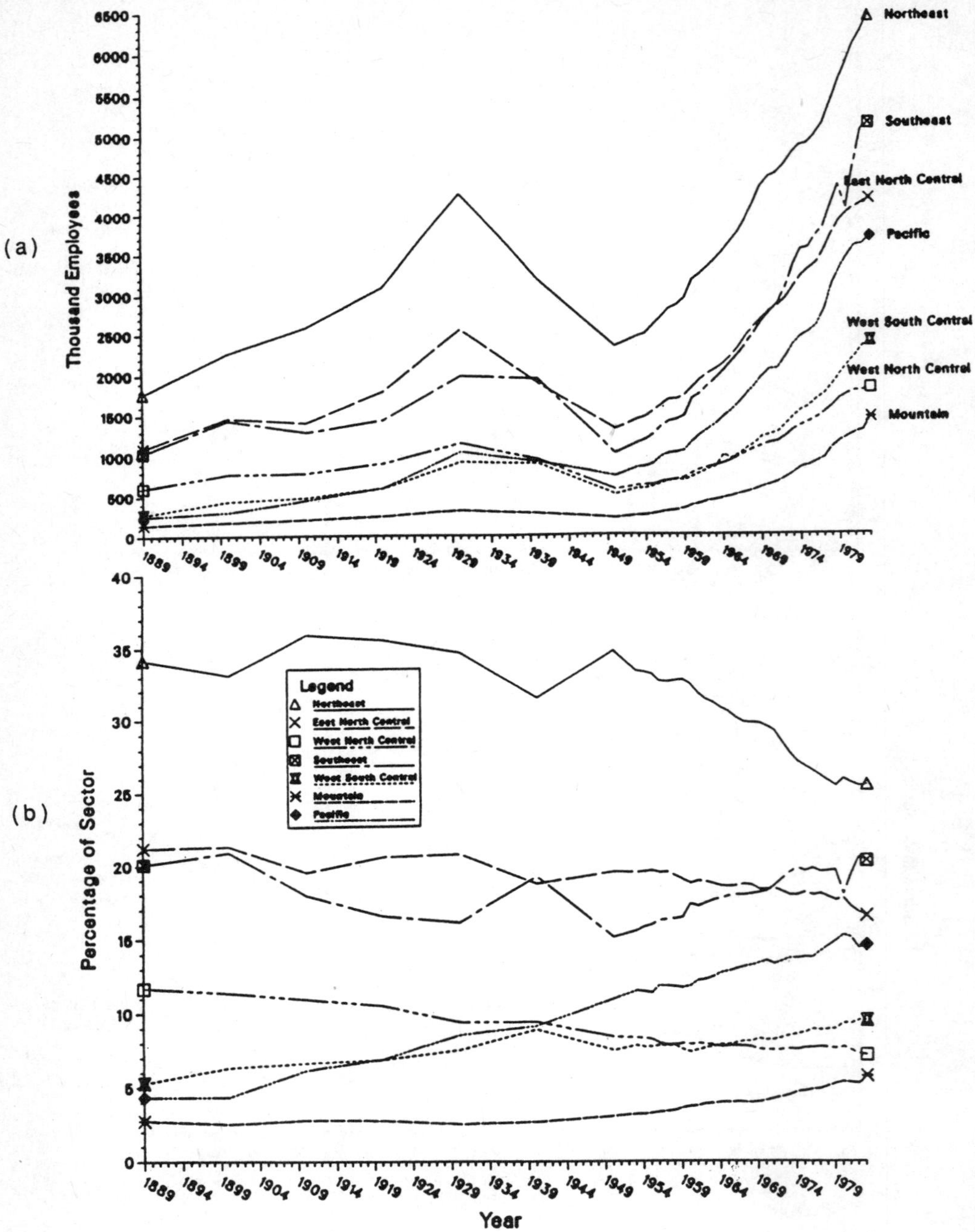

Figure 5. Non-Governmental Services: Regional (a) Employment and (b) Employment Distribution.

Data sources: U.S. economic censuses, various years.

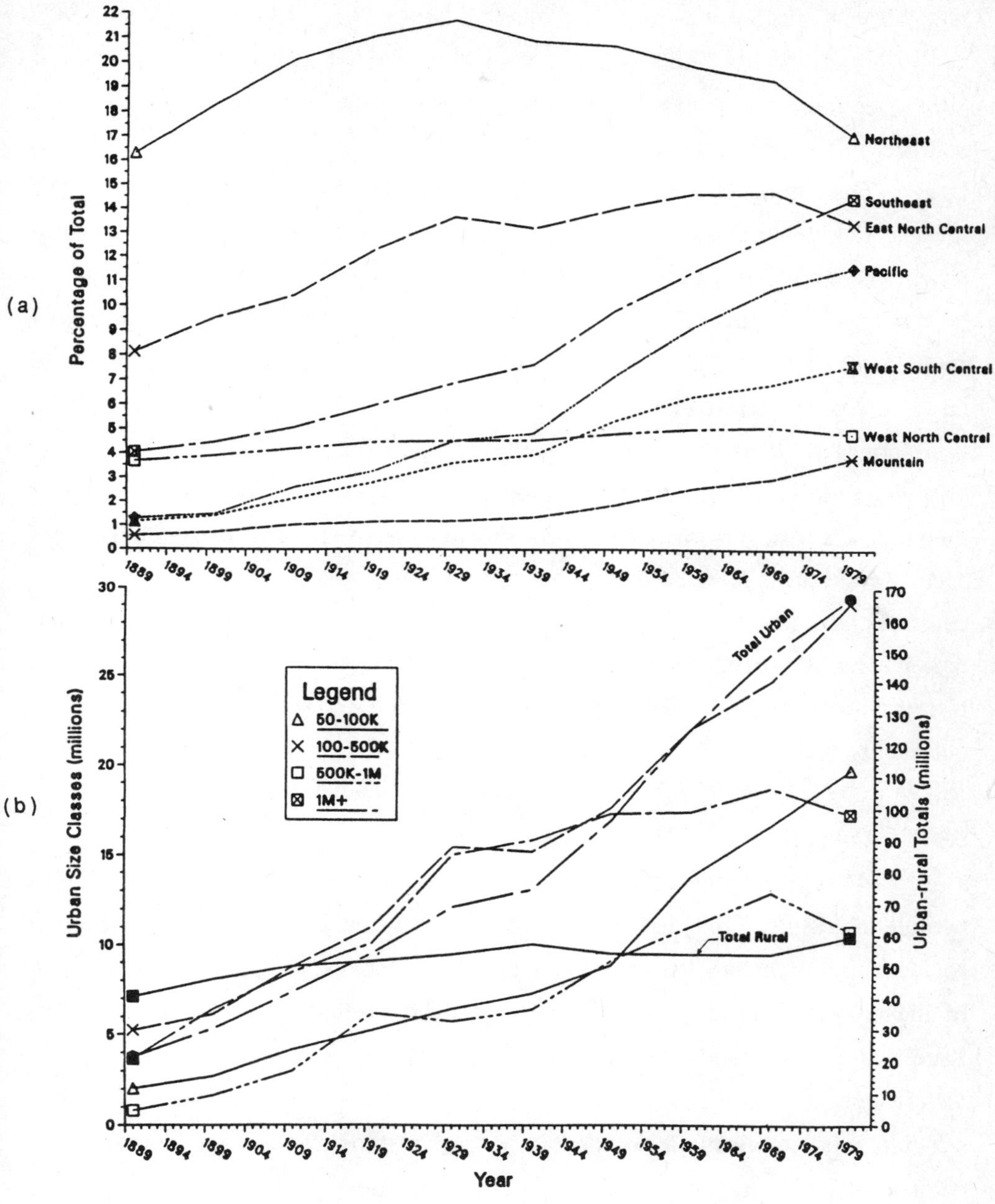

Figure 6. Urban Population: (a) Regional Distribution and (b) Size Categories.

Data sources: U.S. population censuses, various years.

population in the hinterland. The Sunbelt regions' shares thus rise substantially after the late 1930s, reflecting major changes in the regional distribution of urban population and, more deeply, the ongoing transformation in the regional and sectoral division of labor.

The allocation of population among the various urban size classes also reveals the sectoral transformations occurring over the fourth era [Figure 6(b)]. The greater spatial dispersion of the most favored sector, services, with its strong central place functions, ensures faster growth for the small urban size classes (50-100 and 100-500 thousand). In contrast, the population of the larger urban size class (1 million and above) levels off after the middle of the third era (late 1920s), following closely the rise of the services sector and the decline of the Northeastern urban population share. After the late 1960s, the medium size category also peaks, as the long term dynamism of the services sector continues to transform the spatial distribution of population. The contrast between the national urban and rural trends then reflects the nation's rapid urbanization, and the long term impact of the primary sector's least favored condition, as virtually all population growth over the fourth era occurs in cities.

5. Conclusions

As a prospective piece on regional development, this paper has had a twofold objective. First, it has provided a structured overview of an aspect of socioeconomic change that has been much overlooked in regional science and in the spatial literature. There can be little doubt that all the major changes, issues and debates that are of interest today in the area of regional develoment are very much related to a long term process whose parameters, structure and implications we do not yet understand well. It has for too long been a common practice to view spatial development in a very restrictive way, divorcing it from the historical context and longer term dimension of which it is an integral component. Variables, functions and models thus tend to be automatically specified, often without much insight on the actors and the larger forces that set the process in motion and determine its course. This is not to say that such models cannot yield important and powerful insights on the process of development. They have often done so, particularly on very specific aspects of the process. The point is, rather, that they could be much enriched by a better consideration of historical process, and of the human forces and roles that it generates.

A second objective of the paper has been to consider the human factor at the root of the larger process of development. The consideration of this 'human factor', be it in the form of the entrepreneur, the institutional policy-maker, or any other significant actor, and their socioeconomic roles and behavior, must be at the center of our micro-level explorations into the development process. From time to time, regional scientists have called our attention to this much overlooked aspect (see, e.g., Hägerstrand 1969; Alonso 1988). Still, such infrequent calls have all too often been ignored in our paradigms and models.

At the macro level, a understanding of the historical process can also provide a better perspective on the socioeconomic forces determining the outcomes that long wave theorists have attempted to model and explain. All too often these efforts, although very important and significant in their own right, have nevertheless viewed long term changes as the outcome of variables that somehow mechanistically determine the observed outcomes. In such efforts, interest rates, raw material prices, inflation rates, and the like, are taken as the generators of long term change, without considering the deeper historical process and structure that actually determine the postulated 'determinants' of our outcomes. Difficult as it may be to operationalize the historical context into a model of long term change, its consideration cannot fail to enrich our perspectives on the process of development.

At the micro level, a search for insights on behavior and decision-making in the study of the entrepreneur, of policy-makers, and of other key economic actors can further enrich our analytical efforts. The study of risk, and of the individual and collective processes applied to deal with uncertainty, can contribute much to our understanding of the process of development. The location decisions which so influence regional development outcomes are, for example, guided by entrepreneurs' risk-taking propensities, by their perceptions of uncertainty, and by their achievement criteria and objectives, of which the profit motive is merely one among many. All of these considerations also affect the process of innovation, about which we know so very little, in both its broader and more specific aspects. These, crucial as they are, and their medium and long term impacts on the spatial structure of development, are not yet well understood by most spatial analysts. Hopefully, this paper can provide a start that will channel our interests to the study of these aspects in the process of development.

References

Alonso, W. (1988), From Alfred Weber to Max: The Shifting Style of Regional Policy. Paper presented at the 35th North American Meetings, Regional Science Association, Toronto.

Andersson, A.E. (1986), The Four Ligistical Revolutions, Papers of the Regional Science Association 59:1-12.

Ashton, T.S. (1948), The Industrial Revolution, 1760-1830, New York, Oxford University Press.

Batten, D., J. Casti and B. Johansson (eds.) (1987), Economic Evolution and Structural Adjustment, Berlin, Springer-Verlag.

Baumol, W.J. (1968), Entrepreneurship in Economic Theory, American Economic Review 58:64-71.

Baumol, W.J. (1985), Entrepreneurship and the Long Run Productivity Record, Research Report 86-04, New York, C.V. Starr Center for Applied Economics, New York University.

Berg, M. (1985), The Age of Manufactures: 1700-1820, Totowa N.J., Barnes and Noble.

Van den Berg, L., L.S. Burns and L.H. Klaassen (eds.) (1987), Spatial Cycles, Aldershot Engl., Gower.

Boserup, E. (1981), Population and Technological Change: A Study of Long-Term Trends, Chicago, University of Chicago Press.

Chandler, A.D. (1962), Strategy and Structure: Chapters in the History of the Industrial Enterprise, Cambridge, M.I.T. Press.

Chandler, A.D. (1977), The Visible Hand: The Managerial Revolution in American History, Cambridge, Harvard University Press.

Chinitz, B. (1986), The Regional Transformation of the American Economy, American Economic Review 76:300-303.

Cochran, T.C. (1977), Two Hundred Years of American Business, New York, Basic Books.

Cyert, R.M. and H.A. Simon (1983), The Behavioral Approach: With Emphasis on Economics, Behavioral Science 28:95-108.

Davies, S. (1979), The Diffusion of Process Innovations, Cambridge Engl., Cambridge University Press.

Donaldson, G. and J.W. Lorsch (1986), Decision Making at the Top: The Shaping of Strategic Decisions, New York, Basic Books.

Engerman, S.L. and R.E. Gallman (eds.) (1987), Long Term Factors in American Economic Growth, Chicago, University of Chicago Press.

Freeman, E. (ed.) (1987), Design, Innovation and Long Cycles in Economic Development, London, Frances Pinter.

Goodman, D., B. Sorj and J. Wilkinson (1987), From Farming to Biotechnology, New York, Basil Blackwell.

Hägerstrand, T. (1967), Innovation Diffusion as a Spatial Process, Chicago, University of Chicago Press.

Hägerstrand, T. (1969), What About People in Regional Science?, Papers of the Regional Science Association 24:7-21.

Hirschman, A.O. (1958), The Strategy of Economic Development, New Haven, Yale University Press.

Hoenack, S.A. (1983), Economic Behavior Within Organizations, New York, Cambridge University Press.

Jewkes, J., D. Sawers and R. Stillerman (1961), The Sources of Invention, New York, Norton.

Jones, B.G. (1984), Productivity and the Spatial Implications of Structural Change: Empirical Evidence for Simon's Model, Papers of the Regional Science Association 54:1-11.

Kleinknecht, A. (1987), Innovation Patterns in Crisis and Prosperity, London, Macmillan.

Kuhn, T.S. (1970), The Structure of Scientific Revolutions, 2nd ed., Chicago, University of Chicago Press.

Leibenstein, H. (1978), General X-Efficiency Theory and Economic Development, New York, Oxford University Press.

Leibenstein, H. (1987), Inside the Firm: The Inefficiencies of Hierarchy, Cambridge MA, Harvard University Press.

Maddison, A. (1982), Phases of Capitalist Development, New York, Oxford University Press.

Mansfield, E. (1968), Industrial Research and Technological Innovation, New York, Norton.

McDermott, P. and M. Taylor (1982), Industrial Organisation and Location, New York, Cambridge University Press.

Mensch, G.O. (1979), Stalemate in Technology, Cambridge MA, Ballinger.

Mintzberg, H. (1979), The Structuring of Organizations: A Synthesis of Research, Englewood Cliffs N.J., Prentice-Hall.

Molle, W. (1983), Industrial Change, Innovation and Location, Paris, O.E.C.D.

Nelson, R.R. and S.G. Winter (1982), An Evolutionary Theory of Economic Change, Cambridge MA, Belknap.

Nijkamp, P. (ed.) (1987), Handbook of Regional Economics, Amsterdam, North-Holland.

Olson, M. (1982), The Rise and Decline of Nations: Economic Growth, Stagflation, and Social Rigidities, New Haven, Yale University Press.

Ratner, S., J.H. Soltow and R. Sylla (1979), The Evolution of the American Economy, New York, Basic Books.

Rosenberg, N. (1986), Inside the Black Box: Technology and Economics, New York, Cambridge University Press.

Say, J.B. (1803), Traité de Economie Politique, Paris, Deterville.

Scherer, F.M. (1986), Innovation and Growth: Schumpeterian Perspectives, Cambridge MA, M.I.T. Press.

Schumpeter, J.A. (1934), The Theory of Economic Development, Cambridge MA, Harvard University Press (orig. publ. 1912; Leipzig, Duncker and Humblot).

Simon, H.A. (1947), Effects of Increased Productivity upon the Ratio of Urban to Rural Population, Econometrica 15:31-42.

Simon, H.A. (1979), On Parsimonious Explanations of Production Relations, Scandinavian Journal of Economics 81:459-474.

Solomon, S. (1987), Phases of Economic Growth 1850-1973: Kondratieff Waves and Kuznet Swings, New York, Cambridge University Press.

Suarez-Villa, L. (1987), Entrepreneurship and the International Diffusion of Innovations in Manufacturing: A General Approach, Rivista Internazionale di Scienze Economiche e Commerciali 34:369-391.

Suarez-Villa, L. (1988), Innovation, Entrepreneurship, and the Role of Small and Medium Size Industries: A Long Term View, in: M. Giaoutzi, P. Nijkamp and D. Storey (eds.) Small and Medium Size Enterprises in Regional Development, London, Routledge.

Suarez-Villa, L. (1989), The Evolution of Regional Economies: Entrepreneurship and Macroeconomic Change, New York and London, Praeger.

Sweeney, G.P. (1987), Innovation, Entrepreneurs and Regional Development, London, Frances Pinter.

Teece, D.J. (1980), Economies of Scope and the Scope of the Enterprise, Journal of Economic Behavior and Organization 1:223-247.

Thomas, M.D. (1986), Growth and Structural Change: The Role of Technical Innovation, in: A. Amin and J.B. Goddard (eds.), Technological Change, Industrial Restructuring and Regional Development, Boston, Allen and Unwin.

Usher, A.P. (1954), A History of Mechanical Inventions, Cambridge MA, Harvard University Press.

Weiss, J.W. (ed.) (1988), Regional Cultures, Managerial Behavior and Entrepreneurship: An International Perspective, New York, Quorum.

Williamson, J.G. (1965), Regional Inequality and the Process of National Development: A Description of the Patterns, Economic Development and Cultural Change 13:3-45.

Williamson, J.G. (1986), Is Inequality Inevitable under Capitalism?, in: P.L. Berger (ed.), Capitalism and Equality in America: Modern Capitalism, vol. 1, Lanham MD, University Press of America.

Chapter 13

THE EUROPEAN INTERNAL MARKET AND THE INTERNATIONALISATION OF SMALL AND MEDIUM SIZE ENTERPRISES

R. Cappellin

1. Introduction

The completion of the European Internal Market will promote the international mobility of services and of production factors, such as capital, labour, technology and entrepreneurship. Thus, the traditional distinction between national and regional economies, based on the immobility of factors among national economies, has become blurred and this requires a tighter integration between international economics and regional economics.

The completion of the European Internal Market should be analyzed in the framework of the macroeconomic and sectoral changes, which will occur in the next five or more years. These seem to be characterized not only by an increasing internationalization of the firms, but also by faster growth rates than those of the seventies and by the diffusion of pervasive technological innovations, promoting the fast adoption of new product and processes. These factors may facilitate the realization of the European Internal Market in a similar way as the development phase, which characterized the sixties, facilitated the realization of the European Common Market in those years.

The completion of the European Internal Market is not the automatic result of deregulation policies, aiming to remove administrative constraints (Centro Europa Ricerche, 1987). On the contrary, as it deals with politically sensitive issues such as the movements of services and factors, where basic social interests should be preserved and various market imperfections exist, some regulations will be inevitable. Thus, the completion of the European Internal Market requires a long lasting effort aiming to achieve a gradual harmonization of national regulations (for example: OPA, antitrust, fiscal, environmental, consumer protection regulations), to coordinate the various national economic policies and to adopt new European common policies (Padoa Schioppa, 1987). In some cases, when specific regulations are required for an efficient functioning of the market mechanism, even free competition will oblige each country to adapt its regulations to those institutional or organizational innovations which are already adopted in other countries.

The same argument applies to the case of regional policies. In fact, the impact of the European Internal Market on the regional disparities can not be evaluated, under the hypothesis of no regional policies by European or national governments, as regional policies have always been and most probably will be a permanent characteristic of the State intervention in the economy. Therefore, the impact on regional disparities of the European Internal market will most probably depend on the kind of regional policies which will be adopted in the next future at the European and national level.

Then, the problem of evaluating the impact of the European Internal Market on regional and national disparities consists in that of evaluating the effects of the harmonization of national policies and of the adoption of new Community policies in various fields. In particular, the completion of the European Internal Market should be conceived as a phase in the process of the transformation of the European Common Market into an economic and political union, which will require a tighter coordination of national economic policies and the development of new European structural policies aiming to directly affect the allocation of economic resources and to foster economic growth.

Therefore, it is important to evaluate the regional impact of these different policies, which will accompany the construction of the European Internal Market, and the interdependent relationships which exist between the regional policies and the non-regional economic policies. In fact, the relationship between regional policies and the other economic policies is a complex process in which it is possible to distinguish different mechanisms, which may be illustrated with the figure below (Molle and Cappellin, 1988):

		OBJECTIVES OF	
		Community and national	Regional policies
INSTRUMENTS OF	Community and national policies	A	B
	Regional policies	C	D

Figure 1. The interdependence between regional, national and Community policies

Therefore, in agreement with the traditional theory of economic policy, a clear distinction must be drawn between the objectives (columns) and the instruments (rows) of policies. Next, a division can be made into, on the one hand, regional policies (undertaken at the regional level) and on the other hand various non-regional policies (pursued on the national and on the Community level) dealing with specific sectors or domains (industrial policy, international trade policy, transport policy, macroeconomic and exchange rate policy, etc.). The usual assumption is for community or national instruments to aim at Community or national objectives (case A) and regional instruments at regional objectives (case D). However, this does not hold if the complex relationships among regional and national or international factors are taken into account.

In fact, the various national policy measures may have different impacts on the objectives of regional policies, as indicated by case B. This impact will mainly depend on the sectoral composition of the various regional economies. Therefore, a European industrial policy aiming to selfsufficiency in high-tech fields may represent an advantage for the most developed regions, where these sectors are mainly concentrated, unless this policy is not accompanied by the development of new modern sectors also in the less developed regions. Similarly, the lack of an explicit European policy for the small and medium size firms and the delay in implementing an effective antitrust regulation in various countries may hinder the development of the intermediate and peripheral regions, which are characterized by a larger share of small and medium size firms than the more developed central regions.

These effects support the need for an explicit regional policy of the European Community for the less developed regions of the Community (case B), which could compensate the negative impact of other non-regional policies of the European Community. However, the transfers of appropriate financial resources should be oriented in such a way in order to support the endogenous development potential in the regions considered and should allow an active role of the regional and local administrations.

On the other hand, regional policies may contribute to the achievement of the objectives of national industrial policies (case C), as they may promote the diffusion of technological innovations and may help in removing various obstacles which hinder the increase of international competition in the individual regions.

Moreover, the choice of the aims and of the instruments of regional policies (case D) should be adapted to the new macroeconomic, sectoral and technological trends, to the various measures taken at the Community level in order to complete the European Internal Market and in particular to the guidelines defined by the national and European industrial policies. In order to promote the regional endogenous development, regional policy should be capable both to create those local environmental factors and to promote those internal capabilities of the firms, which affect the international competition of these latter in terms of lower costs and of faster adoption of product and process innovations.

Given the uncertainty on the structural policies which will be taken by the European Community, the purpose of this paper can not be that to forecast the changes in regional disparities due to the harmonization of national regulations required by the completion of the European Internal Market. Therefore, this paper will attempt to answer to a more restrictive or methodological question, such as that of the identification of the analytical tools, which may be most appropriate in the evaluation of the impact of the various policy measures, aiming to the completion of the European Internal Market, and it will clarify that the impact of these policies can not be properly analyzed whether we use only the traditional analytical tools, which were elaborated for the analysis of the effects of customs unions. In particular, the paper will analyze the problems created by the completion of the European Internal Market on regional economies characterized by a system of small and medium size firms.

2. The Internationalization Process of SME

The internationalization of the firms may be interpreted as the aim of the process of completing the European Internal Market but also as an external constraint in this process. In particular, the internationalization of firms may be related to four different mechanisms. First of all, the internationalization process of the firms is related to various factors, such as the improvement of communications, the improved knowledge of foreign languages and cultures, the increasing diffusion of information, the adoption of foreign know-how and organizational procedures, the faster international diffusion of consumer fashions. All these factors determine an increasing homogeneity of national markets and a decrease of those national or regional peculiarities in consumer preferences, which were an effective protection for local producers.

A second phenomenon is the increasing articulation or fragmentation of the internal structure of the various national and regional markets, which increasingly require customized products. These transformations of the market structures are tightly connected with long-term factors such as the low demographic increase, the increasing role of older generations, the growth of the female activity rate, the increased fragmentation of the professional profiles, the increase of leisure time, the increase of education levels and, last but not least, the increase of personal income. These factors lead to more articulated and sophisticated consumer patterns and to an increasing importance being attributed by consumers to the quality of the products and services rather than to their costs.

A third factor is the fact that the length of the product cycle becomes shorter and products become obsolete after a shorter delay. In particular, the high costs of research required by the adoption of innovations and the shorter 'lead time' on competitors and imitators compel the firms to look for alliances in research activities and for commercialization agreements with foreign firms, in order to exploit their transitory advantage on all national and regional markets at the same time.

Fourthly, the increasing complexity of the customers needs requires a tighter relationship between the producer and the consumers and an intense exchange of information. Therefore, services should increasingly be integrated with manufactured goods, in forms of consumer assistance, training, financing, advertising, etc. The combination of services with goods allows new forms of product differentiation and of non price competition. Therefore, the exports of goods are increasingly related to the capability to provide to foreign consumers these complementary services. This requires a direct presence of national firms in foreign markets through the establishment of marketing offices in the other countries, of foreign subsidiaries or of commercialization agreements with foreign firms.

Therefore, technological progress and internationalization of regional economies are tightly related. On the one hand, technological progress is an **effect** of the increasing globalization and the increasing customization of national markets, which both imply the need for an increasing specialization of productions of the various firms and thus a faster adoption of product and process innovations. In particular, the creation of a global but segmented market at the European and world scale requires that firms specialize in specific technologies, in order to be recognized by the customers as a 'leader' in specific segments of the market. Therefore, the access to a global and segmented market is a stimulus to innovate.

On the other hand, technological progress is also a **cause** which leads firms to internationalize, in order to diversify their markets and to have access to foreign technology. The high costs of research required by the adoption of innovations and the shorter 'lead time' on competitors and imitators compel the firms to look for alliances with other firms in research activities, in order to be capable to adapt their products to the specific characteristics of new markets. Firms attempt to exploit their transitory advantage on all national and regional markets at the same time. Thus, technological progress promotes a further increase in the internationalization process of national and regional economies.

It should be stressed that a characteristic of the internationalization process is the fact that it does not only imply an **increasing competition** of local firms with foreign firms but it also implies an **increasing cooperation** among firms of different countries. In this respect the internationalization process may differ from the increase of exports, which consists in an exchange of product or services between local firms and foreign consumers. In fact, the internationalization process does not only imply monetary exchanges but also a joint investment of human and financial resources together with foreign firms in order to develop particular collaborations in the fields of research, production, marketing, finance or in any other function of a firm.

The development of collaborations between local and foreign firms faces various obstacles in the various phases which characterize the establishment of these collaboration, such as the collection of information on foreign firms and economies, the understanding of the respective strategies by the various firms, the precise identification of specific fields of collaboration and the adoption of the most appropriate contractual form, which should regulated the collaboration itself.

The internationalization process is a gradual learning process. Thus, it may imply a continuous change in the specific forms of collaboration among the various firms. Through intermediate phases, these relationships may evolve from loose forms of coordination to more effective and binding forms, according to the progress in the reciprocal knowledge and thrust among the firms.

3. The Sectoral Impact on the European Internal Market

The completion of the European Internal Market will lead to the adoption of measures concerning many different fields, among which the following seem particularly important:

1) removal of technical obstacles to the exports of agricultural and industrial products;
2) free movement of capital and right to establish by foreign banks;
3) freedom of sale of insurance services and right to establish by foreign insurance firms;
4) greater integration of transport and communication services, with special respect to air transports and TV networks;
5) liberalization of public procurement policies;
6) harmonization of indirect taxes, of corporate law taxes;
7) increase in the financial resources of the Community budget devoted to common structural policies;
8) adoption of common policies in various fields, such as:
   - antitrust law,
   - research and development,
   - environmental protection.

According to the traditional approach, of the 'theory of customs unions' which has still been used by the Commission of the EC in assessing the impact of the European Internal Market (Cecchini, 1988) the results of measures aiming to eliminate the still existing obstacles to a free movement of goods and factors of productions should be (Commission of the EC, 1988):
1) a greater exploitation of economies of scale;
2) a greater efficiency due to a greater competition.

When these effects are analyzed in an interregional perspective, their impact on the less developed regions of the European Community should consist in:
- an increase of imports from more efficient foreign firms;
- a decrease of production by the less efficient local firms ('trade creation effect');
- an increase of unemployment in the regions which are not capable to sustain foreign competition;
- an increase of regional development disparities;

- an increase of political conflicts in the EEC;
- a need for regional policies by the European institutions.

In fact, the cumulative development process, sustained by unilateral movements of capital and labour toward the most developed areas, should almost certainly determine a crisis of the political institutions of the European Community and a stop or reversal of the unification process, unless the liberalization measures are accompanied by 'compensatory' regional policies. These latter should aim to sustain the income levels of the regions negatively affected, to promote technological change in these regions, to increase the regional production potential or to reconvert local resources toward new sectors.

According to other more 'liberist' scholars, the regional problems could be solved by removing the obstacles which still hinder the movement of labour from the regions where unemployment is increasing to those regions where the productivity of labour is higher and through the adoption of measures increasing the downward flexibility of wages and the intersectoral mobility of labour in the regions of high unemployment. Regions should specialize in the sectors where they have a 'comparative' advantage. Thus, this second policy approach seems to imply the need to decentralize labour intensive industrial and agricultural productions from the more developed regions and countries, where the average productivity and wages are higher, to the less developed regions.

The pessimistic indications of this traditional approach may have the advantage to indicate arguments which support the need for a regional policy directed toward the less developed regions of the Community. However, it may also have the effect to justify measures which will help the largest European firms to grow even further through acquisition and mergers, as they are supposed to be the most capable to exploit the opportunities created by the European Internal Market. This may imply to assign a lower priority to measures aiming to support the small and medium size firms, as it has occurred in the past since these firms have until now not been represented adequately in the national and Community industrial policies.

Certainly, the arguments of the traditional approach still have some truth in the actual situation. However, too many changes have occurred in the last decades in the sectoral structure of the European economies and in the policy strategies of the firms, so that this 'traditional paradigm' is clearly insufficient when not also misleading. In fact, it is based on the same analytical tools which were used more than 30 years ago (Scitovsky, 1958) in analyzing the impact of the Common Market. Thus, it seems to forget the results of the extensive literature on industrial economics, which has been elaborated since those old years. Moreover, the analytical tools should be appropriate not to the objective of investigating the effects of the free movements of goods, which is a phase already achieved in the European Common Market, but rather of investigating the effects of the free movements of services, capital and people, which represent the specific aims of the completion of the European Internal Market.

In particular, various studies indicate that the value of the 'static effects of trade creation' created by the removal of custom tariffs are rather limited (Broecker and Peschel, 1988). Also the 'costs of the Non Europe' (Cecchini, 1988), which are directly related to the organization of various types of control, may be important, but they determine only an once for all gain. On the other hand, the 'dynamic effects' of an economic union, such as the greater efficiency related to the increased competition, the greater exploitation of economies of scale, may be more important, but they are uncertain and ambiguous.

However, an important contribution to this theoretical debate is certainly represented by the results of many empirical analysis of the European development during the last thirty years (Cappellin, 1983 and Camagni and Cappellin, 1984). These studies demonstrate that contrary to the expectations of the traditional theory, many weakly industrialized regions, such as the 'Third Italy' region, have been characterized by a fast development mainly related to the growth of exports to other European and non-European countries. This indicates that the creation of the Common Market for the industrial products has positively affected the growth of small and medium size firms, which represents the most important component of the economy of these regions. A survey of the studies on the regional impact of the European integration (Broecker and Peschel, 1988) indicates that although significant regional impacts of trade liberalization have been observed to affect regions unequally, there is no sign of gains concentrating in the centres and losses in the periphery.

In particular, the expectation that the widening of the market will necessarily lead to an increase of the size of the firms may be historically unverified, while the real outcome may be an increasing specialization of firms. On the one hand, large firms may restructure the location of their productions among their different plants in order to achieve a greater specialization of each plant in a particular production, instead of having the same diversified production range in all countries. On the other hand, the widening of the market and the access to a more advanced final and intermediate demand in foreign countries may decrease the barriers to entry and may promote the birth of new small and medium size firms, specialized in particular productions. These firms may be located in particular areas, endowed with positive location factors, which could not previously be exploited due to the limited size of the local market.

The process of increasing specialization of firms may be demonstrated by analyzing the impact of various effects illustrated by the modern literature in industrial economics, such as the deverticalization of production processes, the existence of network economies, of economies of scope, of changes in the barriers to entry, of economies of specialization, of changes in external economies (Cappellin and Nijkamp, 1990).

First of all, the increased competitive pressure may lead firms to a greater **'deverticalization of production processes'** and this may allow a further decentralization of industrial activities toward the less developed regions.

In particular the increased transparency of the market, created by the removal of various non-tariff obstacles, may increase the advantage of the market (the choice to 'buy' or to externalize productions) with respect to hierarchical organization (the choice to 'make' or to internal integration), as the more appropriate 'governance structure' of transactions (Williamson, 1981). In fact, the completion of the Internal Market will increase the knowledge of potential subcontractors in other countries and it will decrease the transaction costs of purchasing from foreign firms particular components, which have lower production cost and/or a higher quality, thus increasing the deverticalization of production processes.

The Internal Market may facilitate the exploitation of **'network economies'** (Teece, 1982), which may be important especially in the diffusion of the know-how and in the cooperation among firms of different countries in the field of research. Network economies are due to the fact that the advantage to belong to a network increases with the number of the firms belonging to the network. In particular, the competition of a firm increasingly depends on their capability to access to the immaterial networks which link the various firms and through which circulate information on new technological and commercial opportunities. This capability may not be related to firm size but rather to the flexibility of a firm. In particular, small and medium size firms may exploit network economies through the development of bilateral or multilateral agreements, which allow the reciprocal marketing of the respective products by firms in different countries and the collaboration in research.

On the other hand, network economies may be very important in the commercialization functions and this certainly represents an advantage for large multinational firms. In fact, these firms are capable to commercialize the same product in different regional and national markets, thus exploiting the demand of customers who prefer to use products which have a wide diffusion. This is the case of various consumer goods, the demand of which is related to international fashion, or of computers and of other investment goods, which require the compatibility to widely accepted standards.

Network economies are also crucial for various modern service firms, such as advertising, consulting and engineering firms and they explain why these firms are induced to operate in different countries through direct subsidiaries or through agreements with local companies. In fact, this internationalization process of service firms, many of which have a limited size, is required by the need to serve multinational clients and by the aim to better exploit their own internal and tacit know-how by combining it with the access, through agreements with foreign firms, to customers and information in other countries.

In particular, the European Internal Market may facilitate this networking process through improved telecommunications, through the mutual recognition of professional titles of skilled workers and through the free movements of capitals, which will facilitate the creation of joint-ventures, and through public support to joint research projects.

The competition of small and medium size firms with respect to large firms may also be facilitated by the exploitation of **'economies of scope'** in production, as small and medium size firms may successfully compete with the mass production of large firms by adopting an organization based on 'flexible specialization' (Piore and Sabel, 1984). On the other hand, there are limits to the exploitation of economies of scope by large firms, through the mergers and acquisition in various often only loosely related production areas. In fact, there seem to exist diseconomies of scope, as in the eighties many conglomerate firms have disinvested from unrelated business areas and have attempted to concentrate on those areas where they had a particular competitive advantage.

On the other hand, there exist economies of scope or economies of scale in the commercialization of complementary products through the same distribution channels. This may represent an advantage for large multiproduct firms or it may stimulate collaboration and the creation of marketing networks among firms of various countries and sectors, as indicated above.

Economies of scope are often advocated by large firms, which being established in their respective national market may aim to protect themselves from foreign competitors through a further diversification of the scope of the services provided to their customers, as in the case of a greater integration of banking and insurance companies. However, these policies seem to imply a defensive strategy within the national market and may be rather risky in the long term.

The completion of the European Internal Market may also decrease the oligopolistic power which some large firms have on their respective national markets. In fact, it may decrease the **'barriers to entry'** of particular medium size firms in foreign markets, dominated by large national firms, which have the direct and indirect support of the respective national governments, through public procurement and specific national legislation, as it still occurs in the case of both many industrial (automobile and chemical) and of many service sectors (banking and insurance).

In fact, medium size firms, due to their more flexible structure and to their greater autonomy may be capable to expand into new markets which are now practically monopolized by large national firms. On the contrary, large firms, due to their dimensions, will be much less free to undertake aggressive expansionary policies on the markets controlled by other large national firms and will be compelled 'in primis' to resort to agreements and collaboration with local firms, in order to expand into new markets.

In particular, the existing regulations represent an obstacle, which is more important in relative terms for small and medium size firms than for large firms, which have the financial and technical resources through which they can to a large extent even now overcome the existing barriers which separate the various national markets. Therefore, it is possible to argue that the positive, and not only the negative effects, of the completion of the internal market will be more important for small and medium size firms than for large firms.

An opposite factor with respect to economies of scale is represented by the **'economies of specialization'**. These are especially important in the elaboration of new technology and in the fast adoption of innovation, as they are related to economies of scale in research and to the efficient organization of the collection and elaboration of information related to specific productions. Therefore, large firms due to their higher investment in RD may have an advantage. However, large firms are seldom technology leaders in their respective fields and small and medium size firms may have an advantage due to their greater flexibility, the greater autonomy and responsibility of the researchers in these firms, the greater motivation, the greater internal cohesion and the shorter decision process. Moreover, technology becomes more complex and different in the various fields, thus it becomes more difficult to manage the particular complexity of each individual field of application and also small and medium size firms have the possibility to become technological leaders in specific fields and not only larger and more diversified firms.

In particular, for all industrial and service firms and, for example for banks and insurance companies, the completion of the Internal Market requires to make a choice between a strategy of further product diversification, in order to better control the respective national market and a strategy of specialization in particular products and of expansion on new foreign markets. In fact, it is very difficult for firms to diversify their products, while expanding on new markets. On the contrary, as indicated by the case of various English and American financial institutions, specialized firms seem to be capable to operate in a flexible way in various foreign markets, by developing synergies with local partners.

The increasing deverticalization of production processes leads to an increasing integration of the firms with the external environment and to an increase of the importance of '**external economies**' in the competition of firms. In fact, the ratio of intermediate or external costs on total turnover is increasing in all firms. This process may clearly represent a danger for the less developed regions of the Community, where the external environment and the supply of public services, modern infrastructures and modern business services is still inadequate. However, the general increase of costs determined by the agglomeration diseconomies in the more developed areas may determine a crowding out effect of the more traditional production from these regions. In the case of specific industrial districts specialized in particular productions, external economies may compensate the lack of internal economies of scale and the collaboration among local firms may help them in their relationships with external firms (Cappellin, 1988).

Clearly, the effects of all factors described above (economies of scale, deverticalization processes, networks economies, economies of scope, barriers to entry, external economies) have a different importance according to the different sectors and the different regions considered. In particular, it seems that specialized producers, such as small and medium size firms producing goods with high or intermediate technology, may enjoy a positive impact from most of the effects indicated above. On the contrary, small and medium size firms producing non-durable consumer goods with a traditional technology may be negatively affected by most of these effects or may face different obstacles in exploiting positive opportunities.

In particular, it seems important to distinguish between services and industrial firms. In fact, the impact of the Internal Market on **service activities** may be rather different from that on industrial productions, due to the fact that the production of services usually requires a 'face to face' interaction between the producer and the user. Thus, where it is not convenient that the user will move to meet the producer, as in the case of education, health and tourist services, the producer should locate nearby the user, such as in the case of banking, insurance and consulting services. Therefore, the free competition in case of services is tightly related to the freedom to establish in other areas. Thus, the internationalization of services is tightly related to seemingly unrelated issues, such as the integration of national cultures and of the organizational procedures of firms, the recognition of education and professional diplomas, the harmonization of immigration regulations, the free movements of capitals and the harmonization of fiscal regulations. In fact, the European Community is taking measures in these different fields in the perspective of the completion of the Internal Market.

Therefore, the internationalization process is also going to affect service sectors. In particular, the international competition of service firms depends on the faster adoption of process innovations, on the improvement of the quality of the existing services, on the specialization in modern and fast growing services and on the capability to establish collaboration and networks with other foreign service firms.

It seems important to distinguish between the modern producer services and the more traditional services. In fact, new firms of modern services are also developing in the less developed regions, such as the Italian Mezzogiorno, mainly as a spin-off from more traditional firms. Moreover, where local firms are not capable to exploit the increasing local demand for these services, the right of establishment allowed by the completion of the Internal Market will allow the growth of foreign firms in these areas, thus eliminating local oligopolistic situations and providing to local industrial firms those modern services which they need for their growth.

In the long term, a positive factor for the diffusion of modern services in the less developed regions of the EEC is represented by the fact that these activities do not require high capital investments, such as industrial activities. On the contrary, the development of these activities depends on the availability of a qualified labour force with high education levels. This does exist in many less developed regions of the EEC, where the education level of the population is equal to the EEC average, as it is indicated by the emigration of many technicians and other qualified workers from those regions.

On the contrary, the increasing internationalization of services may represent a danger for the less developed regions of the Community with respect to the traditional services, such as many consumer services. The market of these regions may be captured by foreign firms either directly through subsidiary firms or indirectly through acquisitions. In fact, retailing, banking, insurance services are often characterized by a traditional, fragmented and inefficient structure in these regions. Local firms should therefore react to this danger by mergers and by a higher diffusion of cooperation agreements among service firms.

4. The Regional Impact of the European Internal Market

The traditional paradigm of the effects of international integration has several shortcomings, such as:

1. It underlines the role of an increasing competitive pressure but it underestimates that the European Internal Market will also promote various forms of collaboration between the firms of the various countries, thus promoting a diffusion of modern activities. Moreover, the competition among firms is mainly based on product and process innovation and not on production costs, as it is implied by the traditional approach.
2. The removal of non tariff barriers is similar to a reduction of transport costs. Thus, it may have both an effect of concentration or

an effect of diffusion, according to the respective importance of economies of scale and of lower production costs in the less developed regions.

3. The completion of the Internal Market will promote an easier diffusion of know-how in the less developed regions, the adoption of product and process innovations and thus an increase of the competition of small and medium size firms of the less developed regions.
4. The removal of non tariff barriers may allow an easier access to a more sophisticated final and intermediate demand to the small and medium size firms of the less developed regions. This decrease of the barriers to entry may allow the birth of new firms and of new sectors in these regions.
5. The overall increase of income and of demand in the European Community and the reconversion of the firms of the more developed regions toward new sectors, due to the existence of agglomeration diseconomies, may also allow the survival and the growth of many small and medium size firms in the less developed regions, specialized in traditional production. Thus the effects may be different whether a comparative static perspective, which only considers the redistribution of the existing production among the various countries, or a more long-term perspective is adopted. This leads to consider a dynamic process of development, which implies the entrance of new firms, the creation of new high tech products and the constant increase of the gross national products in the various countries and regions.
6. The case of services is different from that of goods, as they can not be directly exported, but they require the establishment of subsidiary firms in other areas. Thus the growth of foreign firms in less developed regions may not affect the level of local employment. Moreover, these subsidiary units may destroy local oligopolistic situations, increase the diffusion of know-how and promote the access of local small and medium size firms to better services.

Therefore, the final effect of the completion of the Internal Market may not be, as indicated by the traditional paradigm:

a) an increased specialization of each area in particular products;

b) an increased spatial concentration of production and an increase of regional disparities.

On the contrary, the completion of the Internal Market will facilitate various forms of collaboration, mergers and acquisitions among the firms of the various countries. This may have a positive effect on the small and medium size firms, which may have an easier access both to modern foreign technologies and to new markets. Moreover, the new opportunities created by the Internal Market will allow to better exploit the potential of local resources, the birth of new firms and the growth of existing firms through a diversification into new products.

Therefore, the completion of the European Internal Market seems to lead to:

a) a greater sectoral diversification of each regional economy, as each region will be represented in almost all economic sectors and almost no region will specialize in just a few sectors;

b) an increase of 'horizontal' or intrasectoral trade among the various countries;

c) to an increased interdependence among the various regional economies, rather than to an increased concentration of production and to an increase of regional disparities.

Thus, the real potential danger of the completion of the European Internal Market seems to consist in the fact that the increased international interdependence will necessarily imply a strengthening of the hierarchical relationships among the firms of the various regions. In fact, hierarchical relationships will not only be determined by mergers and acquisitions but also by the creation of networks and agreements among the firms in the fields of production, finance, marketing or research.

The relationships among the firms within a network are normally characterized by very different degrees of dependence and/or of power. Therefore, the increasing interdependence among the various regions will be accompanied by the strengthening of the hierarchical relationships among the various firms and regions.

We may conclude that the completion of the Internal Market will not necessarily lead to an increase of spatial concentration and of regional development disparities, but it will increase the concentration of economic power. Although the Internal Market will increase the opportunities to the individual regional economies, the increased interdependence will certainly decrease their autonomy.

5. Conclusions

This paper indicates the complexity of the various aspects and effects of the different policies which aim to the completion of the European Internal Market, thus pointing out the need for further research on these issues.

In fact, the completion of the European Internal Market should be interpreted as the policy response of the European Community to the increasing internationalization of the economic activities and not as a dramatic exogenous change with respect to the recent development pattern in the European regions.

Therefore, the impact of the European Internal Market should be evaluated in the framework of the different responses of the various regions to the actual internationalization process of their economies.

In particular, the enlargement of the market will induce firms to customize their products and to specialize in particular market segments, in order to exploit their competitive advantage in the framework of a wider market. The increasing specialization of the individual firms will imply an increased sectoral diversification of the economic structures of the various regions, due to the inevitable differences in the specialization strategies of the individual firms. However, the increased interdependence of the various regional economies and the increased flows of 'horizontal' trade will not necessarily imply 'equal' exchanges among the various regions, but most probably stronger hierarchical relationships among the various firms. These increasing relationships among the various regions will decrease the autonomy of each region on the strategic decisions concerning its long-term development pattern.

Moreover, the impact of the European Internal Market will not be the automatic effect of deregulation policies, but rather the effect of the harmonization of national policies or that of the adoption of new structural policies of the European Community. Therefore, it is necessary to analyze the regional impact of these policies and the special role which will be attributed to the regional policies in this new scenario.

In particular, there is the need to promote a greater articulation of national structural policies according to the different structures of the various regions. A greater decentralization of the implementation of these policies to the regional administrations would allow to take into account the different responses of the different regional environments to the Community policies and could also increase the effectiveness of these policies.

The regional impact of national policies should be evaluated in order to minimize strong negative impacts, which would then require costly compensatory regional policies. On the other hand, regional policies should adjust their strategy to the new scenario and should attribute a greater attention to the problem of finding ways to promote the synergy between endogenous (local) and external (national/international) factors in the regional economy, in order to increase the speed of diffusion of product and process innovation.

Moreover, regional policies may contribute to the achievement of national and Community policy objectives, such as the completion of the Internal Market, as they may eliminate specific local obstacles which hinder the free flows of services and production factors and may also promote a greater coordination among border regions belonging to different countries of the Community.

Finally, the objective of completing the Internal Market should be interpreted as a phase of the more global process aiming to create the European Economic and Political Union. Therefore, a more thorough and, possibly, internationally coordinated research effort may help in the choice of those policies which could allow to achieve this goal without having a negative impact on regional disparities, but promoting their endogenous development potential.

References

Begg, I. (1987), The Regional Impact of Policies Implemented in the Context of Completing the Community's Internal Market by 1992, Final Report to the European Commission, Department of Applied Economics, Cambridge.

Broecker, J. and K. Peschel (1988), Trade, in: W. Molle and R. Cappellin (eds.), Regional Impact of Community Policies in Europe, pp. 127-151, Avebury-Gower, Aldershot.

Camagni, R. and R. Cappellin (1984), European Regional Growth and Policy Issues for the 1980s, Built Environment 3/4, pp

Cappellin, R. (1983), Productivity Growth and Technological Change in a Regional Perspective, Giornale degli Economisti e Annali di Economia, July-August, pp. 459-482.

Cappellin, R. (1988), Transaction Costs and Urban Agglomeration, Revue d'Economie Regionale et Urbaine 2:261-278.

Cappellin, R. and P. Nijkamp (eds.) (1990), The Spatial Context of Technological Development, Gower, Aldershot.

Cecchini et al. (1988), The European Challenge 1992. The Benefits of the Single Market, Gower-Wildwood House, Aldershot.

Centro Europa Ricerche - I.R.S. (1987), Un'industria in Europa, Bologna, Il Mulino.

Commission of EC (1988), The Economics of 1992, European Economy 35.

Molle, W. and R. Cappellin, R. (eds.) (1988), Regional Impact of Community Policies in Europe, Avebury-Gower, Aldershot.

Padoa Schioppa, T. et al. (1987), Efficiency, Stability and Equity: A Strategy for the Evolution of the Economic System of the European Community, Report to the Commission, Brussels.

Piore, M. and C. Sabel (1984), The Second Industrial Divide, New York, Basic Books.

Scitovsky, T. (1958), Economic Theory and Western European Integration, London, Allen and Unwin.

Teece, D.J. (1982), Toward an Economic Theory of the Multiproduct Firm, Journal of Economic Behavior and Organisation 3:39-64.

Williamson, O.E. (1981), The Modern Corporation: Origin, Evolution Attributes, Journal of Economic Literature 19:1537-1568.

Chapter 14

SOUDY 3: A SUPPLY-ORIENTED URBAN DYNAMICS MODEL WITH INNOVATION AND SYNERGY EFFECTS

R.P. Camagni
L. Diappi

1. Introduction

In the last few years a new family of evolutionary models of the urban hierarchy was presented (Camagni, Curti, Gibelli, 1985; Camagni, Diappi, Leonardi, 1986; Diappi and Pompili, 1987), whose main characteristics may be summarized in the following:

i) The aim of linking together different theories and traditions in urban economics, which still remain mainly separated, such as the central place tradition, spatial interaction models, the reflections on 'optimal' city size, the urban life-cycle model, the studies on urbanization and localization economies, intersectoral linkages and synergies in local development.

ii) The aim of substituting the old-fashioned demand side approach, typical of the export base model and of most of part of land use and urban dynamics models, by a new supply-oriented approach; hence the acronym by which we define the new models: SOUDY, supply-oriented urban dynamics models.

iii) The aim of integrating into the framework of urban models the true dynamic driving force of local development that, after Schumpeter, may be synthesized in the term 'innovation'.

iv) The adoption of a methodology based on a 'master equation' approach (Weidlich and Haag, 1982) - through which the probability distribution of the possible states of the urban system is defined at the macroscopic collective level, starting from the properties of the microscopic individual behaviour - and on the dynamic simulation of the growth process of each center of the urban hierarchy. This methodological choice appears nowadays the most appropriate in two principal respects: firstly, it allows a sufficiently easy and efficient handling of complex interactions and synergy effects, and of the consequent non-linearities in the dynamic behaviour of spatial variables (Clarke and Wilson, 1986); secondly, it looks as an effective way for including a stochastic element like innovation in a formalized economic model.

As far as the internal economic logic and the supply-oriented character of SOUDY models is concerned, it rests on the full recognition of the impossibility of understanding spatial development, and in particular the development of small sites in the long run, through demand side models. In fact, the main part of local growth comes not really from an exogenous expansion of existing markets, but from the capability of widening the local share of even poorly expanding markets through continuing increases in local competition; this competition on its turn rest on supply-side elements such as prices, availability and quality of production factors, entrepreneurship and innovation capability.

Also in the location of tertiary activities and in the structuring of a Christaller type hierarchy, the demand side, embedded in the concept of threshold and range, has mainly been stressed by traditional and current literature. On the contrary a supply effect is also implicit in the logic of the model, in the form of agglomeration economies: in the location - exogenously determined - of lower order activities where a higher order activity is centered, and in the opposite logic of the appearance of high order activities only in centers completely endowed with lower order ones (Christaller's 'nesting' principle).

A first model, we may call SOUDY 1, was developed on a supply-oriented and Schumpeterian basis (Camagni, Diappi, Leonardi, 1986), integrating Allen's logic of random appearance of new functions or sectors along the urban hierarchy (Allen and Sanglier, 1981; Allen, 1982) with a logistic model of development à la Wilson driven by the existence of a positive margin between average production benefits and urban locational costs (Wilson, 1983). The simulation results were interesting and gave rise to the emergence and growth of an urban hierarchy mainly of a Christaller type, with a nested structure of centers and market areas.

A second model of urban development was then developed (SOUDY 2: Diappi and Pompili, 1987), reflecting mainly the preoccupation of departing from a nesting principle and allowing a more Loeschian logic to build the production landscape. The previous approach based on the development path of the single center, acquiring through time higher order functions through innovation processes, was substituted for an approach based on the single functions and their locational logic on the production landscape.

In this paper, a further refinement is presented, aiming at bringing the theoretical construction closer to the real pattern of centres distribution. Recent empirical research has shown in fact at least two main departures from the pure theory of central places:

a) the possibility of a presence of some high order functions in centres of relative small size, linked to specialization effects (horizontal synergies) or vertical 'filière' type, intersectoral linkages (vertical synergies) around some local 'vocation productive' (the case of many Third Italy Centres such as Prato, Carpi, Sassuolo, or the case of the silk 'filière' in Como, made up of advanced specialized technological and marketing services, linked to local silk manufacturing) (Camagni and Pompili, 1988); and

b) the presence, at the intermediate and lower levels of the urban hierarchy, of a complex 'network' of mainly functional and complementary linkages among specialized centres, instead of a truly hierarchical structure (De Matteis, 1985; De Matteis et al., 1988).

Traditional central place theory, as well as our first attempts of building a dynamic framework for the urban hierarchy, cannot explain this new reality, due to two main theoretical shortcomings: firstly, the fact that the pattern of industrial location may no more be tackled through a model of non-overlapping and mutually exclusive market areas, and secondly their neglect of intersectoral linkages and externalities, both of a general and sector-specific nature. By these externalities we do not really mean input-output, demand-side linkages, but rather supply-side synergetic effects coming from technological know-how diffusion, customer-supplier cooperations, income effects on price elasticity of local demand which allow to fix higher prices for lower order products in big agglomerations.

In this paper, we present a first attempt to embed these new (empirical and theoretical) elements into a dynamic simulation model of the urban hierarchy, allowing a differentiated locational logic for tertiary and manufacturing functions; while in fact a spatial pattern of market areas still applies to the former, the latter should be better supposed to look for market 'shares' rather than market areas.

2. A Family of Supply-Oriented Dynamic Models

2.1 SOUDY 1: cities as evolutionary systems in a Christallerian space

The first model of the urban hierarchy simulates the growth path of the single centre as a series of structural changes taking the form of urban innovations: creation or attraction of a higher order function, characterized by a superior level of profitability with respect to existing ones. In absence of these qualitative and stochastic 'jumps', the city may well develop through quantitative and mechanistic expansions, but only up to a certain threshold where production benefits are overcome by increasing locational costs, linked with city size (Figure 1).

Each function, typical of a specific urban rank, is characterized on its turn by an appearance threshold, defined by its minimum efficient production size and linked to the urban size through both supply and demand elements: presence of qualified factors in the first case, and presence of an initial local 'captive' demand in the second. Average production benefits of each function are held constant with increasing production (and city) size, and the complete structure of (urban) costs and (production) benefits appear as in Figure 2.

Two important theoretical elements appear from this picture:

* an 'optimal' city size does not exist, but a series of efficient size intervals emerge, each one link to a specific rank of the urban hierarchy, and therefore

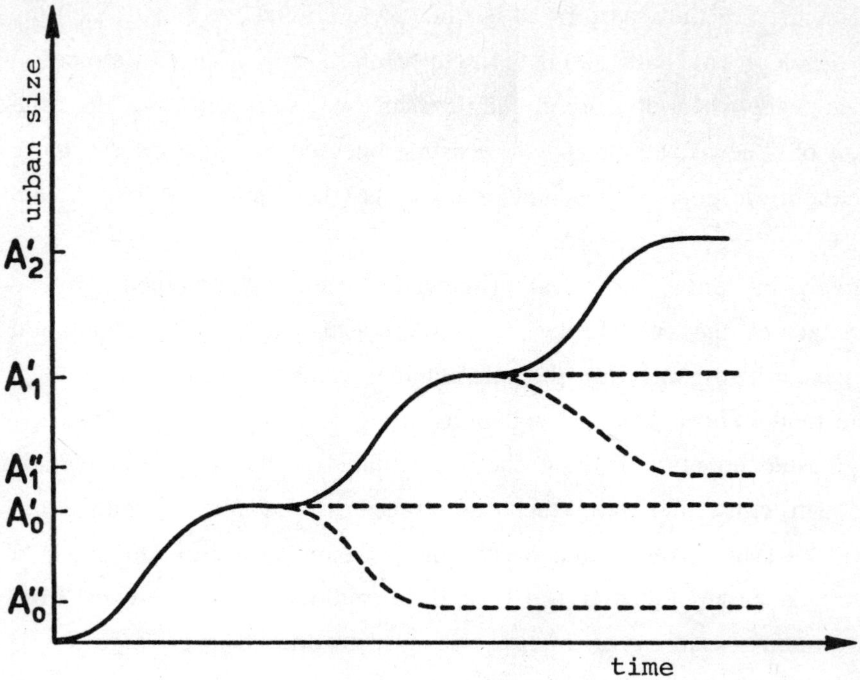

Figure 1. The Development Path of a Single Centre

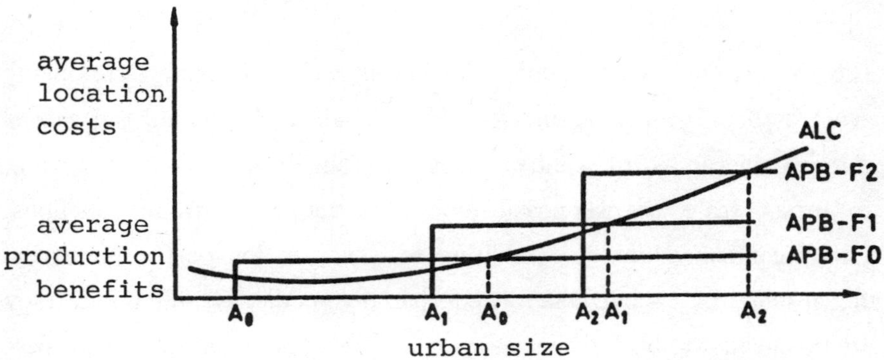

Figure 2. Efficient Urban Size for Different Urban Functions

depending on the functions performed by the single city (A_0-A_0';A_1-A_1',.....);
* in the development path of the single centre, conditions of structural instability appear, when it has already reached a sufficient size for the development of a new function and the existing ones do not provide any more sufficient benefits to cope with actual urban costs (the intervals A_1;A_0';A_2-A_1' in Figure 2).

Driven by both 'structural' (innovative) and 'constrained growth' (multiplier) factors, the size of the single centre expands, draining population through a competitive game with the surrounding urban structure.

The model consists of two equations:

i) a 'master equation' stating the probability of changing urban rank, through entry and exit stochastic processes: the former (entry into rank k+1) is linked to the overcoming of the appearance threshold of the next order function (k+1) and to spillover spatial effects from surrounding centres of bigger size, which decentralize some lower order functions; the latter (exit from rank k into rank k-1) depends on the lack of a sufficient spatial demand for function k, wherever localized. Both processes are activated in the simulation model through a Monte Carlo procedure;

ii) a set of differential equations defining population growth in each centre as a function of the algebraic difference between benefits of the highest function present in the centre and urban location costs, in a spatial setting of centres competing for total population.

The results of the dynamic simulation were interesting in many respects from a theoretical point of view. First of all, they pointed out that a fully developed hierarchy of centres does not show up unless increasing returns to urban scale are supposed along the range of urban functions. Secondly, they confirmed another condition for the creation of the hierarchy, namely the absence of a high generalized spatial impedance, and finally they highlighted the need for bigger centres of speeding up the innovation process if their primacy is to be maintained in presence of technical progress and diffusion tendencies in existing functions towards lower order centres (Figure

3).

2.2 SOUDY 2: cities as interacting networks in a Loeschian approach

While partly building on the previous model, SOUDY 2 shifts its focus to the Loeschian tradition with the aim to formulate a model of urban evolution in more realistic terms. From this point of view, Loesch's theory of spatial allocation of activities, with the superimposition of different game-theoretic market area networks gives us a more general theoretical framework than Christaller's, as far as not only service but also manufacturing activities are considered, and at the same time, once given the supply-side interpretation, it fits empirical evidence better and more robustly. Within this framework, urban population size is disconnected from the rank of the urban functions active in each centre, and Christaller's nesting principle is abandoned.

Soudy 2 introduces a series of innovative features such as synergy effects and agglomeration economies as the driving forces behind the resilience of large centres, and the possibility of the disappearance of activities due to local congestion diseconomies. The dynamic stochastic model - with two variants, for manufacturing and service activities - describes the process of location for each rank of functions, assuming purchasing power (and no more population) as a demand indicator.

The appearance of a new function does not depend strictly on the existence of the activity of immediately lower rank, but on the contrary on synergetic effects with higher order functions. Therefore, the number of activities already existing in each centre is taken as a proxy for diversification, qualified inputs supply and innovation, possibly enhanced by increasing returns to scale.

Moreover, the main locational condition for profit maximization generates a location of supply units that minimizes distance from consumer demand (with the constraint of a commodity-specific minimum threshold) and maximizes distance from competitors.

With regard to the possibility of losing activities, it may occur because a minimum demand requirement is no longer satisfied or because, with the

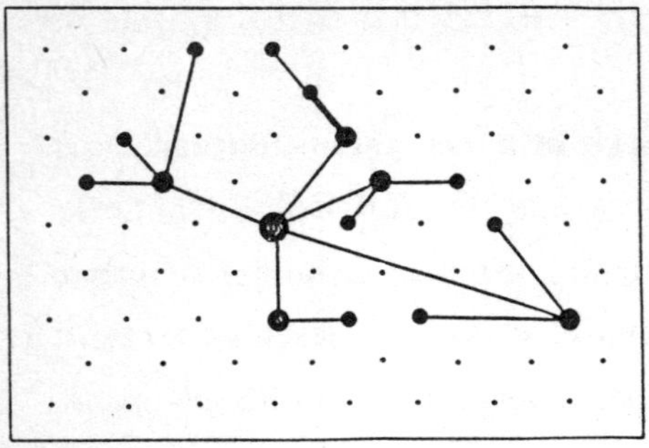

Final spatial structure of centers (rank 5, 6,7) when the initial state of the system is homogenous.

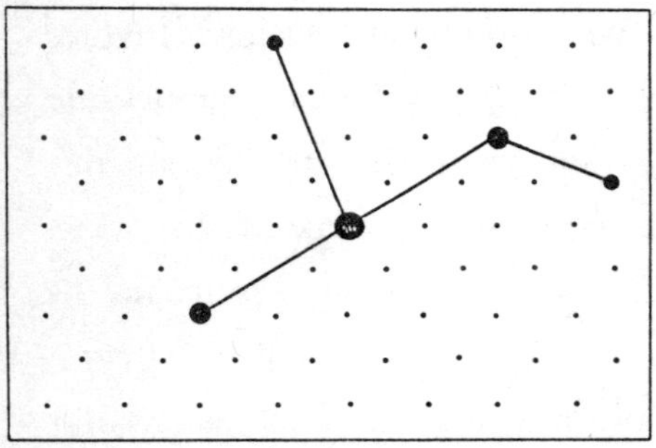

The spatial structure of the urban hierarchy (rank 5,6,7) in case of Zipf-type initial distribution, no technical progress, and linearly increasing returns to urban scale.

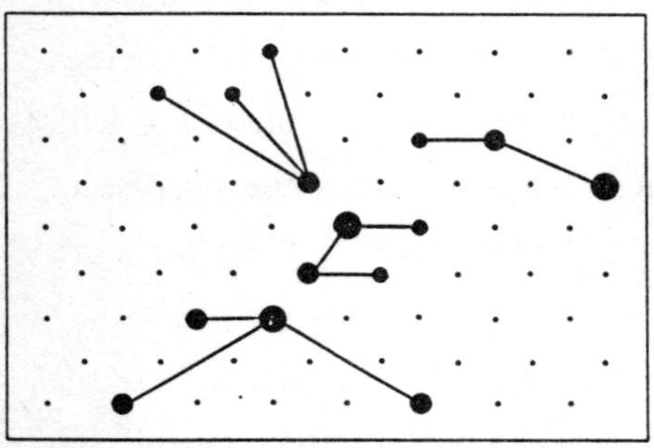

The spatial structure of the urban hierarchy (rank 5,6,7) in case of Zipf-type initial distribution, no technical progress, and exponentially increasing returns to urban scale.

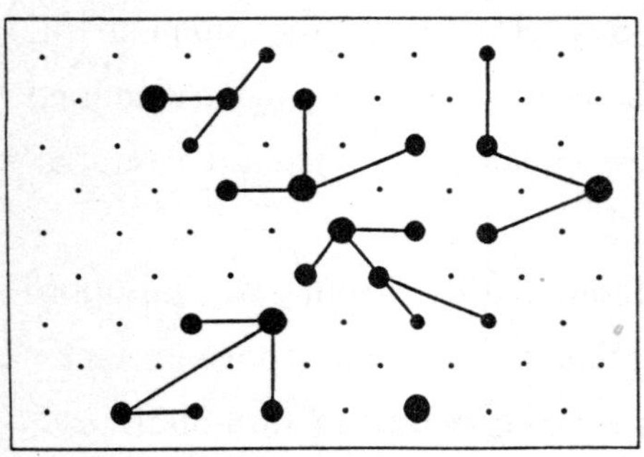

The spatial structure of the urban hierarchy (rank 5,6,7) in case of Zipf-type initial distribution, presence of technical progress, and exponentially increasing returns to urban scale.

Figure 3. Some Results of the SOUDY 1 Simulation Procedure

entry of new higher order activities, locational costs have become unbearable for lower order, less profitable activities.

The models for the location of manufacturing and service activities are held as similar as possible, but they differ basically because of the different nature of demand: in the service case it is assumed that consumers displace themselves to the supply locations, thus bearing directly the cost of movement, whereas in the case of goods it is the producer that displaces output to demand, charging a delivered price which is constant over space.

The randomness of the location process is obtained through a master equation approach for both sectors, governing a birth/death stochastic process.

3. SOUDY 3: Intersectoral Interactions and Local Synergies

3.1 The new hypothesis

Holding the general supply-side orientation of the previous models and the idea of a general urban growth process driven by Schumpeterian innovation and capturing of new and higher order functions, we now want to move on in three main directions:

a) first, to control explicitly for the sectoral structure of the urban economy, directly taking into account:

- intersectoral synergies among sectors (in an exogenous way),
- agglomeration economies (endogenously), and
- a process of intersectoral competition for central locations and 'crowding-out' effects, such as the disappearance of some lower order service and manufacturing activities with rising urban size (also this in an endogenous way);

b) second, to allow for the appearance of higher order functions in centres of relatively low dimension, if local supply and demand conditions generate a sufficient level of intersectoral synergy and economic effectiveness. For tertiary activities, this means that the stochastic structure of the innovation process and in particular the derived probability of acquiring new urban functions when the appropriate urban size is approached, shift upward as a consequence of favourable local supply conditions.

By this we allow a single centre to develop not necessarily the full bundle of functions of the next order rank (which would generate an upward shift in its rank), but only some specific functions within the bundle associated to the next urban rank (or ranks), according to the existence of specific intersectoral synergies with respect to its production structure and its economic tradition or 'vocation'. In order to reach this simulation result, we will consider a wide number of functions and sectors, higher than the number of urban ranks, a hypothesis which enhances the degree of realism of the model.

In the formalized model, we assume, for the sake of simplicity, that synergy effects among sectors and functions develop only with the single centre, but this case could be easily generalized to take into account possible synergies with the surrounding areas through a distance decay function;

c) third, to allow the simultaneous presence of two spatial logics:

- one which holds for services, which requires a minimum level of local know-how and quality of production factors (approximately by city size) for the appearance of each function, and a spatial organization of (only partially overlapping) market areas, and
- one which holds for manufacturing activities, where market areas dissolve and location factors demand favours the reduction of the cost of physical space, and therefore pushes forward a non-metropolitan environment.

3.2 The model

Let us start with:

a) a vector of homogeneous functions $i(i = 1,...,I)$, which:

- are defined in terms of both sectors and type of activity (e.g., a function may be that of manufacturing of textiles, or technological services for chemical productions),
- are ordered in increasing level of average benefits per capita (b_o^i) (an order which, for tertiary activities, coincides with a rising appearance

threshold for each function),

- belong to a limited number of bundles, linked to urban ranks (k<I), where within each bundle the dispersion of sectoral average benefits is limited and supposed close to zero;

b) a set of minimum initial production size X_0^i for each function i;

c) a set of appearance thresholds A_0^i for each function defined in terms of urban size (total urban production X*);

d) a set of sectoral per capita productivities x_0^i;

e) a set of sectoral per capita wages w_0^i;

f) a set of per capita benefits b_0^i (defined as $b_0^i = x_0^i - w_0^i$);

g) productivity, wages, benefits and appearance thresholds of each function are similar within each bundle of functions associated with an urban rank and increase together with urban **ranks**;

h) a matrix (IxI) of intersectoral externalities ϵ^{ij}, expressing the positive or negative effects of the presence of function i on function j in the same area. The externalities are expressed by pure numbers, showing a percentage change in the variables they are linked to (productivity and innovation probabilities, as it will be shown later on);

i) a curve of average location costs c (in terms of average costs per employee), increasing with increasing city size (size in this case is better approximately by total employment N_s^*);

j) a set of possible centres or locations s in space.

In each centre, each **existing** function may develop (or shrink) its production X_s^i according to a logistic path of 'constrained development', driven by a positive (negative) difference between the specific production benefits and the urban location costs (b^i-c). This development may take place both through the entrance of new firms in the same sector or function, attracted by the presence of extra-profits, and through the development of existing firms. When the city grows, thanks to the development of its range of functions, its location costs may overcome the production benefits of some lower order functions, which are therefore destined to decline or disappear.

The role of intersectoral externalities is twofold throughout the model:

a) they allow a higher productivity (and hence a higher benefit or profit, given the constancy of sectoral wages w_o^i in space) to be reached by some functions in a specific spatial setting, and b) determine a higher probability of acquiring some specific higher order functions in centres of limited size.

These externalities may stem from:

i) intersectoral linkages along input-output or technological 'filières', linking different industrial sectors or some industry with some industry-specific services (e.g., textiles manufacturing with textiles equipment makers and fashion designers); these linkages may be considered as a wider form of localization economies;

ii) intersectoral linkages among functions of the same rank, coming mainly from interconnections and spin-off effects on the labour market (a subset of urbanization economies);

iii) income effects on price-elasticity of demand for lower order functions and possibility of raising quality and price of these latter functions when higher order functions are present.

These three main effects generate respectively in the model:

i) the possibility of appearance of higher order functions in centres of low rank;

ii) the widening of the range of functions of the same rank, once one of these is acquired by a centre;

iii) the development of lower order functions once a higher order function is acquired, and the subsistence of some low ranking activities in big centres, where the locational costs conditions are adverse (e.g., the subsistence of retail trade functions in big centres, thanks to higher specialization, higher effectiveness and lower price elasticity of demand).

As a consequence of the presence of these externalities (or idiosyncrasies) in each centre, the general level of sectoral productivity and average benefit change throughout space; in each centre s in fact:

$$x_s^i = x_o^i . (1 + \epsilon_s^i)$$

and consequently

$$b_s^i = x_s^i - w_o^i = x_o^i . (1 + \epsilon_s^i) - w_o^i$$

We may define the externality elements which acts on function i in centre s by the following:

$$\epsilon_s^i = \Sigma_j \epsilon^{ij}$$

for all j active in s, a definition that might be easily expanded to take in consideration also possible synergies coming from functions located in nearby areas, as said before.

The two dynamic equations that are proposed in our study express:

1. the basic mechanism of creation or disappearance of functions i in each centre s; and
2. the related effects of growth or decline induced on urban production size in each centre.

3.3 Master equation: the birth/death stochastic process of urban functions

Assuming a stochastic approach, the urban system evolution corresponds to a dynamic birth/death process of a statistical population of functions displaced in different locations and disaggregated in groups.

Let us consider our abstract urban system with s=1,2,...,S centres and the set I of functions in which two different subsets I' (tertiary functions) and I'' (manufacturing functions) may be distinguished, so that:

$$I' \cup I'' = I$$

The state vector $\mathbf{n}^i$ defines the centres which have an active function i at time t:

$$n^i = n^i_1,\ n^i_2,\ n^i_3,\ldots,n^i_S$$

The macro-state of the urban system at time t will be described by a configuration **n**(t):

$$n(t) = n^1_1,\ n^1_2,\ldots,\ n^1_S,\ n^2_1,\ n^2_2,\ldots,\ n^i_1,\ n^i_2,\ldots,\ n^I_S$$

consisting of all functions active at time t in each centre s.

Over all possible configurations, a probability distribution function P(**n**) may be described, which denotes the probability to find a certain realized configuration at time t:

$$P(n,\ t) = P(n^1_1,\ n^1_2,\ldots,\ n^i_1,\ n^i_2,\ldots,\ n^I_S,\ t)$$

with the normalization condition:

$$\Sigma_n P\ (n,\ t) = 1$$

The evolution of the system may be described assuming a 'master equation' approach (Weidlich, Haag, 1982) where transition probabilities in configurations $\mathbf{n}^i$ have to be formalized by the definition of birth/ death rates of change.

If $ß^i(\mathbf{n}^i,\ \mathbf{n}^i - 1)$ denotes the birth rate of change from state $\mathbf{n}^{i-1,toni}$ and $\mu^i(\mathbf{n}^i - 1,\ \mathbf{n}^i)$ the death rate of change from $\mathbf{n}^i$ to $\mathbf{n}^i - 1$ for activity i, the equation of motion for P(**n**, t) reads:

$$dP(n,\ t)/dt = \Sigma_i ß^i(n^i,n^i - 1).P(n^i - 1,t) + \Sigma_i \mu^i(n^i,n^i + 1).P(n^i+1,t)+$$

$$- [\Sigma_i ß^i(n^i + 1,n^i) + \Sigma_i \mu^i(n^i - 1,\ n^i)].P(n^i,\ t)$$

3.3.1 Transition probabilities for tertiary functions

a) birth rate of change

The birth of a new function of type i' in s can occur if a minimum threshold $A_o^{i'}$ (in terms of total urban production size X*) is overcome, and the probability of this birth may be enhanced by the presence of intersectoral externalities; thus:

$$ß_s^{i'} = (1+\epsilon_s^{i'}) . \exp \alpha_1 (X_s^{*} - A_o^{i'})/A_o^{i'}$$

where α_1 is a parameter and the other symbols have already been presented.

b) death rate of change

An existing function i' in s may disappear due to: a) competition effects on local demand potential (Φ) induced by centres s' which compete with s in the same function i', and b) to excess supply capacity in the entire system with respect to total demand for the output of function i' (X-D). Thus the death probability of the function in centre s is given by:

$$\mu_s^{i'} = \pi_{s'}(\Phi_{s'}^{i'}) \Theta / \Phi_{s'}^{i'} . \exp(X_*^{i'} - D_*^{i'})/D_*^{i'} \qquad s' \neq s$$

where:

$$\Phi_s^{i'} = \Sigma_{s'} X_{s'}^{*} . f(d_{ss'}) . \delta^{i'}$$

(d being the distance or transport cost, and $\delta^{i'}$ being an average input coefficient related to input i'; a superscript * denotes summation over all functions i and a subscript * denotes summation over all centres), and:

$$\Theta = \Theta_{s'}^{i'} = X_{s'}^{i'} / \Sigma_{s'} X_{s'}^{i'}$$

On its turn, total supply and total demand for function i' in the entire urban system may be defined respectively as:

$$X_*^{i'} = \Sigma_s X_s^{i'}$$

and:

$$D_*^{i'} = \Sigma_s \Sigma_i X_s^{i'} . \delta^{i'}$$

3.3.2 Transition probabilities for manufacturing functions

a) birth rate of change

As far as birth rates are concerned, manufacturing activities are treated in the same way as lower order services, with the smallest appearance threshold (A_o) and the smallest level of per capita production benefits. This implies possible ubiquity for these functions, but a decreasing production share in bigger centres due to 'crowding out' processes (as it will be seen in section 3.4). Of course, higher order functions in industrial sectors are treated as tertiary functions, and follow the logic presented in section 3.3.1.

Birth transition probabilities are similar, mutatis mutandis, to those presented for tertiary functions:

$$ß_s^{i"} = (1+\epsilon_s^{i"}) . \exp \alpha_2 (X_s^* - A_o^{i"})/A_o^{i"}$$

b) death rate of change

Due to the fact that manufacturing activities sell worldwide and not to a local market, the only locational condition which may lead to their disappearance is an insufficient share of demand, determined by the overcrowding of producer competitors in the same function i" in the system.

If we define total supply and total demand of good i" as usual, the death transition probability is given by:

$$\mu_s^{i"} = (1/\epsilon_s^{i"}) \exp \alpha_3 [X_*^{i"}(t) - D_*^{i"}(t)]/D_*^{i"}(t)$$

3.4 Second equation: dynamics of urban size

The 'constrained' dynamics of the single centre is generated by the logistic growth process induced by each function active in s at time t, which pushes its production size up to some limiting value corresponding to the equilibrium between per capita benefits and urban costs. Thus, remembering that:

$$X_s^*(t) = \Sigma_i X_s^i(t)$$

then:

$$\dot{X}_s^*(t) = \Sigma_i X_s^i(t)\ \tau(b_s^i - c_s)$$

where:

$$c_s = f(N_s^*) = f[\Sigma_i N_s^i] = f[\Sigma_i (X_s^i / x_s^i)]$$

As the rates of change are a function of both the state of the system and of time, a markovian non-homogeneous growth and decline process is produced.

4. Conclusions

Empirical reality shows a hierarchy of centres that in many respects departs from the traditional Christallerian description: centres may not present the full range of activities associated with their rank and, inversely, may develop some functions that are typical of higher order urban ranks.

On the other side, the urban hierarchy seems to present some stable structural elements of order and sectoral nesting that allow us to associate, perhaps in a lose and sometimes impressionistic way, the single centres to specific ranks, especially in the upper part of the hierarchy; by this, it departs also from a purely random structure of market areas à la Loesch.

In our opinion this empirical urban structure may be captured in theoretical terms by holding the general Christallerian hypotheses for tertiary functions, only releasing some too deterministic and abstract assumptions and allowing for local intersectoral synergies to account for some discrepancies with respect to the pure model in the appearance of some higher order functions in lower order centres. As far as manufacturing functions are concerned, they have to be supposed to behave freely throughout the spatial system, in that they do not require strict urbanization economies and do not sell to well defined, non-overlapping market areas.

In this paper, these two main theoretical assumptions are embedded in a dynamic framework, simulating urban growth as driven by innovative processes of acquisition and birth of new, higher order functions, along a new theoretical research programme based on supply-oriented urban dynamics models.

References

Allen, P.M. (1982), Order by Fluctuation and the Urban System, New Quantitative Techniques for Economic Analysis, London, Academic Press.

Allen, P.M. and M. Sanglier (1981), Urban Evolution, Self-Organisation and Decision Making, Environment and Planning A 13:167-183.

Camagni, R., F. Curti and M.C. Gibelli (1985), Ciclo Urbano: le Città fra sviluppo e declino, in: G. Bianchi and I. Magnani (eds.), Sviluppo Multiregionale: Teorie, Metodi, Problemi, Milano, F. Angeli.

Camagni, R., L. Diappi and G. Leonardi (1986), Urban Growth and Decline in a Hierarchical System: a Supply-Oriented Dynamic Approach, Regional Science and Urban Economics 16:145-160.

Camagni, R. and T. Pompili (1988), Competence, Power and Waves of Urban Development in the Italian City System, paper presented at the Urbinno meeting, Berlin, 10-12 March, forthcoming (P. Nijkamp and J. Ewers, eds.).

Clarke, M. and A.G. Wilson (1986), A Framework for Dynamic Comprehensive Urban Models: the Integration of Accounting and Micro-Simulation Approaches, Sistemi Urbani 2/3:145-177.

Dematteis, G. (1985), Verso Nuove Strutture Urbane Reticolari, in: G. Bianchi and I. Magnani (eds.), Sviluppo Multiregionale: Teorie, Metodi, Problemi, Milano, F. Angeli.

Dematteis, G., A. Fubini and R. Gambino (1988), Le Aree Metropolitane e le Reti Urbane, mimeo, CNR.

Diappi, L. and T. Pompili (1989), Cities as Interacting Networks: a Loeschian Type Dynamic Approach, in: D. Martellato and F. Sforzi (eds.), Studi sui Sistemi Urbani, Milano, F. Angeli.

Weidlich, W. and Haag, G. (1982), Concepts and Models of a Quantitative Sociology: the Dynamics of Interacting Populations, Berlin, Springer Series in Synergetics no. 14.

Wilson, A.G. (1983), Making Urban Models More Realistic: Some Strategy for Future Research, Working Paper no. 358, School of Geography, University of Leeds.

Chapter 15

INTERREGIONAL MIGRATION DYNAMIC THEORY AND COMPARATIVE ANALYSIS

W. Weidlich
M. Sonis

1. Introduction

Migration processes in all their theoretical and empirical aspects are a constitutive part of regional science and demography: No demographic theory and no reliable prognosis of the regional evolution can be made without taking into consideration the migration of populations.

Furthermore, the field of migration is a challenge to develop a fully dynamic theory connecting the micro- and macrolevel of the society. On the microlevel there exist clear individual decisions to maintain or to change the location in a given interval of time, and on the macrolevel a rich quantitative empiric material about migration is available in many countries for a considerably long series of years. These data provide a good empiric basis for the quantitative treatment of the migratory dynamics.

The theoretical model proposed in this paper - and in full detail in a forthcoming book [1] - is designed to serve two purposes:

1. to provide a fully dynamic theory of the migratory process in terms of equations of motion, which link the microlevel of individual decisions with the macrolevel of developing regional populations, and
2. to correlate the migratory trends, being expressed in mobility factors and dynamic regional utilities, to regional timedependent key-factors representing the spatio-temporal socio-economic situation.

2. The Model

After a short discussion of the empirical macrolevel in section 2.1, we turn in section 2.2 to the microlevel of individuals whose migration is described in probabilistic terms. In section 2.3 the dependence of the probability transition rates on mobility factors and 'dynamic regional utilities' is introduced, and in section 2.4 the probability transition rates are used to set up equations of motion for the mean population configuration. Section 2.5 gives a short introduction into the evaluation methods if the model is to be compared with empirical systems.

2.1 The empirical macrolevel

The population is considered to be homogeneous, since the migratory data material in most cases does not distinguish between different subpopulations. The first set of quantities of interest is the population configuration

$$(2.1)\quad n^{(e)} = \{n_1^{(e)}(t)\ ,\ n_2^{(e)}(t) \ldots\ ,\ n_L^{(e)}(t)\}$$

where $n_i^{(e)}(t)$ is the empirical number of people in region i at time t. Index e stands for 'empirical'.

Secondly, one introduces the empiric number of people migrating from region i to region j in the time interval Δt as $\Delta t \cdot w_{ji}^{(e)}(t)$. Hence, $w_{ji}^{(e)}(t)$ is the empiric migration rate per unit of time from i to j. Similarly, $\Delta t w_i^{\beta(e)}(t)$ and $\Delta t w_i^{\mu(e)}(t)$ are the empiric numbers of birth and death, respectively in region i and in time interval Δt. Since the difference $(n_i^{(e)}(t+\Delta t) - n_i^{(e)})$ is obtained by all people migrating into i from any j minus all people moving from i into any j, plus all people born in i minus all people dying in i in the time between t and $t+\Delta t$, the following relation can easily be derived:

$$\text{(2.2)} \quad \frac{n_i^{(e)}(t+\Delta t) - n_i^{(e)}(t)}{\Delta t} = \sum_{j=1}^{L}{}' \, w_{ij}^{(e)}(t) - \sum_{j=1}^{L}{}' \, w_{ji}^{(e)}(t) + w_i^{\beta(e)}(t) - w_i^{\mu(e)}(t)$$

where the bar at the sum excludes summation over $w_{ij}^{(e)}(t)$ with $i=j$.

2.2 Probabilistic description of migratory decisions

The **macrolevel** of an evolving population configuration **n** must now be linked to the **microlevel** of decisions of individuals to migrate from an origin to a destination region. The appropriate description of the individual decision process is a probabilistic one.

Let us introduce the conditional probability

$$\text{(2.3)} \quad p(j,t+\tau|i,t)$$

to find the individual in region j at the time $t+\tau$, given that it was in region i at time t. For obvious reasons the conditional probability fulfils the relation of probability normalization

$$\text{(2.4)} \quad \sum_{j=1}^{L} p(j,t+\tau|i,t) = 1$$

and the initial condition

$$\text{(2.5)} \quad p(j,t|i,t) = \delta_{ji}$$

where δ_{ji} is the Kronecker symbol.

The decisive assumption is now, that the conditional probability for small τ can be expanded in a Taylor series up to first order as follows:

(2.6a) $$p(j,t+\tau|i,t) = \tau p_{ji}(t)+\ldots, \quad j\neq i$$

(2.6b) $$p(i,t+\tau|i,t) = 1 - \tau \sum_{k=1}^{L}{}' p_{ki}(t)+\ldots, \quad j=i$$

In (2.6b) the relation (2.4) has been used. The quantities $p_{ji}(t)$ are denoted as **individual transition rates** (per unit of time) from i to j.

Let us now turn to the **macrolevel**. The individual transitions from i to j induce the following change of the population configuration:

(2.7) $$\mathbf{n} = \{n_1 \ldots n_i \ldots n_j \ldots n_L\} \Rightarrow$$

$$\mathbf{n}^{(ji)} = \{n_1 \ldots (n_i - 1), \ldots (n_j + 1), \ldots n_L\}$$

and an individual birth or death in region i induces the respective transition

(2.8a) $$\mathbf{n} = \{n_1 \ldots n_i \ldots n_L\} \Rightarrow \mathbf{n}^{(i+)} = \{n_1 \ldots (n_i + 1), \ldots n_L\}$$

(2.8b) $$\mathbf{n} = \{n_1 - n_i - n_L\} \Rightarrow \mathbf{n}^{(i-)} = \{n_1 \ldots (n_i - 1), \ldots n_L\}$$

The corresponding **configurational transition rates** (per unit of time) are denoted as $w_{ji}(\mathbf{n},t)$, $w_i^{\beta}(\mathbf{n},t)$ and $w_i^{\mu}(\mathbf{n},t)$, respectively. It is intuitively clear that these rates are linked to the individual transition rate from i to j, and to the individual birth rate $\beta_i(t)$ and individual death rate $\mu_i(t)$ in i as follows:

(2.9a) $$w_{ji}(\mathbf{n};t) = p_{ji}(t)\cdot n_i$$

(2.9b) $$w_i^{\beta}(\mathbf{n};t) = \beta_i(t)\cdot n_i$$

(2.9c) $$w_i^{\mu}(\mathbf{n};t) = \mu_i(t)\cdot n_i$$

The reason for these formulas is, that n_i independent individuals in region i each have the migration rate p_{ji} to region j, produce the birth rate $\beta_i(t)$ and have the death rate $\mu_i(t)$, respectively.

2.3 Probability transition rates as functions of dynamic utilities and mobilities

The migratory trends of the individual must now be taken into account by giving his transition rate $p_{ji}(t)$ a special functional form in terms of these trends.

We choose the general form

$$(2.10)\quad p_{ji}(t) = \nu_{ji}(t) \cdot G_{ji}(t)$$

where the symmetric **mobility factor** $\nu_{ji}(t)$ (with $\nu_{ji} = \nu_{ij}$) includes the geographic, economic and social influence of an **effective distance** between i and j, whereas the nonsymmetric factor $G_{ji}(t)$ (with $G_{ji} \neq G_{ij}$) includes **push/pull-effects.**

It is plausible and consistent with empirical data to put the mobility factor equal to

$$(2.11)\quad \nu_{ji}(t) = \nu_0(t) f_{ji}$$

with the global timedependent mobility

$$(2.12)\quad \nu_0(t) = \frac{1}{L(L-1)} \sum_{i,j}{}' \nu_{ji}(t)$$

and a time-independent factor

(2.13) $$f_{ij} = f_{ji} = L(L-1)\exp(-D_{ij})$$

which is normalized as

(2.14) $$\frac{1}{L(L-1)} \sum_{i,j}^{L}{}' f_{ij} = 1$$

and depends on the 'effective distance' D_{ij} between the regions i and j.
The **push/pull factor** $G_{ji}(t)$ on the other hand depend on the attractiveness or 'utility' of the origin region i and destination region j for each individual in view of his migration decision. Introducing a **'dynamic utility'** $u_i(t)$ for every region i (not to be confused with the static concept of utility used in economics), the push/pull factor can be written in the plausible and empirically verifiable form:

(2.15) $$G_{ji}(t) = \exp(u_j(t) - u_i(t)) = e^{u_j(t)} \cdot e^{-u_i(t)}$$

The first factor in (2.15) describes the pulling of the destination region j with increasing utility $u_j(t)$, and the second factor the pushing of the origin region i with decreasing utility $u_i(t)$.
Inserting (2.10) with (2.11) and (2.15) into (2.9a) one obtains the formula of central importance for the configurational probability transition rate:

(2.16) $$\begin{aligned} w_{ji}(n,t) &= n_i \nu_{ji}(t)\exp(u_j(t)-u_i(t)) \\ &= n_i \nu_o(t) f_{ji} \exp(u_j(t)-u_i(t)) \end{aligned}$$

Since only the differences of the utilities enter the transition rates (2.16), the utilities are only determined up to a common additive constant which can be chosen so that

$$(2.17)\quad \sum_{i=1}^{L} u_i(t) = 0$$

is fulfilled.

2.4 The migratory equations of motion

Equations of motion for population configuration **n** can be derived on two interconnected levels: The first level is the **stochastic level.** It is fully consistent with the probabilistic description of individual decisions and contains fluctuations on the macrolevel too. The fundamental equation on this level is the master equation. The second level is the **quasideterministic description.** It consists of a set of equations of motion for the mean values of the population configuration. These equations can be derived from the master equation.

Let us begin with the stochastic level! At first one introduces the configurational probability

$$(2.18)\quad P(n_1,n_2...n_L;t) \equiv P(\mathbf{n};t)$$

to find realized configuration **n** at time t.

Evidently, the normalization condition

$$(2.19)\quad \sum_{\mathbf{n}} P(\mathbf{n};t) = 1$$

must hold, where the sum extends over all possible configurations. If migratory transitions as well as birth and death processes are taken into account, the configurational probability satisfies the following **fundamental master equation:**

$$(2.20)\quad \frac{dP(\mathbf{n};t)}{dt} = \sum_{i,j}^{L}{}' w_{ji}(\mathbf{n}^{(ij)};t)P(\mathbf{n}^{(ij)};t) - \sum_{i,j}^{L}{}' w_{ji}(\mathbf{n},t)P(\mathbf{n};t)$$

$$+ \sum_{i=1}^{L} w_i^{\beta}(\mathbf{n}^{(i-)},t)P(\mathbf{n}^{(i-)};t) - \sum_{i=1}^{L} w_i^{\beta}(\mathbf{n},t)P(\mathbf{n};t)$$

$$+ \sum_{i=1}^{L} w_i^{\mu}(\mathbf{n}^{(i+)},t)P(\mathbf{n}^{(i+)};t) - \sum_{i=1}^{L} w_i^{\mu}(\mathbf{n},t)P(\mathbf{n};t)$$

It is easy to understand the meaning of this dynamic probability balance equation: The left hand side is the change with time of the probability of configuration **n**. The three lines on the right hand side show how this change is accomplished by migration, birth and death, respectively. The first term of this first line contains the probability fluxes $w_{ji}(\mathbf{n}^{(ij)};t)\cdot P(\mathbf{n}^{(ij)};t)$ from the neighbouring configurations $\mathbf{n}^{(ij)}$ into the configuration **n**, thus increasing P(**n**;t), whereas the second term contains the probability fluxes $w_{ji}(\mathbf{n};t)P(\mathbf{n};t)$ from configuration **n** into neighbouring configurations $\mathbf{n}_{(ji)}$, which diminish the probability P(**n**,t) of **n**. The structure of the birth and death process terms is analogous.
Obviously solution P(**n**;t) of (2.20) not only tells us which configurations $\hat{\mathbf{n}}(t)$ are the most probable ones at time t, but also how probable it is to find deviations from $\hat{\mathbf{n}}(t)$.

Let us now turn to the **quasideterministic level of meanvalues** $\bar{\mathbf{n}}(t)$ defined by

$$(2.21)\quad \bar{n}_k(t) = \sum_{\mathbf{n}} n_k\, P(\mathbf{n};t)$$

It can easily be derived from the master equation that the meanvalues obey the equations of motion

$$\frac{d\bar{n}_k(t)}{dt} = \sum_i^{L}{}' \overline{w_{ki}(\mathbf{n},t)} - \sum_j^{L}{}' \overline{w_{jk}(\mathbf{n},t)} + \overline{w_k^{\beta}(\mathbf{n},t)} - \overline{w_k^{\mu}(\mathbf{n},t)} \tag{2.22}$$

where on the right hand side appear the meanvalues of the configurational transition rates.

If the probability P(**n**,t) is a sharp-peaked unimodal distribution, the approximate relation holds:

$$f(\mathbf{n},t) \cong f(\bar{\mathbf{n}}(t),t) \tag{2.23}$$

which transforms eqs. (2.22) into a set of selfcontained - in general nonlinear - equations of motion for the meanvalues:

$$\frac{d\bar{n}_k}{dt} = \sum_i^{L}{}' w_{ki}(\bar{\mathbf{n}}(t),t) - \sum_j^{L}{}' w_{jk}(\bar{\mathbf{n}}(t),t) + w_k^{\beta}(\bar{\mathbf{n}}(t),t) - w_k^{\mu}(\bar{\mathbf{n}}(t),t) \tag{2.24}$$

They take the fully explicit form after inserting (2.9) and (2.16)

$$\frac{d\bar{n}_k}{dt} = \sum_i^{L}{}' \bar{n}_i \nu_{ki} \exp(u_k - u_i) - \sum_i^{L}{}' \bar{n}_k \nu_{ik} \exp(u_i - u_k) + (\beta_k - \mu_k)\bar{n}_k \tag{2.25}$$

Identifying the theoretical meanvalues $\bar{\mathbf{n}}(t)$ with the empirical configuration $\mathbf{n}^{(e)}(t)$, it can be seen by comparison of equations (2.24) and (2.2) that the theoretical mean configurational transition rates

$w_{ki}(\bar{\mathbf{n}},t)$, $w_k^{\beta}(\bar{\mathbf{n}},t)$ and $w_k^{\mu}(\bar{\mathbf{n}},t)$ must agree with the empirical rates $w_{ki}^{(e)}(t)$, $w_k^{\beta(e)}(t)$ and $w_k^{\mu(e)}(t)$, respectively, if one wants the theory to yield results in agreement with empiric data. This observation is the starting point of the regression analysis of the next section.

The dynamic utilities $u_i(t)$ may - or may not - themselves depend on the mean population configuration $\bar{\mathbf{n}}(t)$. This alternative is decisive for the mathematical structure of eqs. (2.24). In the case when u_i depends on $\bar{\mathbf{n}}(t)$, they are **nonlinear**, and in the simpler case of u_i being independent of $\bar{\mathbf{n}}(t)$ they are **linear**.

In the simple - but not always realized - case that the rate of natural increase ρ_k does not depend on the region:

$$\rho_k(t) \equiv (\beta_k(t) - \mu_k(t)) \cong \rho(t) \tag{2.26}$$

it is easy to separate off the birth/death-processes from the migratory process by introducing the relative population shares through

$$N(t) = \sum_{k=1}^{L} \bar{n}_k(t) \ ; \ \bar{n}_k(t) = N(t) \cdot x_k(t) \tag{2.27}$$

Making use of the equation for the total population growth

$$\frac{dN}{dt} = \rho(t)N \ ; \ N(t) = N(0)\exp\left(\int_0^t \rho(t')dt'\right) \tag{2.28}$$

which easily follows from (2.25) and (2.27), one obtains the purely migratory equations for the population shares $x_k(t)$:

(2.29) $$\frac{dx_k}{dt} = \sum_{i=1}^{L}{}' x_i \, p_{ki}(t) - \sum_{j=1}^{L}{}' x_k \, p_{jk}(t)$$
$$= \sum_{i=1}^{L}{}' x_i \nu_{ki}(t)\exp[u_k(t)-u_i(t)] - \sum_{j=1}^{L}{}' x_k \nu_{jk}(t)\exp[u_j(t)-u_k(t)]$$

If mobilities ν_{ki} and dynamic utilities u_i are constant with time, the **stationary solution** x_k of (2.29) can immediately be obtained and verified by inspection:

(2.30) $$\hat{x}_k = c \exp[2u_k] ,$$
$$\text{with } c = \{ \sum_{i=1}^{L} \exp[2u_i] \}^{-1}$$

2.5 The estimation of parameters

The application of the theory to concrete migratory systems consists in two main steps:

- In a **first** step all trend parameters of the theory directly relating to the migratory process - namely mobilities and regional utilities - are estimated by comparison with empirical data.
- In a **second** step the migratory trend parameters are correlated to socio-economic spatio-temporal key factors, in order to provide insight into the possible causal relation between the socio-economic situation and the migratory dynamics.

The separation of the analysis into these two steps proves to be very convenient. If migratory data are available, the migration can be analyzed in terms of mobilities and regional utilities, independent of their socio-economic interpretation. The second step, the socio-economic analysis, depends on the available set of spatio-temporal variables and may then go through several and even alternative runs without detriment to the determination of the migratory trends in the first step.

The first step: Regression Analysis for Trend Parameters.

Assuming that we have separated off the birth/death-process, we can restrict the consideration to the migratory process only. As already mentioned in section 2.4, the agreement between empirics and theory will be better, the higher the agreement is between the theoretical rate $w_{ki}^{th}(t)$ given by (2.16) and the empiric rate $w_{ki}^{(e)}(t)$. In this spirit we make use of the least square method trying to minimize the difference between $w_{ki}^{th}(t)$ and $w_{ki}^{(e)}(t)$. It turns out that the mathematically simplest procedure is the log-linear estimation consisting in minimizing the expression

$$(2.31)\quad F[\nu,u] = \sum_{t=1}^{T} \sum_{k,l=1}^{L} \{\ln w_{kl}^{(e)}(t) - \ln[n_l^{(e)}(t)\nu_{kl}(t)\exp(u_k(t) - u_l(t))]\}^2$$

by variation of the mobilities $v_{kl}(t) = v_{lk}(t)$ and the regional utilities $u_k(t)$. The calculation is straightforward and yields explicit expressions for the optimal utilities and mobilities:

$$(2.32)\quad u_i(t) = \frac{1}{2L} \sum_{k=1}^{L}{}' \ln \frac{p_{ik}^{(e)}(t)}{p_{ki}^{(e)}(t)}$$

where

$$(2.33)\quad p_{ji}^{(e)}(t) = \frac{w_{ji}^{(e)}(t)}{n_i^{(e)}(t)} .$$

is the empiric individual transition rate, and

$$(2.34)\quad \nu_{ji}(t) = \nu_{ij}(t) = (p_{ji}^{(e)}(t) . p_{ij}^{(e)}(t))^{1/2}$$

If $\nu_{ji}(t)$ is assumed to be of the form (2.11), the optimization of the factors f_{ij} and $\nu_o(t)$ yields:

$$(2.35)\quad f_{ji} = C \cdot \prod_{t=1}^{T} \left(p_{ji}^{(e)}(t)\,p_{ij}^{(e)}(t)\right)^{\frac{1}{2T}}$$

and

$$(2.36)\quad \nu_0(t) = \frac{1}{c}\;\frac{\prod'_{k,l}\left(p_{lk}^{(e)}(t)\,p_{kl}^{(e)}(t)\right)^{\frac{1}{2L(L-1)}}}{\prod'^{L}_{k,l}\;\prod_{t=1}^{T}\left(p_{lk}^{(e)}(t)\,p_{kl}^{(e)}(t)\right)^{\frac{1}{2TL(L-1)}}}$$

and factor c follows from the normalization (2.14).

It should be mentioned that instead of the log-linear also a nonlinear estimation can be performed, based on the minimization of the expression:

$$(2.37)\quad F[\nu,u] = \sum_{t=1}^{T}\sum_{k,l}^{L}{}' \left\{w_{kl}^{(e)}(t) - n_l^{(e)}(t)\,\nu_{kl}(t)\exp(u_k(t)-u_l(t))\right\}^2$$

The results of this optimation are slightly better, but cannot be written in analytical form. A direct survey of the quality of the fitting procedure can be given by plotting the fitted theoretical transition rates $w_{ji}^{th}(t)$ against the empirical rates $w_{ji}^{(e)}(t)$. (For the ideal fit all points are on the line y=x). Figures 1a and 2a show the fit of $w_{ji}^{th}(t)$ to $w_{ji}^{(e)}(t)$ in the case of Germany (L=11 regions and T=27 years, giving 2970 matrix elements $w_{ji}^{(e)}(t)$) in the log-linear and non-linear estimation, respectively, if a general mobility $\nu_{ji} = \nu_{ij}$ is allowed, and Figure 1b and 2b show the results of the same fit, if the mobility is restricted to the form $\nu_{ji} = \nu_{ij} = \nu o(t) f_{ij}$.

The second step: The Dependence of Trend Parameters on Socio-Economic Key-Factors

The important question must now be solved, whether there exists an influence

The important question must now be solved, whether there exists an influence of the socio-economic situation - expressed by spatio-temporal proxy-variables $\tilde{\Omega}(i,t)$ - on the migratory trends. Since direct causalanalyses are very difficult if not unfeasible, an indirect measure of a possible causal relationship must be introduced: The degree of the spatio-temporal correlation of socio-economic variables $\tilde{\Omega}(i,t)$ to trend parameters like the regional utility $u_i(t) \equiv u(i,t)$, which is also a spatio-temporal function. Such a high degree of correlation between the regional utility u(i,t) and certain determinants $\tilde{\Omega}_\alpha(i,t)$ of the socio-economic situation would exist, if u(i,t) should turn out to be representable as

$$(2.38) \quad u(i,t) = \sum_{\alpha=1}^{r} b_\alpha \Omega^\alpha(i,t) + \Delta(i,t)$$

where $\Delta(i,t)$ is a small rest-term only, and where

$$(2.39) \quad \Omega^\alpha(i,t) = \frac{\tilde{\Omega}^\alpha(i,t) - \bar{\tilde{\Omega}}^\alpha(t)}{\bar{\tilde{\Omega}}^\alpha(t)}$$

with

$$(2.40) \quad \bar{\tilde{\Omega}}^\alpha(t) = \frac{1}{L} \sum_{i=1}^{L} \tilde{\Omega}^\alpha(i,t)$$

is the scaled dimensionless form of the variable $\tilde{\Omega}^\alpha(i,t)$.

Formally, one can consider the spatio-temporal functions u(i,t) and $\Omega^\alpha(i,t)$ as vectors $|u\rangle$ and $|\Omega^\alpha\rangle$ in an $L \cdot T$-dimensional space, writing (2.38) in the form

$$(2.41) \quad |u\rangle = \sum_{\alpha=1}^{r} b_\alpha |\Omega^\alpha\rangle + |\Delta\rangle$$

The conventional regression analysis then consists in taking all available $|\Omega\alpha\rangle$

$|\Delta>$:

(2.42) $||\Delta||^2 = <\Delta|\Delta> =$ Minimum

In this notation one makes use of the scalarproduct defined as:

(2.43) $$<u|v> = \sum_{i=1}^{L} \sum_{t=1}^{T} u(i,t)v(i,t)$$

This procedure, however, has disadvantages for the following reasons:

a) The number of socio-economic variables at disposal can be very large, in particular if time-delayed variables are also included.

b) Variables of almost the same spatio-temporal functional form are simultaneously considered, although it would suffice to take only one proxy of them.

Hence, it is desirable to introduce an alternative procedure having the advantage

a) that an automatic ranking of relevance of the socio-economic variables takes place, which allows to terminate the representation of $|u>$ according to equation (2.40) by taking the most important r keyfactors $|\Omega^\alpha>, \alpha = 1,2,\ldots,r$ only and

b) that automatically only one out of a group of variables $\Omega^\alpha(i,t)$ of almost the same spatio-temporal functional form is taken into account.

The main steps of the alternative procedure, denoted as ranking regression analysis, are only sketched here, since it will be treated in detail elsewhere [1,2].

In a first step, the utility vector $|u>$ is tentatively represented as

(2.44) $$|u\rangle = a_1|\hat{\Omega}^1\rangle + |u_{\perp 1}\rangle$$

$$\text{with } a_1 = \langle\hat{\Omega}^1|u\rangle$$

where $|\hat{\Omega}^1\rangle$ is that normalized ($||\hat{\Omega}^1|| = 1$) socio-economic variable out of all $|\hat{\Omega}^\alpha\rangle$, which leads to the minimal rest $|u_\perp\rangle$ of the utility vector. This rest utility is orthogonal to $|\hat{\Omega}^1\rangle$

(2.45) $$\langle\hat{\Omega}^1|u_{\perp 1}\rangle = 0$$

expressing the fact, that $|u_{\perp 1}\rangle$ is that part of $|u\rangle$, whose functional dependence is not at all captured by $|\hat{\Omega}^1\rangle$.

In a second step, all the remaining socio-economic variables $|\Omega^\alpha\rangle$ $\alpha=2,3,...$ are decomposed into a term parallel (linearly dependent) to $|\hat{\Omega}^1\rangle$ and a remaining term, which is orthogonal to $|\hat{\Omega}^1\rangle$:

(2.46) $$|\Omega^\alpha\rangle = c_{1\alpha}|\hat{\Omega}^1\rangle + |\Omega^\alpha_{\perp 1}\rangle$$

$$\text{with } c_{1\alpha} = \langle\hat{\Omega}^1|\Omega^\alpha\rangle \;;\; \langle\hat{\Omega}^1|\Omega^\alpha_{\perp 1}\rangle = 0 \;;\; \alpha=2,3,...$$

The part $|\Omega^\alpha_{\perp 1}\rangle$ of $|\Omega^\alpha\rangle$ contains that spatio-temporal functional dependence, which is not already represented by $c_{1\alpha}|\hat{\Omega}^1\rangle$.
Now the whole procedure can be iterated with respect to $|u_{\perp 1}\rangle$: This rest vector is now represented by

(2.47) $$|u_{\perp 1}\rangle = a_2|\hat{\Omega}^2_{\perp 1}\rangle + |u_{\perp 2}\rangle$$

$$\text{with } a_2 = \langle\hat{\Omega}^2_{\perp 1}|u_{\perp 1}\rangle$$

where $|\hat{\Omega}_{21}\rangle$ is that (normalized) $|\hat{\Omega}^{\alpha}_{\perp 1}\rangle$ among all $\alpha=2,3,...$ which is the best in representing $|u_{\perp 1}\rangle$, that means which leads to the rest term of second order $|u_{\perp 2}\rangle$ with minimal norm. The latter rest term now fulfils

$$(2.48) \quad \langle\hat{\Omega}^{1}|u_{\perp 2}\rangle = \langle\hat{\Omega}^{2}_{\perp 1}|u_{\perp 2}\rangle = 0$$

The remaining variables $|\Omega^{\alpha}_{\perp 1}\rangle$, $\alpha=3,4,...$ can again be decomposed into a term parallel to $|\hat{\Omega}^{2}_{\perp 1}\rangle$ and a remaining term:

$$(2.49) \quad |\Omega^{\alpha}_{\perp 1}\rangle = c_{2\alpha}|\hat{\Omega}^{2}_{\perp 1}\rangle + |\Omega^{\alpha}_{\perp 2}\rangle$$

$$\text{with} \quad c_{2\alpha} = \langle\hat{\Omega}^{2}_{\perp 1}|\Omega^{\alpha}_{\perp 1}\rangle \quad ; \quad \langle\hat{\Omega}^{2}_{\perp 1}|\Omega^{\alpha}_{\perp 2}\rangle = 0$$

Therefore, the whole procedure can be repeated with the rest term of second order $|u_{\perp 2}\rangle$, and so on. The ranking regression procedure ends with

$$(2.50) \quad |u\rangle = \sum_{\alpha=1}^{r} a_{\alpha}|\hat{\Omega}^{\alpha}_{\perp\alpha-1}\rangle \Rightarrow \sum_{\alpha=1}^{r} b_{\alpha}|\Omega^{\alpha}\rangle$$

if the rest term $|u_{\perp r}\rangle$ of r-th order is small enough to be neglected. In practical cases the ranking regression leads to the systematic selection of very few key-factors representing $|u\rangle$ via (2.50) out of dozens of socio-economic variables, whose majority proves to be irrelevant with respect to the migratory trends.

3. Evaluation for Individual Countries, Selected Results and Comparative Analysis

In the forthcoming book [1] a systematic investigation of interregional migration has been implemented by the authors A. Andersson, G. Haag, I. Holmberg, J. Ledent, M. Munz, D. Pumain, G. Rabino, R. Reiner, N. Sarafoglou, M. Sonis and W. Weidlich, for the six countries Canada,

Federal Republic of Germany, France, Israel, Italy and Sweden, on the basis of the model of section 2. Here we present and compare selected results out of this investigation. The analysis proceeded along the lines indicated in section 2.5.

3.1 Determination of the global mobility and of regional utilities

At first the global mobility and the regional utilities have been determined for the migration between the regions within each country in a considerable period of time by fitting the theoretical rates

$$(3.1)\quad w_{ji}^{th}(t) = \nu_o(t) f_{ji} \exp(u_j(t) - u_i(t)) n_i(t)$$

to the empirical rates $w_{ji}^{(e)}(t)$.

Let us begin with the global mobility $\nu_0(t)$. The temporal evolution of the global mobilities (scaled to meanvalue 1) of the countries FRG, Canada, France, and Israel, Italy, Sweden is depicted in Figures 3a and 3b, respectively.

Remarkable differences can be observed between the countries with respect to this global measure of the intensity of interregional migration. Some 'old' European countries - Germany, Italy and Sweden - exhibit a permanent long-term decrease of the global mobility beginning after 1960 in Germany and after 1970 in Sweden and Italy. France, on the other hand, has a prolonged increase of the global mobility until 1975 due to a somewhat delayed urbanisation process. In Canada and Israel, whose histories differ characteristically from those of European countries, practically no decline of the mobility takes place. Canada even showed a small but long prevailing trend to increasing mobility, only reversed after the 1973 oil crisis, and Israel had strong initial fluctuations due to immigration waves before reaching a more stationary internal migration after 1970.

Secondly, the regional utilities have been determined. Here the ranking regression analysis had lead to the remarkable result that in all six countries the regional population $n_i(t)$ and its square, $n_i^2(t)$, are the two most important key-factors in the representation of u (t). This leads to the decomposition of the $u_i(t)$ into two terms:

$$(3.2) \quad u_i(t) = s_i(t) + \delta_i(t)$$

where s (t) is the part of u (t) depending on the regional size - that means on $x_i(t) = \frac{n_i(t)}{\bar{n}(t)}$ and on $x^2(t)$ - and where δ (t) are size-independent residual terms describing regional preferences due to the socio-economic situation. The size-term in all six countries turns out to have the form in the scaled variable $x_i(t)$

$$(3.3) \quad s_i(t) = \varkappa(x_i(t)-1) + \sigma(x_i^2(t) - \frac{\overline{n^2(t)}}{\overline{n(t)}^2})$$

where the agglomeration parameters and the (negative) saturation parameters σ for the six countries, together with their t-values, are listed in Table 1.

We now show as an example the regional utilities (including size-effects) and the regional preferences (excluding size-effects) for a European country, the Federal Republic of Germany, and a non-European country, namely Canada.

Figures 4a and 4b show the regional utilities and preferences, respectively, of the Federal Republic of Germany, and Figures 5a and 5b the regional utilities and preferences of Canada. In both cases the spread of the regional utilities expressed by their variance

$$(3.4) \quad \sigma_u^2(t) = \frac{1}{L-1} \sum_{i=1}^{L} u_i^2(t)$$

is larger than the spread of regional preferences with variance

(3.5) $$\sigma_{\delta}^{2}(t) = \frac{1}{L-1} \sum_{i=1}^{L} \delta_{i}^{2}(t)$$

because the utilities also reflect the different sizes of the regions, which is not the case for the preferences. The time-dependence of the utilities and preferences of Canada is more pronounced in comparison to the smoother behaviour of the same quantities of the FRG. This may reflect in a global sense the more equilibrated situation in the 'old' country Germany and the less uniform and developing situation in the 'young' country Canada.

The more detailed inspection of regional preferences shows that in Germany exist regions (federal states) with slowly increasing preferences and regions with a long-term decline of preferences. This development seems to be connected with regional economic structures, for instance ageing industries (steel, shipyards) in the case of declining regions, and prospering industries in the case of increasing regional preferences. The more dramatic variations of preferences of regions in Canada, for instance the peaks 1964, 1969 and 1980 of Alberta, British Columbia and Ontario also coincide with booms of economic prosperity in these regions, for instance by oil prospecting in the case of Alberta.

3.2 The representation of mobilities and utilities in terms of key-factors

Let us now go over to the second part of the analysis, the even more important representation of the global mobility and the regional utilities in the six countries by socio-economic key-factors. The problem of the choice of comparable socio-economic factors is not trivial, since the data sets available in the different countries are not directly comparable in the strict sense. The way out of this difficulty consists in considering the individual socio-economic variables as a proxy of some more general class of influences, on which migration may depend. In this sense the variables available in every country belong to one of the following classes:

Size and Population Structure (SP)

Labour Market (LM)

Housing Market (HM)
Industry and Investment (II)
Public Sector, Politics (PP)
Living Standard (LS)
Climate (C)

The results of these investigations within the six countries are summarized in Table 2 giving the representation of the global mobility in terms of key-factors, and in Table 3 giving the analogous representation of the regional preferences.

It is satisfactorily interpretable and not too surprising, that the few relevant key-factors selected by the ranking regression analysis out of many variables for all six countries belong to the following classes, in sequence of their importance:

Labour Market (LM)
Living Standard (LS)
Public Sector (PP)

Further variables of Industrial Structure (II) and of Housing Market (HM) are of complementary but not so prominent importance for describing the regional differentiation of migration in the six case studies under consideration.

3.3 The migratory stress

In section 2.4 we have seen that for a given set of constant regional utilities $\{u_1,u_2,...U_L\}$ the regional population shares $\mathbf{x}(t) = \{x_i(t),...x_L(t)\}$ eventually reach the stationary values

(3.6) $\hat{\mathbf{x}} = \{\hat{x}_1,\hat{x}_2,...\hat{x}_L\}$ given by

$$\hat{x}_i = c\exp(2u_i)$$

with $c = [\sum_{i=1}^{L} \exp(2u_i)]^{-1}$

The values $\hat{x}_i$ are the relative population shares into which the actual population would evolve, given that the utilities would remain constant for ever. The actually realized population shares x(t) in general differ from the virtual equilibrium values $\hat{\mathbf{x}}$, and the 'migratory stress' is introduced to be a measure for the 'distance' between the actual and the equilibrium distribution belonging to the given $\{u_1 \ldots u_L\}$.
One introduces the migratory stress as

(3.7) $s(x,\hat{x}) = \frac{1}{2}(1-r(x,\hat{x})) \quad ; \quad 0 \leq s(x,\hat{x}) \leq 1$

where $r(\mathbf{x},\hat{\mathbf{x}})$ is the correlation coefficient

(3.8) $$r(\mathbf{x},\hat{\mathbf{x}}) = \frac{\sum_{i=1}^{L}(x_i-\bar{x})\cdot(\hat{x}_i-\bar{x})}{\sqrt{\sum_{i=1}^{L}(x_i-\bar{x})^2 \cdot \sum_{j=1}^{L}(\hat{x}_j-\bar{x})^2}}$$

varying between $-1 \leq r(\mathbf{x},\hat{\mathbf{x}}) \leq +1$. Because of $r(\mathbf{x},\hat{\mathbf{x}}) \to 1$ for $\mathbf{x} \to \hat{\mathbf{x}}$, the migratory stress is - by definition - the smaller the more the actual population distribution approaches its virtual equilibrium.

In Figure 6 the timedependent correlation coefficient $r(\mathbf{x},\hat{\mathbf{x}})$ is depicted for the six countries. One sees that some countries are very near to the virtual equilibrium. These are the cases, where the regional utilities are almost stationary over a long period of time, so that the population distribution had sufficient time to adapt to these utilities. On the other hand there also exist countries developing away from the virtual equilibrium, because changes of preferences and a reorientation of migratory trends have led to new regional utilities $\{u_1,u_2 \ldots u_L\}$ and a new virtual equilibrium $\{\hat{x}_1,\hat{x}_2 \ldots \hat{x}_L\}$ so that the actual distribution $\{x_1 \ldots x_L\}$ is now farther away from the new equilibrium $\{\hat{x}_1 \ldots \hat{x}_L\}$than from the original equilibrium $\{\hat{x}_1 \ldots \hat{x}_L\}$

Acknowledgements

The authors are very grateful for the financial support provided by the German-Israeli Foundation.

Notes

[1] W. Weidlich and G. Haag (eds),
'Interregional Migration - Dynamic Theory and Comparative Analysis'
(387 pages with 129 figures and 64 tables) (1988)
ISBN 3-540-18441-4
Springer Verlag Berlin Heidelberg New York

[2] R. Reiner and M. Munz
'Ranking Regression Analysis of Spatio-Temporal Variables'
Environment and Planning A. Vol 21, in print

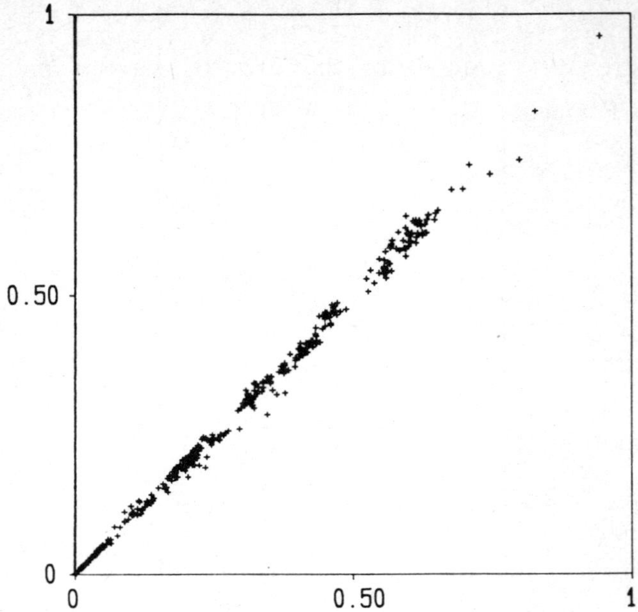

Figure 1a and 1b

Theoretical versus empirical migration matrices $w_{ji}(t)$ with log-linear estimation for a) undecomposed $\nu_{ji}(t)$, and b) for $\nu_{ji}(t) = \nu_0(t) f_{ji}$

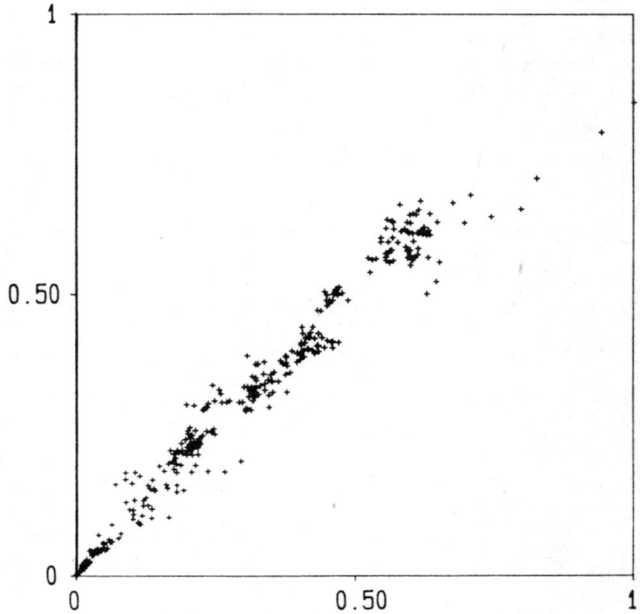

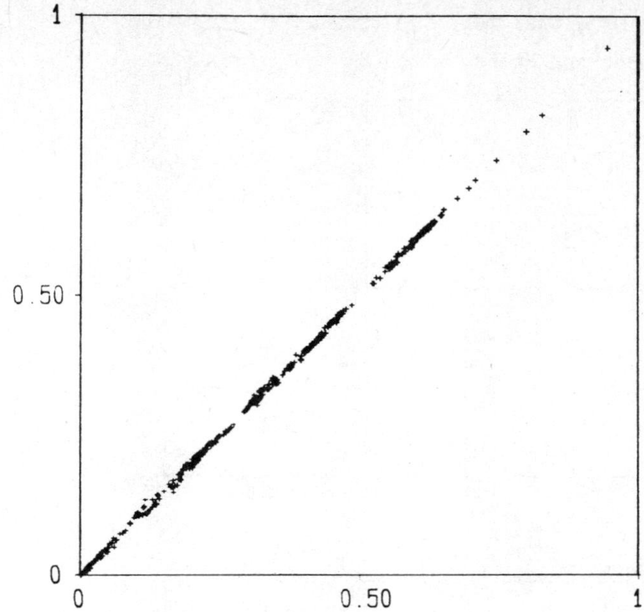

Figure 2a and 2b

Theoretical versus empirical migration matrices $w_{ji}(t)$ with non-linear estimation for a) undecomposed $\nu_{ji}(t)$, and b) for $\nu_{ji}(t) = \nu_0(t)f_{ji}$

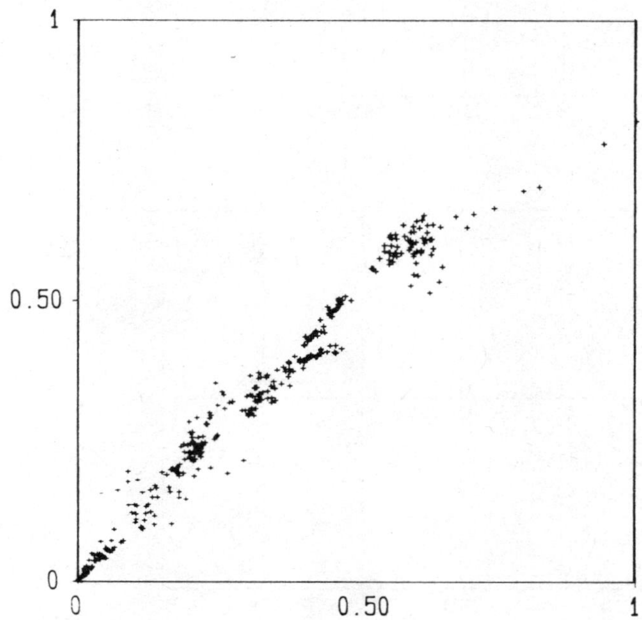

Figure 3a The global mobility (scaled to meanvalue 1) of FRG, Canada and France

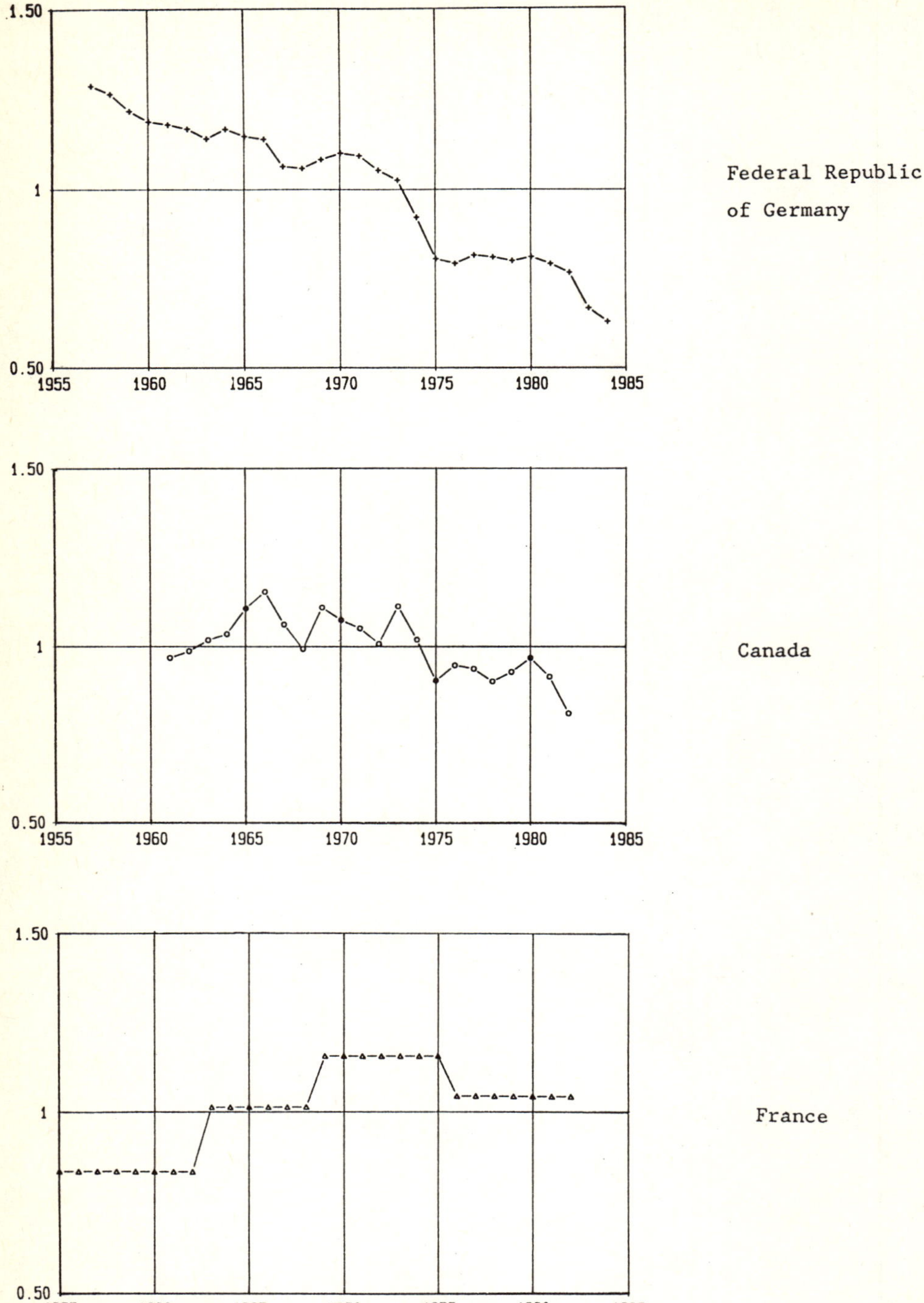

Figure 3b The global mobility (scaled to meanvalue 1) of Israel, Italy and Sweden

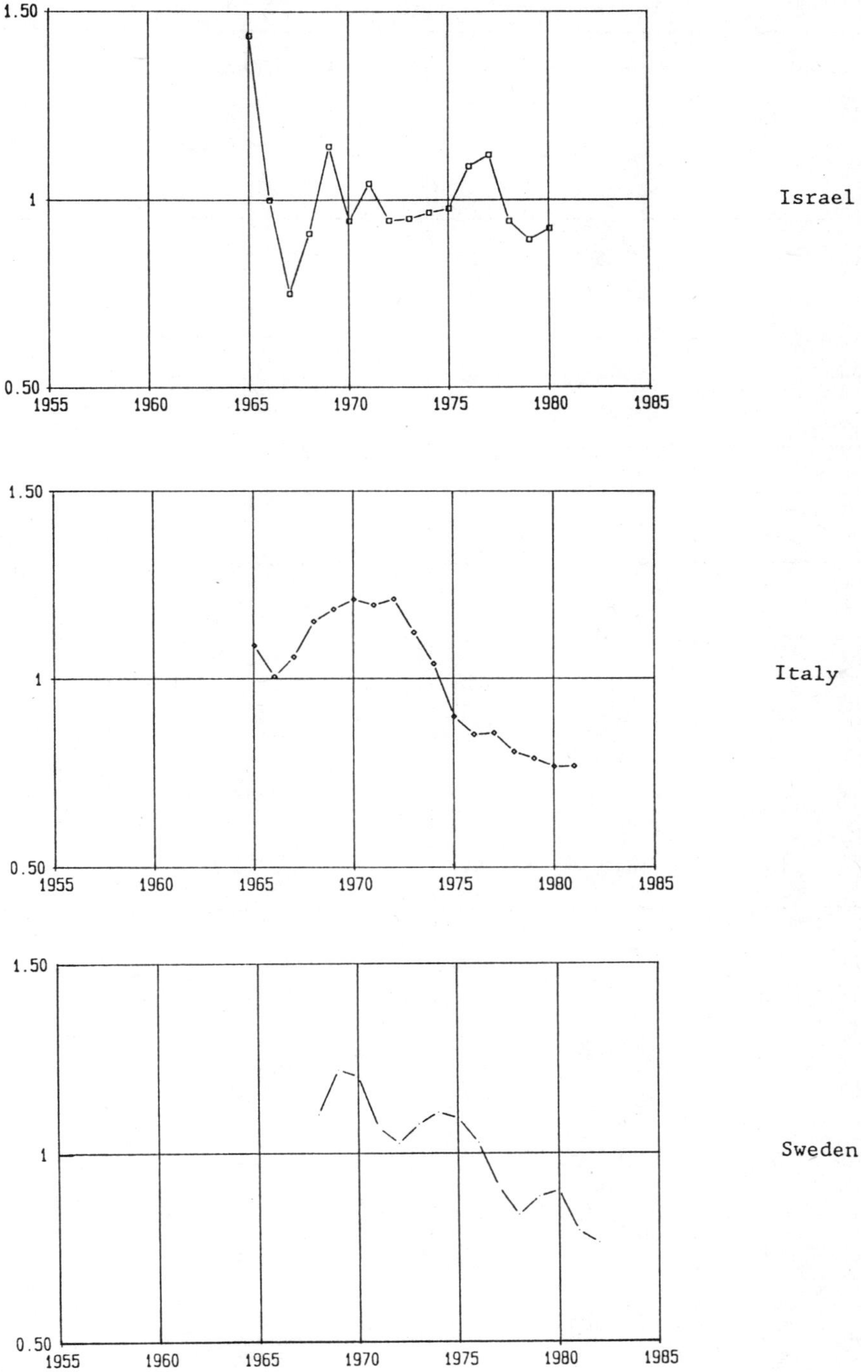

Table 1

Agglomeration parameter $\varkappa$ and saturation parameter σ for six countries

country	κ	σ	stability of homogeneous distribution (s=stable, u=unstable)	Symbols
Federal Republic of Germany	1.300 (+151)	-0.258 (-27)	u	+
Canada	1.546 (+105)	-0.313 (-15)	u	○
France	0.831 (+90)	-0.124 (-15)	u	△
Israel	1.250 (+92)	-0.179 (-16)	u	□
Italy	1.285 (+73)	-0.245 (-14)	u	◊
Sweden	1.106 (+66)	-0.295 (-5)	u	

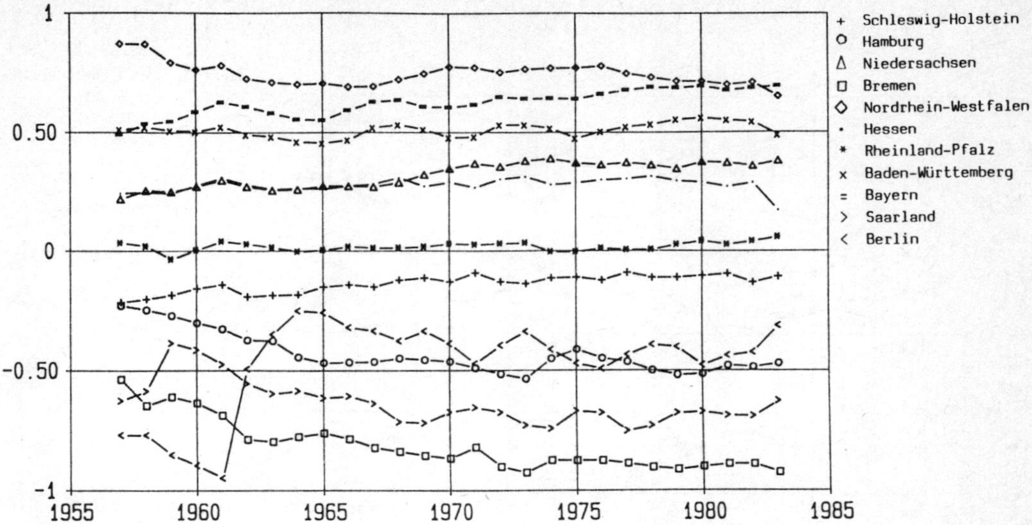

Figure 4a

The utilities of the Federal States of the FRG

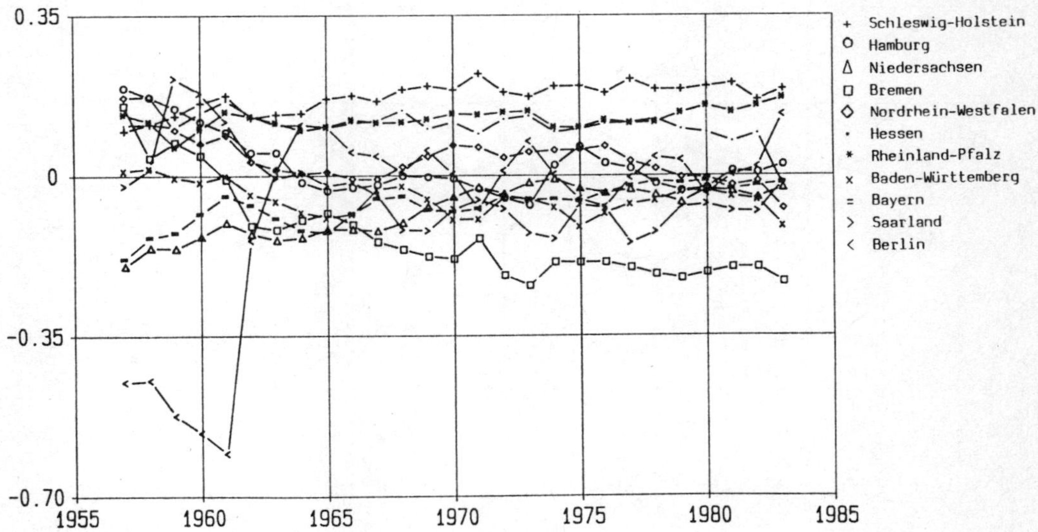

Figure 4b

The preferences of the Federal States of the FRG

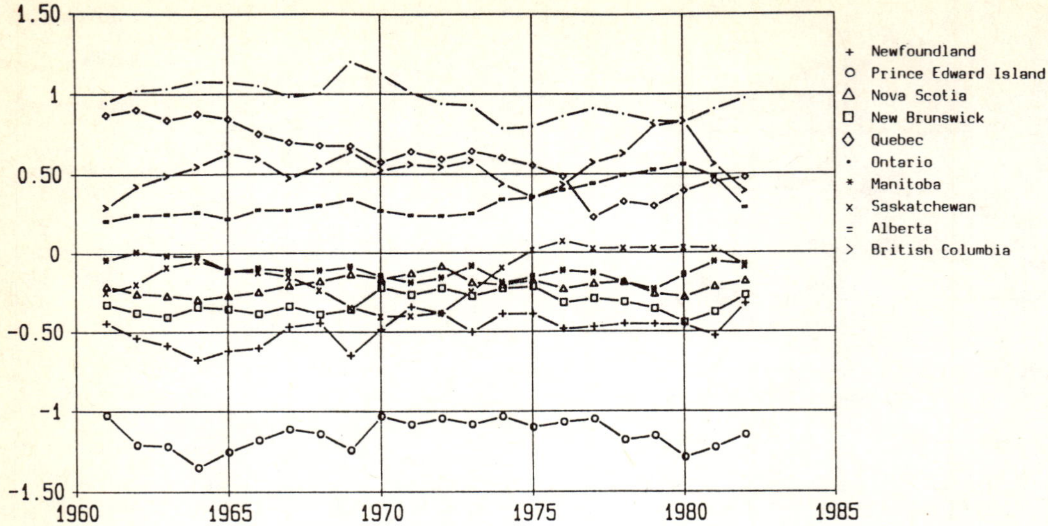

Figure 5a

The utilities of the regions of Canada

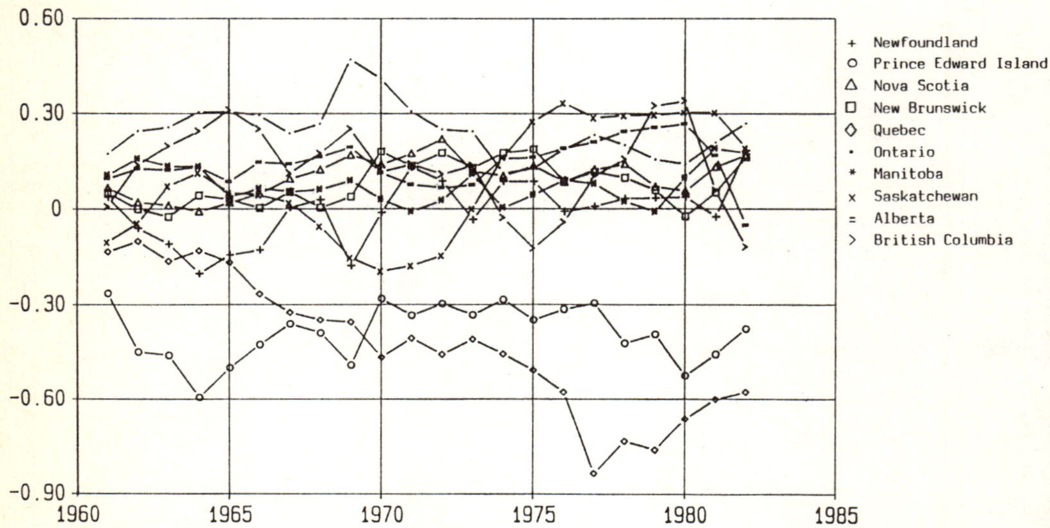

Figure 5b

The preferences of the regions of Canada

Table 2

Key - factors of the global mobility for different countries

country	1. key factor	2. key factor	3. key factor	4. key factor
Federal Republic of Germany	(LS) real income (-49.3)	(LM) vacancies (12.7)	(II) investment structure index (8.4)	(LM) employment (-7.3)
Canada	(LS) income per capita (-34.8)	(LM) employment index (6.2)	(LM) unemployment rate (2.3)	-
France	(HM) housing construction (34.5)	-	-	-
Israel	-	-	-	-
Italy	(SP) population (-86.3)	(LM) agricultural employment (-9.2)	(LS) gross national product (7.0)	-
Sweden	(HM) housing constructions (20.6)	(LM) rate of vacancies (10.7)	-	-

Table 3

Key-factors of the regional preferences for different countries

country	1. key factor	2. key factor	3. key factor	4. key factor
Federal Republic of Germany	(LS) (C) overnight stays per capita (17.7)	(II) export structure index (9.2)	(LM) unemployment rate (-5.7)	(LM) (PP) tertiary sector employment rate (5.0)
Canada	(LM) labour income per capita (12.3)	(LS) (PP) benefit payments (-5.6)	-	-
France*	(LM) unsatisfied job demand per capita (-26.2)	(HM) housing constructions per capita (18.9)	(C) temperature (15.5)	(LM) (PP) tertiary sector employment rate (11.1)
Israel	(SP) age structure (-3.2)	(LS) auto ownership (9.1)	(LS) persons per room (2.7)	-
Italy	(LM) employment index (12.4)	(HM) housing constructions (9.4)	-	-
Sweden	(LM) labour market variable (-7.8)	(PP) (LM) public sector employment (5.4)	(LM) rate of vacancies (3.2)	-

*as the fifth key-factor turned out: qualification (3.1)

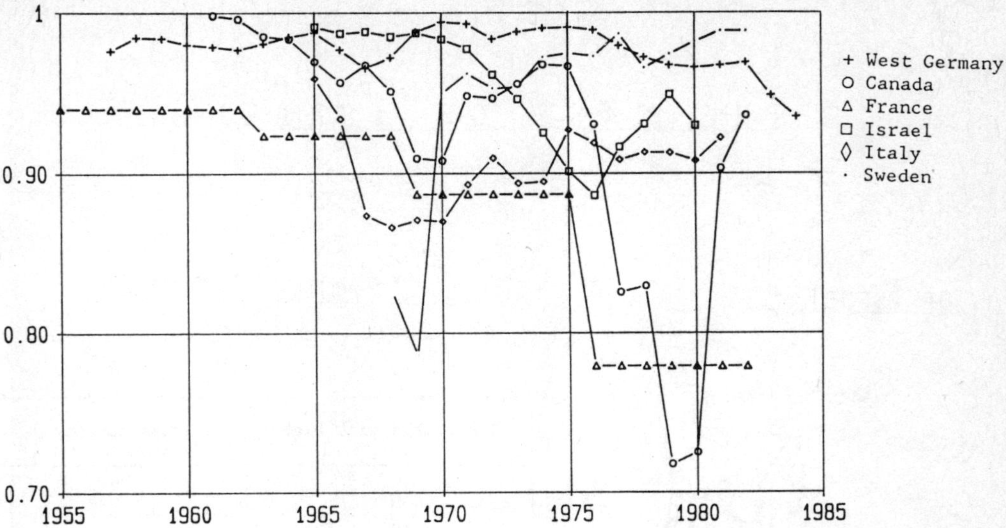

Figure 6

The correlation coefficient $r(x,\hat{x})$ for six countries

Fig. 7

THE CASE OF ISRAEL

Map of Israel

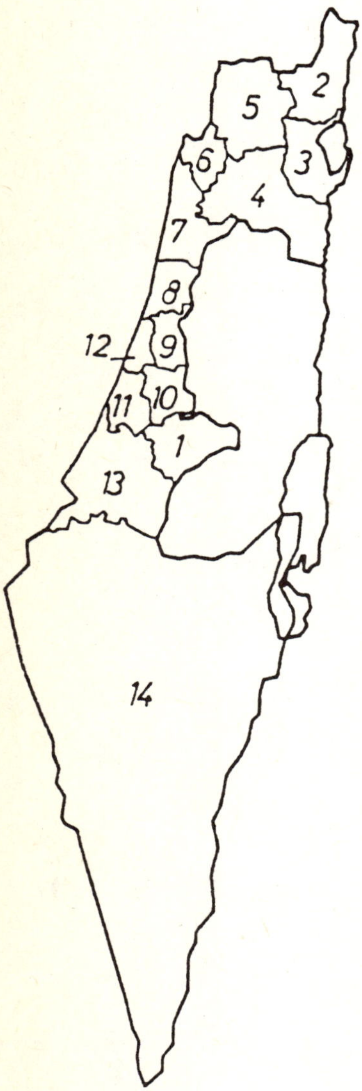

List of Regions

1. Regions with higher population density		
symbol	region	name
+	1	Jerusalem
o	6	Haifa
Δ	12	Tel Aviv
□	14	Beer Sheva

2. Regions with lower population density		
symbol	region	name
+	2	Zefat
o	3	Kinneret
Δ	4	Yizreel
□	5	Akko
◊	7	Hadera
·	8	Sharon
*	9	Petah Tiqva
x	10	Ramla
-	11	Rehovot
>	13	Ashqelon

Fig. 8

ORDER OF MAGNITUDE of the INTERNAL MIGRATION STREAMS in Israel

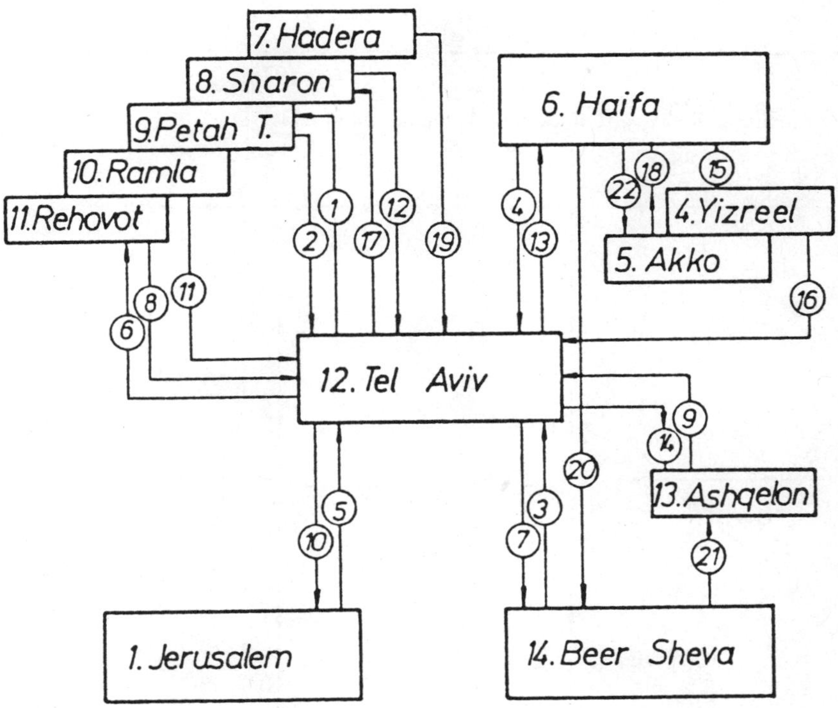

Graphical presentation of spatio-temporal stability of Israeli internal migration streams, 1965–1980

Fig. 9

The Global Mobility $v_0(t)$ of Israel

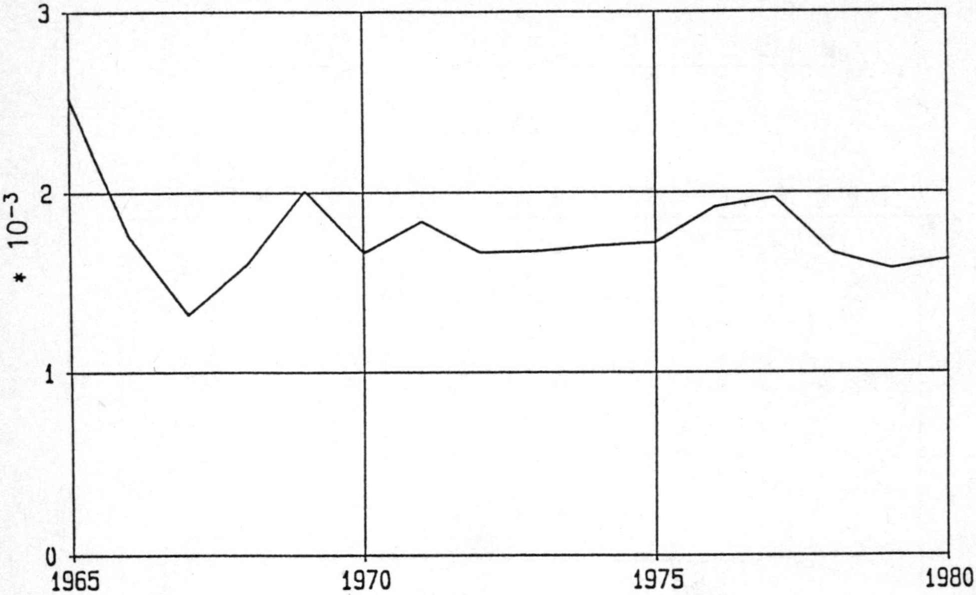

The global mobility $v_0(t)$ of the Jewish population in Israel

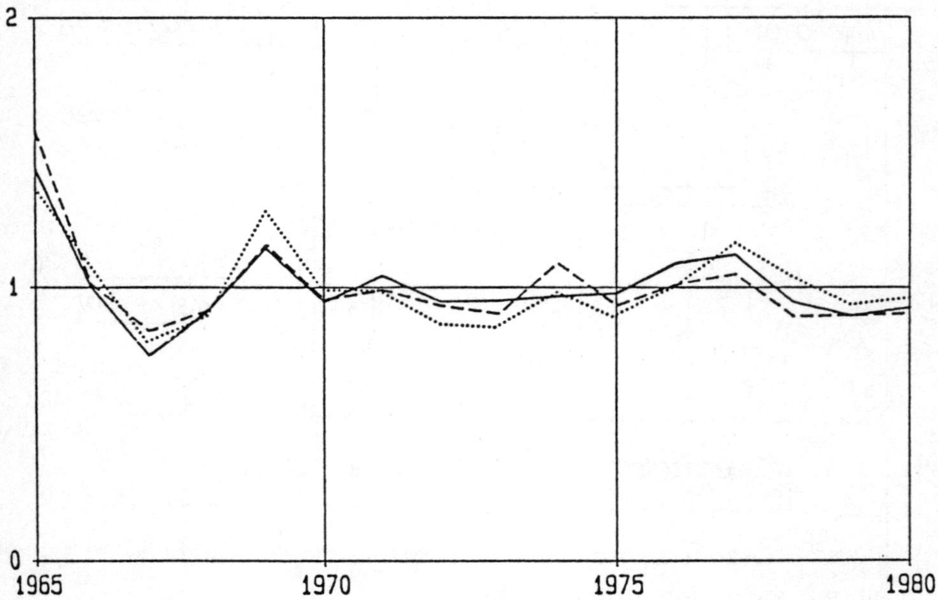

Interdependence between the global mobility $v_0(t)$ (solid line), the rate of interregional migration (dashed line) and the rate of internal migration (dotted line).

Fig. 10

The Utilities of the Regions of Israel

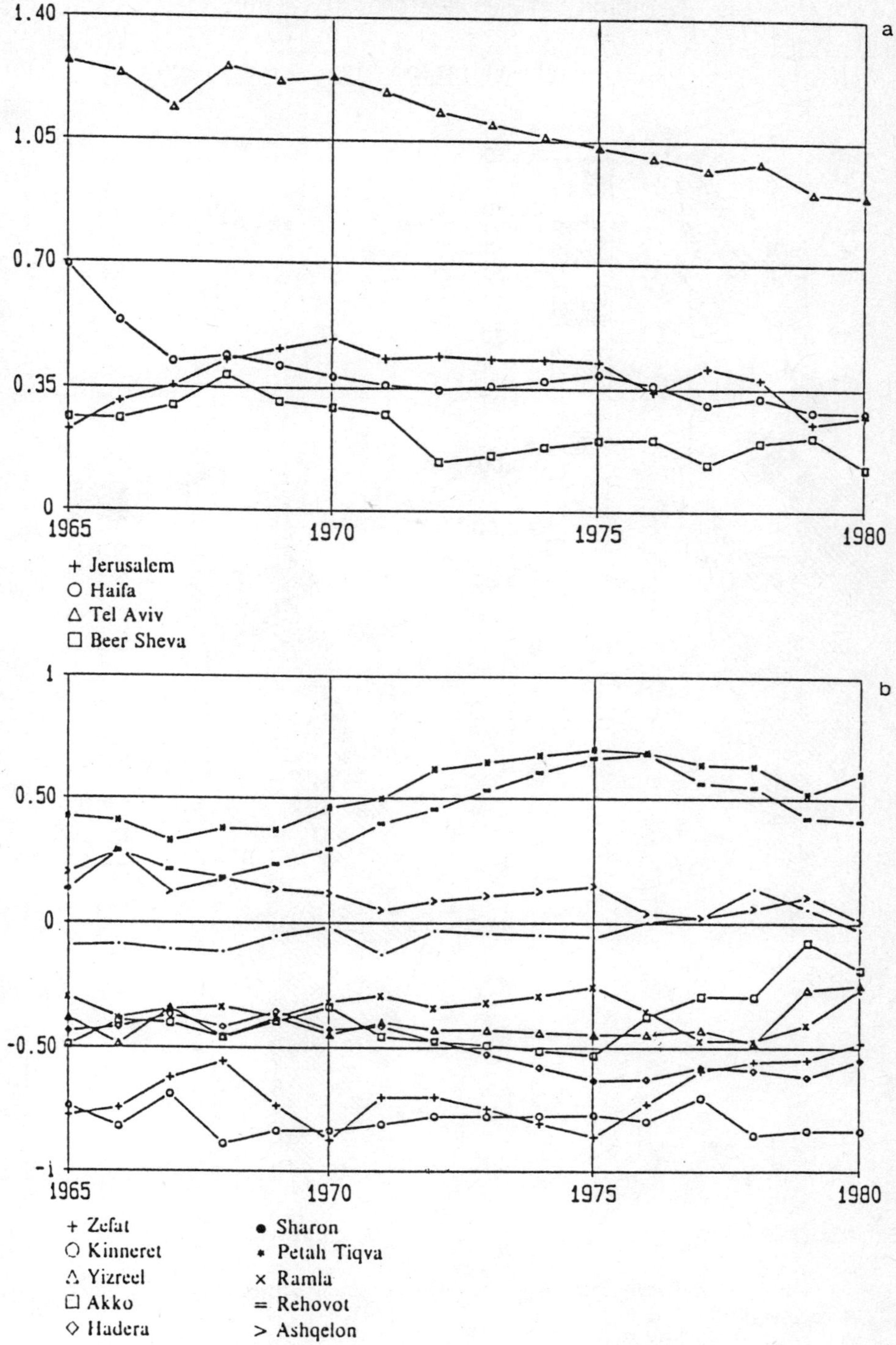

Fig. 11

Relation between Regional Utilities and Population Size

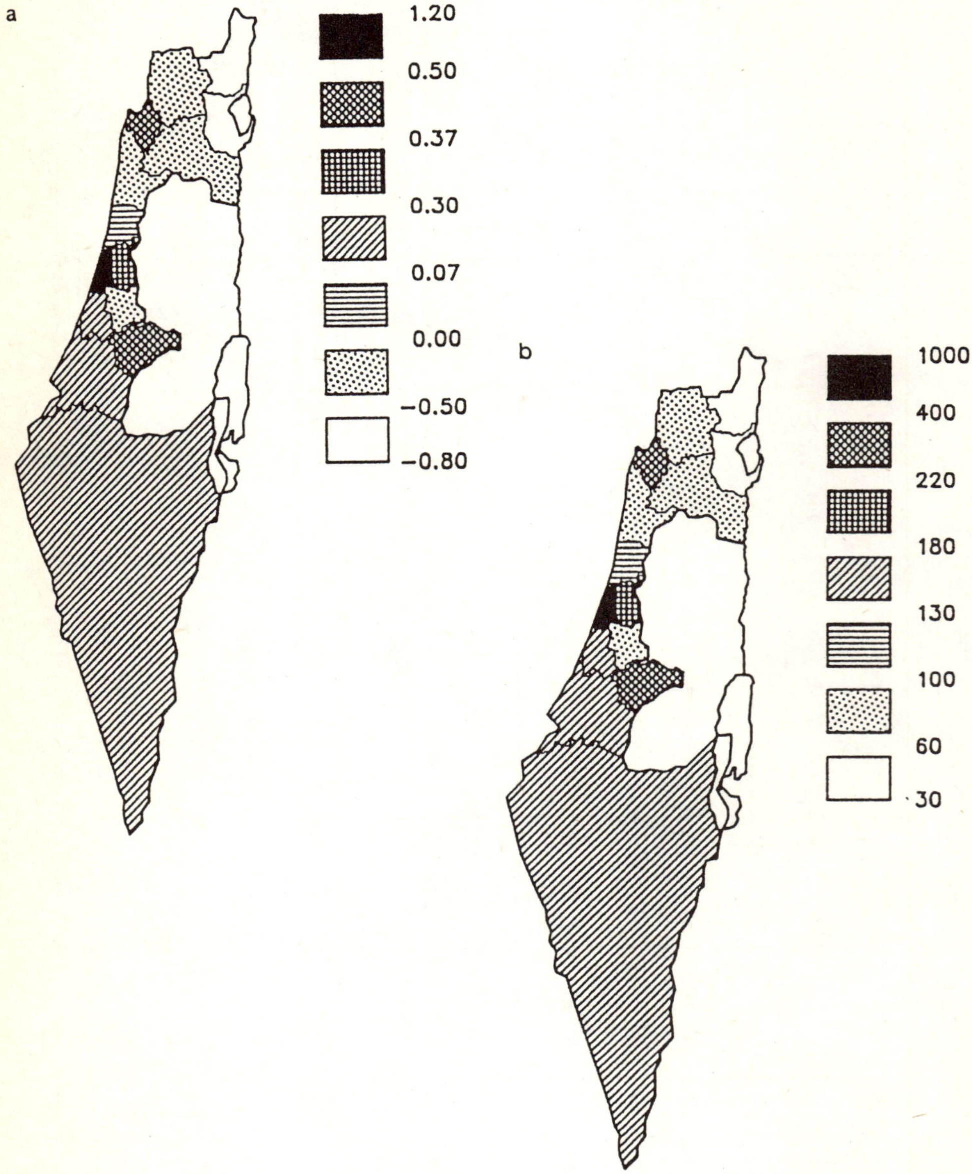

Cartographical representation of the interconnections between regional utilities (a) and population size (b)

Fig. 12

The Preferences of the Regions of Israel

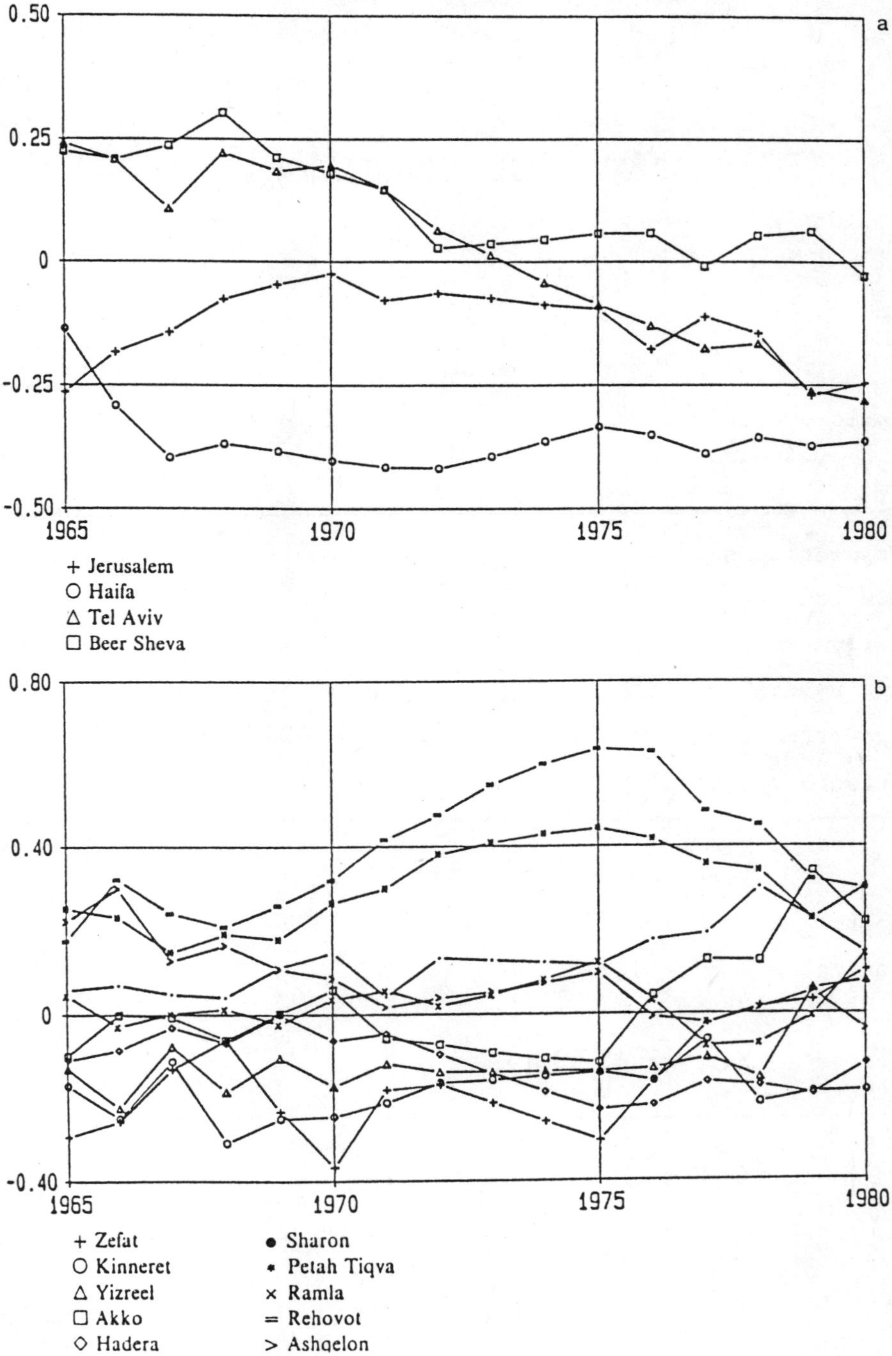

Fig. 13

KEY FACTORS

of Regional Utilities of Israel

Socio-economic key factors	Timelag	coefficient	t - value
Population size	0	0.596	21.5
squared population size	0	-0.079	-3.5
car ownership	0	1.755	9.1
age structure	0	-0.423	-3.2
persons per room	0	1.172	2.7

Correlation R^2 - 0.974
Corrected R^2 - 0.958

Fisher's test value: 60.4

Fig. 14

ROBUSTNESS OF THE EVALUATION

OF THE GLOBAL MOBILITY

a) Straight Line:

Global Mobility estimated from All Interregional Migration Streams $w_{ij}(t)$

b) Dotted Line:

Global Mobility estimated from Regional In- and Out-Migration only

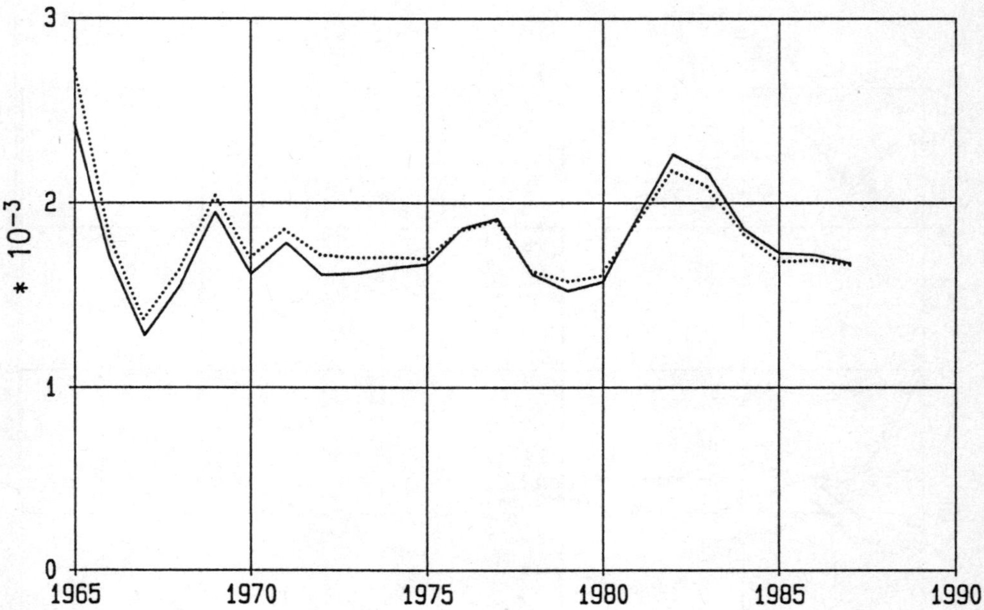

Fig. 15

ROBUSTNESS OF THE EVALUATION OF THE REGIONAL UTILITIES

a) Lines with Symbols:

Utilities estimated from All Interregional Migration Streams $w_{ij}(t)$

b) Dotted Lines:

Utilities estimated from Regional In- and Out-Migration only

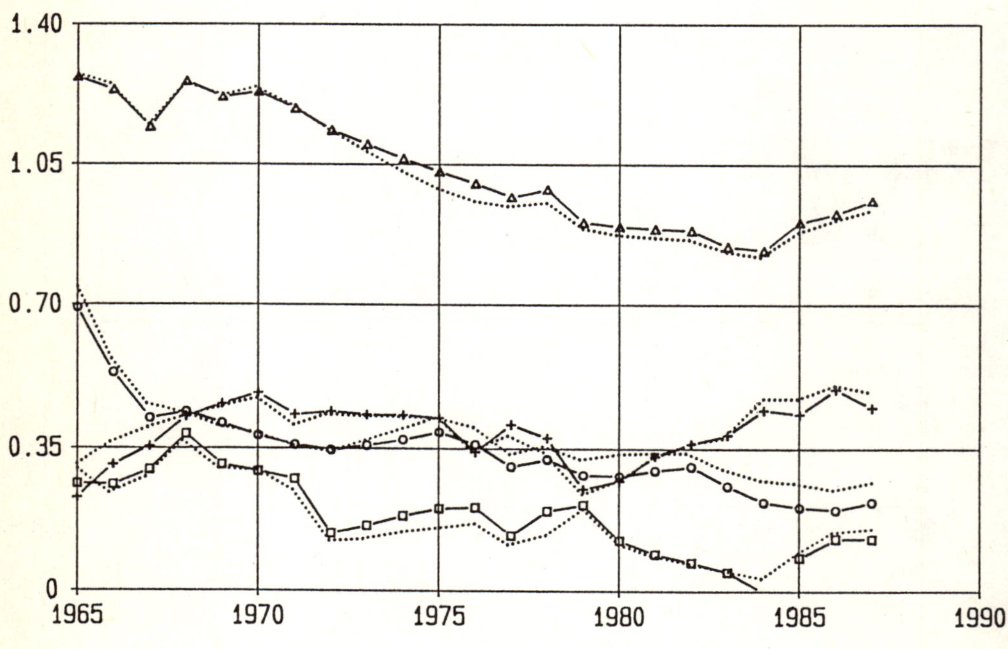

Part D. ADVANCES IN REGIONAL SCIENCE METHODS

Chapter 16

QUANTITATIVE METHODS IN REGIONAL SCIENCE: PERSPECTIVES ON RESEARCH DIRECTIONS

L. Anselin

1. Introduction

The use of formal mathematical concepts and expressions, and the application of statistics, optimization and other quantitative methods constitutes the most common characteristic of research in regional science as it has become known over the past thirty years. To address the current state of the art and to outline potential future developments for quantitative methods in regional science would therefore be a rather formidable task. In order to focus the discussion, and to limit the overlap with the other papers in the series that address substantive research areas (e.g., spatial interaction and migration, location, regional development, energy and the environment), I will limit the scope of my remarks to two specific areas. I will focus on operational models of the urban and regional economy, and on methods for the analysis of spatial data. Both of these are areas in which recently some important new developments have come to the fore. In addition, these are also areas in which the quantitative methods have made the transition into the practice of planners and policy makers. Even though my choice of topic is admittedly narrow, it allows a focus on those methods that have emphasized and dealt with the central role of space in regional science models. Arguably, this is less the case with some other, and equally important sets of techniques, such as optimization methods and decision theory. In addition to these conceptual motivations, there is also a practical constraint on the length of the paper, which precluded me from covering a more comprehensive range of topics.

After a summary review of the state of the art in general and the progress achieved during the last five years in particular, the emphasis of my comments will be on identifying important shortcomings and on outlining some potentially fruitful directions for future work. Clearly, these remarks do not claim to be comprehensive and reflect my personal biases about what holds the most promise. The common theme for the discussion is my perception of growing discrepancies between theory and practice, and between methods and data. These discrepancies are becoming more obvious with the recent acceleration of advances in computing technology, and its greatly increased accessibility. In my opinion, these issues form an important challenge to the overall societal relevance of quantitative methods in regional science.

2. Where We Are: A Summary Review of the State of the Art

In this section, I will formulate some remarks on the application of quantitative methods to models of the macro-economy of cities and regions and to the analysis of spatially referenced data. For the former, the particular emphasis will be on general economic models, i.e., models that integrate a range of actors and markets. I will therefore not deal with specific urban markets, such as urban housing and transportation markets (for recent reviews of issues in urban modeling, see, e.g., Anas, 1987; Kain, 1987), nor with specialized demographic, spatial interaction, or other models which are treated in other papers in this series. For the analysis of spatial data, I will focus on some recent advances in the fields of spatial econometrics and spatial statistics.

The current state of the art in regional modeling has been assessed in a number of publications, e.g., in Nijkamp, Rietveld and Snickars (1986), Batey and Madden (1986), and Harrigan and McGregor (1988a). Similarly, the progress in spatial data analysis is reviewed in Anselin and Griffith (1988), Anselin (1988a), Griffith (1988), and Odland, Golledge and Rogerson (1989). It is not my intent to present still another review of the literature. Instead I will focus on a limited number of salient issues in areas where the application of quantitative methods in regional science has achieved important progress, and particularly where significant advances have occurred in recent years.

2.1 From regional accounts to regional social accounting matrices

In many of the early approaches to regional modeling in the 1960s and 1970s, there was an underlying assumption that a model had to be based on a solid and consistent system of regional accounts (e.g., as typified by the work of Czamanski, 1973). More recently, this is rarely mentioned any more, and there even is some evidence of an overall wariness with the idea, in light of the tremendous data limitations encountered in practice (e.g., the lack of data on capital flows or on detailed personal consumption expenditures at the regional level, and problems with the definition of regional closure).

Instead, the research emphasis has shifted to the application of a more encompassing social accounting matrix (SAM) to regional economies, partly as an extension of input-output analysis (see, e.g., the review by Batey and Rose, 1988), and partly as a line of inquiry in its own right. The latter is represented most recently by the work of Round (1986, 1988), and Sonis and Hewings (1988a) (see also the discussion in Stone and Weale, 1986).

The SAM approach allows for a detailed description of the interaction between sectors in the regional economy in the broadest sense, i.e., between firms, factors of production, and institutions, and often has a particularly detailed focus on the role of different income classes. In addition to providing this detailed description, the SAM also allows for some interesting decompositions of various types of multipliers, e.g., into components related to internal linkage effects (within one region), open-loop and closed-loop interregional feedbacks, as well as into various functional components (between production sectors, factors and institutions).

The application of a SAM at the regional scale, and particularly at the multiregional scale is not without problems, however. Clearly, the SAM framework is extremely data hungry and the necessary information is often not available at the regional scale. There are problems with sectoral classification and in particular with the designation of income classes to track distributional impacts. In a multiregional model, there is a difficulty with incorporating transfer costs and factor mobility in a consistent fashion, issues which are still largely unresolved to date. Furthermore, and maybe most importantly, the transition from an accounting system to a 'model' is not without problems, and is typically based on very restrictive assumptions of fixed proportional relationships. To date, the application of SAM frameworks in regional modeling has therefore been mostly limited to 'large' regions in developing countries, and to theoretical discussions of measurement issues and multiplier decompositions.

2.2 From regional input-output to computable general equilibrium models

The application of Leontief's input-output approach at the regional scale has come a long way since its initial popularization in Isard's Methods of Regional Analysis (1960). Recent reviews of this progress can be found in, e.g., Richardson (1985), Hewings and Jensen (1986, 1988), Miller, Polenske and Rose (1989), and Jensen (1988).

An early focus of research on regional input-output analysis was on the methodology of regionalization (e.g., survey vs. non-survey, hybrid or short-cut techniques, the 'reconciliation' debates), and its extension to multiple regions (interregional vs. multiregional) and multiple time periods (dynamic I/O models). Also, substantial attention was paid to the incorporation of energy sectors and environmental effects, and to methodological issues associated with the application of activity analysis.

Many of these issues have been more or less resolved, or, maybe more precisely, have resulted in a consensus of informed opinion on what approaches are most suitable in particular situations. In addition, most of the 'accepted' techniques have found their way into the practice of planning and consulting, facilitated by the availability of software (e.g., the discussion in a recent issue of the Review of Regional Studies, and particularly in Brucker, Hastings and Latham, 1987).

On the other hand, some recent research attention has started to focus again on issues associated with the input output modeling framework as such. Various extentions of the model have been suggested, e.g., to take into account labor markets and migration (e.g., Batey, 1985), or income distributional effects (Rose and Beaumont, 1988). Also, the role of a supply side perspective in input-output analysis has been re-examined, and less rigid frameworks have been suggested (e.g., Bon, 1988; Oosterhaven, 1988). These extensions are similar to and often intertwined with the development of a SAM approach alluded to earlier.

More interestingly, in recent years some fundamental issues related to the input output model in a regional context have begun to be addressed (again) as well. Three areas in particular seem to be receiving increased attention: measurement and errors; the use of the input output framework as a tool to understand regional development; and the development of I/O into more encompassing Computable General Equilibrium (CGE) models.

A focus on the structure of the linkages in the model (e.g., Szyrmer, 1985) and the associated measurement errors and error propagation has generated a renewed interest in probabilistic approaches to input-output modeling (e.g., West, 1986; Jackson, 1986). It also has led to the introduction of new concepts, such as 'fields of influence', and summary measures of the most important sectors, such as 'inverse important coefficients' (e.g., Sonis and Hewings, 1988b). This has obvious implications for empirical work on updating and constructing regional input-output tables, but it has also been extended to make a link with regional development theory in general. Here, the notion of a 'fundamental economic structure' (FES) of Jensen, West and Hewings (1988) may open up an interesting new path of analysis.

A more fundamental development is the extension of the input output model into a Computable General Equilibrium framework (or, alternatively, the consideration of I/O as a limiting case of a CGE). The CGE approach has received considerable interest as a tool to analyze national public policy (taxation) and international trade, but much less attention as a model for a regional economy (e.g., recent discussions in Higgs, Parmenter, Rimmer, 1988; Spencer, 1988; and Harrigan and McGregor, 1988b). Although conceptually more encompassing than the input output model, the CGE is difficult to implement in practice at the regional scale, due to the emphasis on prices and the problems in dealing with factor mobility. Nevertheless, a CGE provides a powerful framework to carry out simulations and other numerical exercises and merits more attention from researchers in regional science.

2.3 From reduced forms to structural regional econometric models

The econometric modeling of a regional macro economy is another instance where the application of quantitative methods in regional science has found widespread acceptance in the practice of planning and policy analysis. The early limited efforts have evolved into ambitious and detailed structural models for a wide range of regions and metropolitan areas, as recently reviewed in Bolton (1985) and Bennett and Hordijk (1986).

For the most part, regional econometric models are the poor cousins of their national counterparts, due to data limitations, but also because of a lack of a distinct 'regional' growth theory that can be implemented empirically. Indeed, data on capital and related variables that are the focus of recent developments in the theory of regional dynamics are hardly available at the regional or metropolitan scale. Most regional econometric models are Keynesian in nature and consist of extensions of the economic base principle, although a few are grounded in neo-classical ideas. All too often however, operational models tend to be specified in an ad hoc fashion, in function of available data.

Overall, the 'theory' of regional econometric model building seems to be stagnating somewhat. The interest in model error and model validation in the early 1980s (e.g., the work of Taylor, 1982a, 1982b) has not led to a more critical approach to model specification or to a more careful assessment of predictive performance. Also, the 'method' used in these models tends to lag behind theoretical developments in the econometrics of estimation and forecasting. Most models are still of the standard structural form and are estimated by means of ordinary least squares, in spite of well known problems with this approach, both conceptual as well as methodological (see, e.g., Anselin, 1988b). Few alternative techniques from 'national' modeling, such as ARIMA modeling or vector autoregressive approaches have been introduced at the regional scale (for an exception, see Kinal and Ratner, 1986).

2.4 From single region - single model approaches to multiregion integrated frameworks

Since the path breaking suggestions for integrating various regional science models were formulated in Isard's (1960) 'Channels of Synthesis', many efforts have been made towards the further development and application of these ideas. Integrated or linked models have ranged from fairly straight-forward two- or three-model combinations for a single region, to more ambitious encompassing multiregion and multi-model frameworks (for a review and discussion, see, e.g., Isard and Anselin, 1982; Isard, 1986; Batey and Madden, 1986); Anselin and Madden, 1990).

A precise definition of an 'integrated' model does not really exist. The integration can pertain to the combination of different methodologies (e.g., input output and econometric), to the consideration of different substantive areas (e.g., economics, demography and ecology), or even to the combination of national and regional models at various spatial scales. From a methodological standpoint two different approaches to integration have been developed (e.g., the discussion in Wegener, 1986; and Anselin, 1988c). On the one hand, all parts can be encompassed within a single overall model, with a unifying methodology, the so-called 'embedding' or 'unified' approach (e.g., an entropy model for production and shipment of goods). Alternatively, separate modules can be linked in terms of their inputs and outputs, in the sense that results from one module can be used as inputs into other modules. This is the so-called 'linking', 'modular', or 'composite' approach. Typically, the linkage is not implemented at the model estimation and calibration stage, nor is it necessarily simultaneous.

Several partial attempts at the integration of models have been implemented in the 1980s, primarily between econometric and input output models. However, a complete and operational application of the more ambitious framework does not yet exist. In part, this is due to the lack of data, but it is also a result of the lack of a clear solution to more conceptual problems on how to deal with built-in redundancies, and with issues of convergence and consistency. More recently, there has been a renewed vigor in the research associated with such large scale models, in part stimulated by the availability of supercomputing technology and parallel processors (e.g., Boyce, 1988).

2.5 From spatial autocorrelation indices to a geographical analysis machine

A great majority of empirical work in regional science deals with spatially referenced data, i.e., with information associated with points, lines or areas in space. This organization often leads to spatial dependence and spatial heterogeneity, which invalidate many of the properties of standard statistical and econometric methods. As pointed out in the recent review by Anselin and Griffith (1988), the early concern with testing for spatial autocorrelation and spatial pattern (e.g., Cliff and Ord, 1973; Getis and Boots, 1978; Paelinck and Klaassen, 1979) has evolved into a body of methods to deal with a wide range of spatial effects and spatial process models, encompassed in the specialized fields of spatial statistics and spatial econometrics (e.g., reviewed in Anselin 1988a; Griffith, 1988; Odland, Golledge and Rogerson, 1989).

In recent years the research in these fields has started to deal with some of the complexities associated with realistic spatial data: new estimators that transcend the traditional maximum likelihood framework (e.g., nonlinear approaches, instrumental variables techniques and the bootstrap); diagnostics for spatial effects based on standard regression results (e.g., tests based on the Lagrange Multiplier principle, or on the randomization approach); robust estimators and tests (e.g., based on permutation and randomization); techniques to deal with limited dependent, categorical and latent variables (e.g., spatial LISREL, Poisson regression models); spatial perspectives on model fit and outliers (e.g., adjusted measures of fit and leverage); and a general interest in specification testing and model validation.

However, this flurry of results on methods and techniques has not been accompanied by a wide dissemination into the practice of empirical research. To a large extent, this is due to the lack of readily available software that incorporates the spatial tests and estimators. Currently, none of the popular statistical or econometric packages for either mainframe or personal computers include any techniques for spatial analysis. The same holds to a large extent for the commercial geographic information systems (GIS), which have seen an explosive growth in the last few years. It should be worrisome to regional scientists that the implementations of this new technology, which according to some has the potential to develop the ideal model of spatial information, lack features to carry out any but the most rudimentary forms of spatial analysis (e.g., Goodchild, 1987; Openshaw, 1987).

Efforts to counter this sorry state of affairs have started to develop. For example, in a collaborative effort between Anselin and Griffith, a number of macros and specialized software are being written to carry out spatial analysis within standard statistical and econometric packages (e.g., Griffith, 1989; Anselin, 1989). Also, the current popularity and commercial success of GIS has led to a call for increased academic research in this area, e.g., as exemplified by the institution of national and regional research centers in the U.S. and U.K. (e.g., Abler, 1987).

An interesting recent development in this respect is the creation of a so-called 'geographical analysis machine' (GAM), as a mixture of a GIS, a spatial statistical analysis and expert system to carry out an automated spatial data analysis (Openshaw, et al., 1987; Openshaw, 1988). In its current form, the GAM is still much too rudimentary and specialized to gain widespread acceptance or to make a convincing case for its superiority to other approaches. Although one can take issue with this form of a-theoretical spatial analysis, the idea of a GAM constitutes a fascinating development which merits serious attention as a means to stimulate further progress in research on the relation between spatial theory and spatial methods.

3. Problems, Limitations and Challenges

3.1 Fundamental theoretical framework

In terms of the fundamental theoretical framework that underlies the application of quantitative methods in regional science, a number of dichotomies and apparent conflicts can be distinguished in the current state-of-the-art. Six of these come to mind readily: equilibrium vs. disequilibrium; statics vs. dynamics; micro vs. macro; model driven vs. data driven analysis; linear vs. nonlinear specifications; and single model vs. integrated model.

The input output methodology and its developments (SAM, CGE) are based on the concept of a long run general equilibrium. In contrast, most regional econometric models do not portray a general equilibrium, but a partial equilibrium at best. It is hard to imagine that the traditional notion of equilibrium would be valid for the small open economy that constitutes a region. Nevertheless, the implementation of an appropriate notion of regional or spatial disequilibrium is a much ignored topic. This aside, the general equilibrium approach in regional models is not without problems in and of itself. Although important theoretical advances have been made in the analysis of a spatial price equilibrium based on transportation networks, these are typically ignored in the portrayal and specification of interregional flows in regional models.

The input output model is static and focuses on the long run, whereas the econometric models are dynamic, and aimed at the short run. In essence these two are irreconciable, and all but the most simplistic efforts to link an econometric model with an input output framework run into formidable problems in this respect. Most solutions that have been suggested for this are ad hoc adjustment processes, without a formal conceptual framework for the timing of the various feed-backs and multiplier effects in the model.

A further major dichotomy is the choice between a micro and a macro approach, or the determination of the proper behavioral unit of analysis. This is not unique to regional science, but in a spatial context it takes on an additional aspect, in that the relevant spatial scale at which the behavioral unit operates needs to be specified. This interaction between individual spatial behavior and the region as an organizing principle can be tackled by means of micro-simulation approaches, although much remains to be done to yield satisfactory and realistic results.

Two opposite approaches are also evident in spatial data analysis. One is model-driven and lets spatial theory determine which specifications need to be empirically validated (e.g., spatial econometrics). The other is data driven and is geared towards elucidating theory from the data (e.g., spatial statistics). Similar techniques have been developed in both approaches, with very little cross-reference between the two (Anselin, 1986). In addition to the loss in efficiency from such parallel developments, is should be clear that the two approaches are not mutually exclusive and competitive, but complementary. With the recent gain in power and popularity of GIS, and a growing acceptance of the viewpoint of 'letting the data speak for themselves' (e.g., in the GAM alluded to earlier) it is crucial to keep an important role for spatial theory in spatial data analysis.

Finally, two issues are more of a technical nature. Most regional models are (still) primarily linear specifications, while nonlinearities prevail in reality. Models that include nonlinear specifications (or still better, chaotic change and bifurcations) are still very aggregate and mostly conceptual. Similarly, most current operational models are single region and not of an integrated structure. The extent to which the multiregional and/or integrated approach indeed does provide a superior framework remains a largely unresolved issue.

3.2 Measurement, spatial and sectoral scale and aggregation

Most data used with quantitative methods in regional science is (still) notoriously bad: the scope of available information is limited, it is often not collected in a consistent fashion, and is only loosely connected to the concepts that underlie regional theory. Unfortunately, this data is the only basis for the operational implementation of the various models.

The data problem in regional modeling has led to a large amount of effort being spent on constructing 'observations' at the regional level for variables that are only observed at the national or supra-regional scale. Examples are the use of location quotients and purchase coefficient to regionalize input output tables, and the use of artificial or synthetic regional sectoral output measures in econometric models (e.g., by means of the Kendrick-Jaycox and similar approaches). However, the artificial nature of the 'data' is typically ignored in the estimation or calibration of the model. Moreover, there is often no way to assess the extent to which this 'data' corresponds with reality (e.g., the problem of comparing various non-survey methods to construct regional input-output tables).

In addition to these well-known issues, there are measurement problems that follow from the nature of the models themselves, and in particular from the sectoral and spatial organization that they impose on the analysis. It is somewhat ironic that 'regional' models essentially ignore the dynamics of the region, both over time as well as across space. There is little connection between the administrative regions for which data are collected or constructed and the spatial market and supply areas that are supposed to be reflected in the models. Ideally, a GIS with data available at a disaggregate level of organization will allow the determination of the 'region' to become endogenous to the analysis. However, in addition to technical issues related to the organization of such a data base (e.g., the polygon overlay problem and spatial interpolation issues), legal, ethical and cost considerations are likely to keep that ideal state from being attained for quite some time to come.

There is also little relation between the various sectoral classifications that are used in regional models, and the reality of business and political decision making. For example, corporate structure and re-structuring, horizontal and vertical integration, or the institutional complexities of locational decision making are poorly implemented in current operational models. Furthermore, there is no obvious mechanism to incorporate the development and diffusion of new technologies and products, or the change in political priorities and laws. These aspects of the current state-of-the-art of regional modeling severely limit their use as tools to analyze the effects of realistic policy problems.

A final measurement issue pertains to the extent to which regional differentiation or regional heterogeneity can or should be incorporated in the models. For example, the well known TFS approach to regional modeling (Treyz and Stevens, 1985), which has also had considerable commercial application imposes the assumption of homogeneity on the regional system. In other words, relations between purchase coefficients and various explanatory variables (e.g., area, population) that are observed for one set of regions (e.g., U.S. states) are assumed to hold for all regions (e.g., for each county to which the model may be applied). In reality, such an assumption may be hard to maintain, as indicated by a substantial body of empirical work of estimates on labor market models for various regions. The implication of reigonal heterogeneity should not be taken lightly: if each region needs its own unique set of variables and relations, what will remain to be explained by a 'general' regional theory?

3.3 Estimation methodology and model validation

The poor quality of the data is mostly ignored in the practice of estimating or calibrating models in regional science. The incompatibility between the spatial units of observation and the spatial scale of the various phenomena that are measured is likely to result in spatial dependence and/or spatial heterogeneity. These spatial effects lead to errors in the statistical inference, misleading indications of model validity, and other problems. As pointed out earlier, they necessitate the application of the specialized methodology of spatial statistics and spatial econometrics.

This methodology is far from having reached a satisfactory state of the art, and several important problems remain unresolved. A major issue is related to the specification of the structure of spatial dependence (the spatial weights problem), and, to a lesser extent, of spatial heterogeneity. The spatial weights matrix which is used in most analyses of spatial dependence introduces a degree of arbitrariness in the modeling process, since it incorporates a prior notion of dependence which may be inappropriate. Tests and estimators are clearly sensitive to the choice of this weight matrix, but very little progress has been made towards determining unambiguous criteria for its specification.

In addition, most methods of spatial econometrics and spatial statistics are geared towards a model of normality in the data (or error term), and are not very robust to deviations from this assumption. They also tend to focus on linear regression models. Methods that explicitly take into account spatial dependence in models with limited dependent variables or count data remain to be developed. Similarly, the design of methods to deal with space-time processes is still in its infancy. It is mostly limited to the application of a time series perspective to spatial data, which is not always appropriate due to the two-dimensional nature of the dependence and the lack of homogeneity.

Spatially referenced data are also subject to edge or boundary effects, which are considerably more complex than in time series. Although tests and estimators have been shown to be sensitive to these effects, a satisfactory method to deal with them has yet to be found. The scope for application of spatial statistics and spatial econometrics is limited by the size of realistic data sets as well. When a small number of observations is available, a conceptual problem results. Although practical data sets in regional science are mostly of small to moderate size, most formal properties of tests and estimators are based on asymptotic derivations, i.e., for infinitely large data sets. These asymptotic properties often no longer hold in finite samples, and for many situations encountered in practice the proper adjustments and techniques have not yet been derived. On the other hand, for large data sets numerical considerations become limiting, since most estimators are based on a non-linear optimization, in which a matrix of dimension equal to the number of observations needs to be inverted at each iteration. Although considerably less stringent, even on supercomputers there is still a limit on the size of matrices for which this can be satisfactorily carried out. In the more mundane micro computer context this numerical problem severely limits the types of spatial process models that can be analyzed.

In addition to these more technical issues, 'errors in variables' are the rule rather than the exception in regional science. This aspect of the data is also typically ignored in the estimation and calibration of the models. Again, this may lead to the wrong conclusions, but more importantly, it projects a false sense of precision which is not matched by the quality of the data or the models. This is aggravated by the almost total lack of rigorous model validation in the practice of the use of quantitative methods in regional science. For example, the presence of spatial dependence and/or spatial heterogeneity invalidates the indications of model fit or forecast error that are based on standard (a-spatial) methods. Some specialized spatial measures of fit have been developed, but much remains to be done. This applies also in the use of GIS, where means need to be found to adequately portray the error in not only the models, but in the data as well.

3.4 Use in practice: buyer beware!

Finally, it is a rather lamentable state of affairs that new perspectives on quantitative methods have only very slowly made the transition to the practice of regional planning. This practice is still too often characterized by the uncritical application of canned programs or simplistic approaches, without much insight into the limitations and reliability of data, models and projections. The recent proliferation of 'user-friendly' microcomputer software and GIS packages has provided the practitioner with a feeling of power of quantitative analysis, which is often much beyond the reliability of the underlying methods.

In addition, quantitative methods are undergoing some sort of credibility crisis in the real world of planning practice. When each party to a conflict has its own expert with associated model to prove its point, there is often a reaction to reject the quantitative results altogether. The lack of importance that is given in public hearings and planning commissions to projections and forecasts of impacts that follow from regional models is not something that the field of regional science can be proud of.

There is a large agenda of research left before the quantitative methods can actually deliver what they promise. It is crucial for the social relevance of the discipline that it continues to make progress in this direction.

4. Possible Directions for Future Research

4.1 Integration of theory and models

Further significant progress in the development of quantitative methods in regional science is conditional upon advances in spatial theory. However, this alone is not enough. It is also imperative that the theory be successfully incorporated into the models and techniques. For example, models of the economy of cities and regions should reflect insights into the spatial and locational behavior of individuals, households and firms, and include the interface with policy and politics. The concept of a region should not be an arbitrary framework for the collection of data, but should become an endogenous part of the models. Similarly, space-time dynamics should be part of the structure of the models, and not an artifact of data collection.

Maybe the progress in the GIS technology will facilitate such advances in the theoretical validity of the methods of regional analysis. However, totally new theoretical perspectives may be necessary as well. The recent attention to chaotic change, fractals and structural dynamics, a focus on disequilibrium modeling, and the interest in computational process modeling and artificial intelligence may all form the seeds of fundamental changes in future research on quantitative methods in regional science.

4.2 Realistic perspectives on data

Equally important to the link with theory is the need for quantitative methods to keep a realistic perspective on the availability and quality of data. This does not only constrain the scope of regional models, but should also drive the research efforts in spatial data analysis. Here again, the advent of superior GIS technology may provide a means to achieve these goals, but it is unlikely to be sufficient in and of itself.

Research in spatial statistics and spatial econometrics needs to focus on realistic frameworks for data encountered in empirical regional science. This means that the dominating role for the normality assumption needs to be relaxed, and more attention should be paid to specification diagnostics, robust approaches, categorical data, exploratory data analysis, and errors in variables perspectives. In general, an assessment of the relative merits of the alternative paradigms for the analysis of spatial data is needed, especially given the potential role of spatial analysis in the future of GIS. The advent of cheap and plentiful computational capacity will undoubtedly broaden the range of computation intensive approaches to the estimation and validation (sensitivity analysis) of spatial models.

An essential by-product of methodological developments in spatial data analysis has to be the dissemination of user-friendly software. In many respects this constitutes the necessary pre-condition for the effective diffusion of the improved methodology to the practice of regional science.

4.3 Truth in packaging

Finally, in order for advanced quantitative methods to become more accepted in the practice of regional science, it will be necessary to develop ways to present a realistic perspective on the validity of a model without running the risk of it being rejected as unuseful (or irrelevant). This implies that research should focus not just on the models and methods per se, but also on their interpretation, use and abuse in practice. It may very well be that the false sense of precision that the quantitative methods provide to a policy maker will ultimately turn out to be scientifically untenable, even though the use of the models is undoubtedly politically attractive if their results are sympathetic to a policy maker's cause. In the end, it is mandatory that a sound methodology for regional analysis be able to provide useful, realistic and ethical tools for policy making. It may be that one of the major challenges to research in this area will be to develop a solid framework through which the insights of quantitative methods can be translated into qualitative notions that are meaningful and relevant to the non-specialist.

5. Conclusions

In all, considerable progress has been made in recent years in the development of quantitative methods in regional science. In my comments I wanted to stress the potential danger of the tools running ahead of concepts and theory, particularly with the recent explosive growth in GIS and cheap and available computing technology. There is a need for a vigorous development of theoretical insights as well as for a solid bridge to practice and policy making. Moreover, it is crucial that a continued effort is made to convince researchers in other social sciences that the special role of space which underlies the essence of regional science merits its own set of appropriate quantitative methods.

References

Abler, R.F. (1987), The National Science Foundation National Center for Geographic Information and Analysis, International Journal of Geographical Information 1:303-326.

Anas, A. (1987), Modeling in Urban and Regional Economics, New York, Harwood Academic Publishers.

Anselin, L. (1986), Some Further Notes on Spatial Models and Regional Science, Journal of Regional Science 26:799-802.

Anselin, L. (1988a), Spatial Econometrics: Methods and Models, Dordrecht, Kluwer Academic Publishers.

Anselin, L. (1988b), Model Validation in Spatial Econometrics: A Review and Evaluation of Alternative Approaches, International Regional Science Review 11:279-316.

Anselin, L. (1988c), An Integrated Multiregional Approach Toward Water Policy Analysis at the State and Sub-State Level. Paper presented at the International Conference on the Construction and Use of Regional Input-Output Models, Alpine Lake Resort, West Virginia.

Anselin, L. (1989), Spatial Regression Analysis on the PC: Spatial Econometrics Using GAUSS. Institute for Mathematical Geography Discussion Paper #2. Ann Arbor, IMaGE (forthcoming).

Anselin, L. and D.A. Griffith (1988), Do Spatial Effects Really Matter in Regression Analysis? Papers, Regional Science Association 65:11-34.

Anselin, L. and M. Madden (1990), New Directions in Regional Analysis: Integrated and Multiregional Approaches, London, Belhaven Press.

Batey, P. (1985), Input-Output Models for Regional Demographic-Economic Analysis: Some Structural Comparisons, Environment and Planning A 17:73-99.

Batey, P. and M. Madden (1986), Integrated Analysis of Regional Systems, Lonson, Pion.

Batey, P. and A. Rose (1988), Extended Input-Output Models and Social Accounting Matrices: Progress and Potential. Paper Presented at the International Conference on the Construction and Use of Regional Input-Output Models, Alpine Lake Resort, West Virginia.

Bennett R. and L. Hordijk (1986), Regional Econometric and Dynamic Models, in: P. Nijkamp (ed.), Handbook of Regional and Urban Economics I, pp. 407-441, Amsterdam, North Holland.

Bolton, R. (1985), Regional Econometric Models, Journal of Regional Science 25:495-520.

Bon, R. (1988), Supply-Side Multiregional Input-Output Models, Journal of Regional Science 28:41-50.

Boyce, D. (1988), Renaissance of Large-Scale Models, Papers, Regional Science Association 65:1-10.

Brucker, S., S. Hastings and W.R. Latham III (1987), Regional Input-Output Analysis: A Comparison of Five 'Ready-Made' Model Systems, The Review of Regional Studies 17:1-16.

Cliff, A.D. and J.K. Ord (1973), Spatial Autocorrelation, London, Pion.

Czamanski, S. (1973), Regional and Interregional Social Accounting, Lexington MA, Lexington Books.

Getis, A. and B. Boots (1978), Models of Spatial Process, London, Cambridge University Press.

Goodchild, M.F. (1987), A Spatial Analytical Perspective on Geographical Information Systems, International Journal of Geographical Information Systems 1:327-334.

Griffith, D.A. (1988), Advanced Spatial Statistics, Dordrecht, Kluwer Academic Publishers.

Griffith, D.A. (1989), Spatial Regression Analysis on the PC: Spatial Statistics Using Minitab. Institute for Mathematical Geography Discussion Paper #1. Ann Arbor, IMaGe.

Harrigan, F. and P.G. McGregor (1988a), Recent Advances in Regional Economic Modelling, London, Pion.

Harrigan, F. and P.G. McGregor (1988b), Price and Quantity Interaction in Regional Economic Models: The Importance of 'Openness' and 'Closures', in: F. Harrigan and P. McGregor (eds.), Recent Advances in Regional Economic Modelling, pp. 178-207, London, Pion.

Hewings, G.J.D. and R. Jensen (1986), Regional, Interregional and Multi-regional Input-Output Analysis, in: P. Nijkamp (ed.), Handbook of Regional and Urban Economics I, pp. 295-355, Amsterdam, North Holland.

Hewings, G.J.D. and R. Jensen (1988), Emerging Challenges in Regional Input-Output Analysis, The Annals of Regional Science 22:43-53.

Higgs, P., B. Parmenter and R. Rimmer (1988), A Hybrid Top-Down, Bottom-

Up Regional Computable General Equilibrium Model, International Regional Science Review 11:317-328.

Isard, W. (1960), Methods of Regional Analysis, Cambridge MA, MIT Press.

Isard, W. (1986), Reflections on the Relevance of Integrated Multiregion Models, Regional Science and Urban Economics 16:165-180.

Isard, W. and L. Anselin (1982), Integration of Multiregional Models for Policy Analysis, Environment and Planning A 14:359-376.

Jackson, R. (1986), The Full Description Approach to the Aggregate Representation in the Input-Output Modelling Framework, Journal of Regional Science 26:515-530.

Jensen, R.C. (1988), Construction and Use of Regional Input-Output: Progress and Prospects. Discussion Paper, Regional Research Institute, West Virginia University.

Jensen, R.C., G.R. West and G.J.D. Hewings (1988), The Study of Regional Economic Structure using Input-Output Tables, Regional Studies 22:209-220.

Kain, J. (1987), Computer Simulation Models of Urban Location, in: E. Mills (ed.), Handbook of Regional and Urban Economics II, pp. 847-875, Amsterdam, North Holland.

Kinal, T. and J. Ratner (1986), A VAR Forecasting Model of A Regional Economy: Its Construction and Comparative Accuracy, International Regional Science Review 10:113-126.

Miller, R., K. Polenske, A. Rose (1989), Frontiers of Input-Output Analysis, New York, Oxford University Press.

Nijkamp, P., P. Rietveld, and F. Snickars (1986), Regional and Multiregional Economic Models: A Survey, in: P. Nijkamp (ed.), Handbook of Regional and Urban Economics I, pp. 257-294, Amsterdam, North Holland.

Odland, J., R. Golledge, P. Rogerson (1989), Recent Developments in Mathematical and Statistical Analysis in Human Geography, in: G. Gaile and C. Wilmott (ed.), Geography in America, pp. 719-745, Columbus OH, Merril.

Oosterhaven, J. (1988), On the plausibility of the supply-driven input-output model, Journal of Regional Science 28:203-217.

Openshaw, S. (1987), An Automated Geographical Analysis System, Environment and Planning A 19:431-436.

Openshaw, S. (1988), Building an Automated Modeling System to Explore a Universe of Spatial Interaction Models, Geographical Analysis 20:31-46.

Openshaw, S., M. Charlton, C. Wymer and A. Craft (1987), A Mark 1 Geographical Analysis Machine for the Automated Analysis of Point Data Sets, International Journal of Geographical Information Systems 1:335-358.

Paelinck, J. and L. Klaassen, (1979), Spatial Econometrics, Farnborough, Saxon House.

Richardson, H. (1985), Input-Output and Economic Base Multipliers: Looking Backward and Forward, Journal of Regional Science 25:607-661.

Round, J.I. (1986), Social Accounting for Regional Economic Systems, in: P. Batey and M. Madden (eds.), Integrated Analysis of Regional Systems, pp. 90-106, London, Pion.

Round, J.I. (1988), Incorporating the Interregional, Regional, and Spatial Dimension into a SAM: Some Methods and Applications, in: F. Harrigan and P. McGregor (eds.), Recent Advances in Regional Economic Modelling, pp. 24-45, London, Pion.

Rose, A. and P. Beaumont (1988), Interregional Income-Distribution Multipliers for the West Virginia Economy, Journal of Regional Science 28:461-475.

Sonis, M. and G. Hewings (1988a), Superposition and Decomposition Principles in Hierarchical Social Accounting and Input-Output Analysis, in: F. Harrigan and P. McGregor (eds.), Recent Advances in Regional Economic Modelling, pp. 46-65, London, Pion.

Sonis, M. and G. Hewings (1988b), Error and Sensitivity in Input-Output Analysis: A New Approach, in: Miller et al. (eds.), Frontiers of Input-Output Analysis, New York, Oxford University Press.

Spencer, J.E. (1988), Computable General Equilibrium, Trade, Factor Mobility and the Regions, in: F. Harrigan and P. McGregor (eds.), Recent Advances in Regional Economic Modelling, pp. 140-152, London, Pion.

Stone, J.R.N. and M. Weale (1986), Two Populations and their Economies, in: P. Batey and M. Madden (eds.), Integrated Analysis of Regional Systems, pp. 74-89, London, Pion.

Szyrmer, J. (1985), Measuring Connectedness of Input-Output Models: 1. Survey of the Measures, Environment and Planning A, 17:1591-1612.

Taylor, C. (1982a), Econometric Modeling of Urban and Other Substate Areas: An Analysis of Alternative Methodologies, Regional Science and Urban Economics, 12:425-448.

Taylor, C. (1982b), Regional Econometric Model Comparisons: What Do They

Mean?, The Annals of Regional Science 18:1-15.

Treyz, G.I. and B.H. Stevens (1985), The TFS Regional Modelling Methodology, Regional Studies 19:547-562.

Wegener, M. (1986), Integrated Forecasting Models of Urban and Regional Systems, in: P. Batey and M. Madden (eds.), Integrated Analysis of Regional Systems pp. 9-24, London, Pion.

West, G. (1986), A Stochastic Analysis of an Input-Output Model, Econometrica 54:363-374.

Chapter 17

A MULTIREGIONAL INPUT-OUTPUT MODEL FOR ISRAEL AND EXTENSIONS: METHODOLOGY AND EXPERIENCE

D. Freeman
H. Talpaz
A. Fleischer
O. Laufman

1. Introduction

Ever since the initial development of Input-Output (I-O) models by Wassily Leontief (1951), they have been given a great deal of attention by economists and national and regional planning authorities throughout the world. These authorities have recognized the potential usefulness of this tool in simplifying and guiding the planning process. Indeed, the increased interest brought about significant advances, both methodologically and empirically, which extend the range of areas and problems that could be handled by I-O models far beyond those originally intended by Leontief.

One area in which an I-O model was successfully implemented is multi-regional modeling. This extension is primarily due to Isard (1953, 1960) and Moses (1955), who have recognized that observed structural dissimilarities in production and consumption among regions call for the construction of a disaggregated multi-regional I-O model. Among the principal reasons for such disaggregation is the desire to capture multiplier effects of multi-regional linkage between producers of commodities in one region and users of these commodities in other regions. By showing how autonomous changes in one region or industry generate impulses that are transmitted to other regions and industries, and by producing high quality measures of these impacts, multi-regional I-O models are particularly useful for analysis of regional development planning aiming at the restructure of the spatial distribution of population and production. With the aid of this tool, plans for regional growth can be evaluated for both feasibility and for embedded regional and national growth rates.

Despite the theoretical merits of these models, they are rarely implemented and many economists remain skeptical of their empirical validity. Criticisms often center around two fundamental and multi-related difficulties. Firstly, the construction of an multi-regional I-O model involves many practical difficulties stemming from the need to identify the origins and destinations of multi-regional commodity flows. Such data are usually not available and their collection is expensive and their use requires several simplifying assumptions. Secondly, and more important, applying such models for prediction purposes raises some doubts since, unlike the production coefficients, the trade coefficients may not be very stable over time (Freeman et al., 1985; Weksler et al., 1986). This is particularly important for developing regions where structural changes in trade patterns are an integral part of the development process. Multi-regional I-O models are particularly useful for analysis of regional development planning, aiming at the restructure of spatial distribution of population and production.

We have been using versions of Multi-regional Input-Output (MRIO) models for the last decade, for the evaluation of projects in peripheral regions, because they are particularly useful in regional planning. They respond to the need for an analytical tool, capable of generating estimates of the impacts of different regional development investment and policies, simultaneously at four spatial levels: project-region-country-world. This makes it readily available to evaluate multi-branch and multi-regional impacts of alternative policies. To date, twenty major regional projects have been evaluated, and the outcome presented as multipliers of output value, number of employed, value added, and multipliers thereof.

The presentation of this paper has two objectives:

1. To present the application of the MRIO models as a policy evaluative tool, and to illustrate how we do it in a relatively short period of time per each project;
2. To evaluate what we have learned from the projects that we have studied.

The structure of the model is composed of two multi-regional direct column coefficient matrices according to K.R. Polenske (1970):a) an input-output matrix; b) a Trade Flow (TF) Matrix. We have so far experimented with seven types of variations in the model, for the purpose of representing and evaluating different policies: 1) Inclusion of a special 'Rest of the World' region devoid of direct I-O coefficients, in which the margins are the import and export values at the ports and/or borders of the country (Freeman et al., 1985); 2) A Markov Chain adjustment of trade flow matrices, to represent the impact of middlemen (see Weksler et al., 1986, for a detailed description of this method); 3) Changes of the TF matrices, to represent locating new industries in development regions (Alperovich et al., 1987); 4) Substitution of columns representing new technologies for those of current technologies (Freeman et al., 1984); 5) Extending matrix sizes to represent the coefficients new project(s) (Freeman D., 1984; Freeman and Talpaz, 1987; Freeman et al., 1988). 6) Integrating a Quadratic Programming algorithm with the MRIO model (Fleischer et al., 1987), for the purpose of analyzing the regional impacts of economic policy options such as prices, subsidies, taxes, maximum capacity, etc. 7) Integrating into the MRIO model a matrix of the proportion of skilled workers by branch in each region for the purpose of evaluating and predicting employment impacts by profession. Further development of this version into a full-scale multi-regional Social Accounting Matrix (SAM) is under way.

The construction of MRIO models is regularly restricted by data difficulties. However, they are essential for the evaluation of comprehensive impacts of economic activities in peripheral regions. Thus our team is continuing to devise ways and means to simplify the process and do it relatively quickly.

The basic idea is to reduce to a minimum the need for statistical surveys. The ensuing basic strategy is double pronged:

(1) Survey of the gross production and final uses of the region (or regions) of interest, and construction of its (or their) I-O table (or tables).

(2) Insert into the margins of the TF tables: as sums of columns the sums of I-O rows, and as sums of TF rows the sums of the I-O columns. Then, insert

into TF rows the actual surveyed flows of outputs of the region (or regions). These are deducted from the column totals and the residuals are inserted into the row of 'Rest of the Country' region (see Freeman et al., 1981). The purpose of this paper is to present the ways and means used by the team in applying the model to different evaluative needs each within a relatively short period of time.

Our first model (Freeman et al., 1985) was applied to two development regions in Israel: Galil (north), and Negev (south). The 'Rest of Israel' as a third region and the 'Rest of the World' as a fourth region. This application focused on the calculation and analysis of multipliers and their regional breakdown. To achieve this goal two different versions of the model were estimated: an **'open'** model where **household** consumption is assumed to be **exogenous**, and a **'closed'** model where **household** consumption is assumed to be an **endogenous** sector. The results obtained were quite encouraging, since they were similar in magnitude to findings of similar USA studies. Later studies focused only on the Negev (the south) as the development area, subdivided into several regions (Alperovich et al., 1987), versus the 'Rest of Israel' and the 'Rest of the World' as the other two regions. The discussion is presented in the following sections.

2. The Model

While the principal objective of a national I-O model is to identify empirically and analyze the prevailing inter-industry structure of the national economy, the main objective of the multi-regional I-O model is to give empirical content to the spatial elements of the national production and distribution structure, and to utilize this information for studying the structures and linkages of individual regions. As such, multi-regional I-O models are most useful for analyzing and evaluating different development strategies, and their direct and indirect impacts on the economy of the investigated region and on other regions. It is this goal that led to the formulation and implementation of the present model. In this section we illustrate the model used in the empirical implementation.

Being a natural extension of the standard national I-O model, the derivation of multi-regional I-O is quite straightforward.

Let:

i,j - be indices of industries ($i,j=1,2...N$).

g,h - be indices of regions ($g,h=1,2...R$).

X_i^{og} - be total supply, from all regions of commodity i in region g.

X_i^{go} - be total output produced by industry i in region g.

X_i^{gh} - be the flow of commodity i from region g to region h.

X_{ij}^{g} - be total output of i used as intermediate input in the production of commodity j in region g.

Y_i - be total output of i consumed by final users in region g.

The basic structural equation on which our multi-regional I-O model is based can be written:

$$(1) \qquad X_i^{og} = \sum_{j=1}^{N} x_{ij}^{go} + y_i^{g}, \qquad ; \quad i,j = 1,2..., N \ ; g = 1,...,R$$

Equation 1 is an accounting equation which simply states that total supply, from all regions of commodity i in region g must be equal to total demand for i in region g. The latter being composed of both intermediate demand used in production and of demand by final users. Accepting the basic assumption of I-O models, namely, that there is a fixed technical relationship describing the amount of commodity i required per unit of commodity j produced in region g. This equation can also be rewritten:

$$(2) \qquad X_i^{og} = \sum_{j=1}^{N} \alpha_{ij}^{g} \, x_i^{go} + y_i^{g} \qquad ; \quad i,j = 1,2...,N \ ; \ g = 1,...,R$$

where α_{ij}^{g} is an input-output coefficient describing the amount of input i, wherever located, required per unit of output of j in region g. This is the basic equation used in all multi-regional I-O models. However, as one may see, equation 2 does not indicate anything about the spatial inter-relationhsips of production. In order to introduce such a dimension, there is a need for an additional assumption concerning regional trade patterns and its stability. An obvious and ideal formulation of the regional trade pattern is due to Isard (1951), who assumed that the trade pattern is stable and required that there will be a fixed proportion relationship describing total amount of i, from region g needed per unit of output of j in region h.

Integrating the spatial input coefficients $ß_{ij}^{gh}$ into equation 1, the ideal model can be rewritten:

$$(3) \qquad x_i^{og} = \sum_{j=1}^{N}\sum_{h=1}^{R} \beta_{ij}^{gh} x_i^{go} + y_i^g \qquad ; \qquad g = 1,...,N$$

This model is ideal both from a descriptive and analytical viewpoint. However, its empirical application is restricted by the lack of adequate multi-regional flow data needed to compute the coefficients $ß_{ij}^{gh}$.

In general there are several possible approaches to make an empirically applicable simplification of equation 3, with each partially sacrificing some of the descriptive and analytical contents of this model. The main differences of the various approaches lie in the way multi-regional trade coefficients are calculated. The approach adopted in our application is known in the literature as a column coefficient model. The basic assumption of this approach is that commodity i is shipped between regions g and h according to the fixed proportion of the total amount of commodity i purchased by region h. According to this assumption multi-regional trade coefficients are calculated from:

$$(4) \qquad c_i^{gh} = \frac{x_i^{gh}}{x_i^{oh}} \qquad ; \qquad i,g = 1,...,N$$

It might be interesting to note that Isard's spatial input-output coefficients can be obtained as the product of the technical coefficients - α_{ij}^{h} and the trade coefficients c_i^{gh} is

(5) $$b_{ij}^{gh} = c_i^{gh} \alpha_{ij}^{h}$$

(6) $$x_i^{og} = \sum_{j=1}^{N} \sum_{h=1}^{R} c_i^{gh} a_{ij}^{h} x_j^{go} + \sum_{h=1}^{R} c_i^{gh} y_i^{g} \quad ; \quad g = 1,\dots,N$$

which is the basic multi-regional I-O model used in this paper. The equivalent presentation of the model in terms of matrix notation is:

(7) $$X = CAX + CY$$

where,

X is an nm column vector describing the gross outputs of i produced in every region;

C is an (nm,nm) matrix of the multi-regional trade coefficients;

A is an (nm,nm) block diagonal matrix describing the I-O technical production coefficients; and

Y is an nm column vector representing the total final demand of i in each region.

Solving this system of equations for X yields:

(8) $$\mathbf{X} = (\mathbf{I}-\mathbf{CA})^{-1}\,\mathbf{CY}$$

where I is an nm-order identity matrix.

The following discussion will present a concise description of the variations that were applied in our analysis so far, and which we go on using in our work.

3. Variations

The first variation was the introduction of The Rest of the World Region. This region is devoid of I-O coefficients. The margins of its I-O table are the CIF value of imports and FOB values of exports at the harbors and borders of Israel (Freeman et al., 1985). The objective for the inclusion of such a region was to separate the share of exports and imports in the total multiple impact. This gives the regional planner the ability to find out the intra-country multi-regional impacts by branch, simultaneously exclusive from the share of foreign goods in the total.

The structure of the **I-O tables** has the following specific characteristics:

- Values of the imports (M) per branch are in each domestic region, while the Rest of the World (W) region has a row of zeros.
- Each domestic region has an export (E) column of zeros, while region w has a column of exports per branch and no other values in it.
- The structure of a **TF Table** for each domestic branch, has three groups of elements:
- The value of transactions between regions X_i^{hg}
- The value shipped as exports E_i, and
- A row of zeroes for imports in region W.

The structure of the **TF table** for region W has only a row of imports (M) per region, while all other values are zeroes (including the W'th column as well).

3.1 Derivation and adjustment of trade flows

Trade Flows are derived in four steps:

1. Survey of flows of output of businesses and enterprises from each of the peripheral regions to the Rest of Israel region.
2. Deduction of these flows from the column totals (which are also the totals of the I-O rows).
3. The residuals per column, are the estimate of the row of the Rest of Israel region.

4. An adjustment by a Markov Chain to account for shipments to middlemen in the Rest of Israel region and reshipment to the region of origin and all other regions.

Some clarifications seem to be in order. Each branch has, in general, its own TF coefficients between the regions. That facilitates the use of this approach for either developed or underdeveloped regions, since the shipments may be different in magnitude between any two regions. One of our studies (Weksler et al., 1986) aimed at defining the magnitude of reshipments. The data used were milk products from Tnuva, the largest marketing organization of agricultural products in Israel. The data composed of two streams of flows: (1) from producers (creameries) to middlemen; (2) from middlemen to retail.

According to the Markov-Chain procedure, these two flows were converted into probabilities by division of row elements into the total of each row. The product of their multiplication became a matrix of conditional probabilities. It then was premultiplied by a diagonal matrix of output per region. The product is a matrix of the **net multiregional trade flows.** It was compared then with a matrix of the **observed gross flows.** The mean percent deviation between the elements in these two matrices has been found to be 14%. This might be considered the lower bound for such differences since the milk products are perishable and are sold quickly. It is reasonable to expect that the role of middlemen will be larger in non-perishable goods, with greater mean deviation.

On the basis of the foregoing discussion we found it necessary to use this method to adjust for reshipments since data about middlemen were either not available or too costly to collect.

3.2 Policy simulation through variation of trade flows

Impact analysis of alternative policies for promoting regional growth in the peripheral region of interest (R) can in general be conducted in terms of output, income and employment multipliers.

The first level of analysis is a 'micro' nature and refers to situations where the government is interested in promoting the growth of a particular industry j in the region R, and does so by focusing on the target industries without paying attention to other industries. A much greater impact can be achieved if, at the same time, other industries which are important suppliers of inputs to industry j are also attracted to region R.

The second level of analysis is 'Macro' in nature and describes a situation in which the government is pursuing indiscriminate measures aiming to encourage mobility of industries into the region under consideration. From the economic point of view, such a policy is always less desirable. Results obtained for the 'macro' analysis provide further insight into the issues involved.

In one of our papers (Alperovich et al., 1987), the government policy is represented by a parameter Θ_J showing the proportion of the target industry's j product to be imported from other regions, while $1-\Theta_J$ is the proportion of it planned to be produced in the region, for both intermediate and final consumption. The off-diagonal elements of the jth column are multiplied by Θ_J, $(0<\Theta_J<1)$ and the augmented production is reflected in the diagonal elements of the TF matrix, C as:

$$(9) \qquad C_J^{RR*} = C_J^{RR} + (1 - \theta_J) \sum_{i=1}^{R} C_J^{RG} \quad ; \qquad R=G$$

If industry k, which is not located in the region, is closely related to industry j as a major supplier of its intermediate resources, a policy directive might be to increase its production in the region. Thus, an identical Θ_J might be applied to it.

We have found this kind of analysis instructive and illuminating, but not sufficient for our clients. The regular question is, can one express the total impact in terms of total added outputs, employment and product of all the branches together. The procedure needed to do that by applying these Θ_J is somewhat too complicated, due to the following needed changes in the I-O transactions:

1. Increase in output of j in region R, must be inserted into the total of its jth column of the transaction table.

2. This must be simultaneously deducted from the identical branch in the current supplier region g.

4. Extensions

4.1 Policy simulation through implanting of new columns

In the foregoing discussion, a method to demonstrate an increasing intra-regional impact is presented. It is an acceptable method if one is satisfied with estimates of the multiplier impact of a branch or several branches only. However, it has been pointed out that it is much more difficult to apply it if the **total impact** is required. Then, one needs to change the sums of rows and columns of I-O and TF tables in at least two regions (if the flow is only between two regions).

These difficulties enhanced the need for a less complicated procedure. A possible method has been hinted at by Hewings (1972), and the difficulties it entailed in defining the row coefficients by Greytak (1972). The method is elucidated by Katz and Burford (1981). It requires data only for the new column(s), that will augment the matrix. It does not require row coefficients since they are zeroes. Thus, for a MRIO model the procedure involves several steps:

1. Determine a project's (or an industry's) intermediate purchases from all branches $X_{i,n+1}$ and convert these values into coefficients: $A_{i,n+1}$.
2. Insert this column into the A matrix of the region of interest, and add to it an $n^{th} + 1$ row of zeroes. Add columns and rows of zeroes in all other rows. This will augment the matrix, and the augmented A* is illustrated in the following figure:

$$A^{*} = \left[\begin{array}{c|c} & a_{1,n+1} \\ A & \vdots \\ & a_{n,n+1} \\ \hline o \ldots\ldots\ldots o & 1 \end{array}\right]$$

3. Distribute the output of the new branch as a final demand of the regions that require it. This procedure permits the planner to estimate the partial impact of a project, a firm, a group of firms, etc. It has the disadvantage that is does not estimate at the same time the substitution effect that might occur due to the introduction of a new branch that might have cross-elasticities with other branches. The advantages of the procedure stem from its ability to represent new technologies that are not yet present in the region. Its disadvantages are the missing values of its contribution to intermediate transactions into other branches, and thus it lacks the representation of the changes due to the substitution effect.

The obtained results have shown that the final demand in a development region, generate in the 'Rest of Israel' region, about one-half to two-thirds of the multiplier's impact, which means that products demanded by the region due to a development expenditure create more output and employment in other parts of the country. The ability to estimate such impacts by branch, could give planners a tool for evaluating the necessary development investments simultaneously with extra investments necessary in the rest of the country, to supply the needs of the region.

Thus far the model has two drawbacks: 1) It is a static model and prediction of each year has to be undertaken separately. 2) The estimate of the required extra resources due to an investment, does not indicate whether they are idle, or has to draw them away from other production. Thus, we hope to develop a dynamic model with an optimal path of resource use in order to be able to respond to these weaknesses. Despite the foregoing weaknesses we find that this model has a definite advantage over single region models, which estimate only the intraregional impacts, and not the interregional ones.

4.2 An optimal equilibrium model

The Quadratic Multiregional Input-Output (QMRIO) model evaluates the effects of economic policy on the various geographical regions of the economy. However, a brief discussion follows a detailed description and model estimation can be found in Flescher et al., 1987. It takes the market price dynamics into consideration and provides a meaningful evaluation criterion and by this overcomes drawbacks of the MRIO model. In the MRIO model there are two strong implicit assumptions concerning the elasticities of supply and demand: (i) Final demands have infinite price elasticities, and no cross elasticities; and (ii) supplies of primary inputs are infinitely elastic until the point of full capacity and possess zero elasticities from there on. Another problem is a lack of general implied criteria for comparison and evaluation of the different policy alternatives.

The QMRIO model has a quadratic target function representing the consumers' and producers' surpluses as expressed in equation (10):

(10) $Z = H'Q + .5Q'FQ$

(11) $Q' = [R' : X' : Y']$

(12) $H' = [-H' : O' : H']$

(13) $F = \begin{vmatrix} -G : & O : & O \\ O : & O : & O \\ O : & O : & B \end{vmatrix}$

Where:

H = a(g + 2mn)x1 vector

H_s = a gx1 vector of intercepts in the supply curves

H_d = a mnx1 vector of intercept in the demand curves

O = matrices of zeroes with the appropriate dimensions

F = a(g + 2mn) x (g + 2mn) matrix

G = a g x g matrix of slopes in the supply curves

B = a mn x mn matrix of slopes in the demand curves

Z is brought to maximum under two sets of constraints:

The **first set** of constraints means that the demand for primary inputs can not exceed its supply, i.e., equilibrium in the input markets.

(14) $A_r X +_r Y \leq R$

The **second set** of constraints means that the intermediate use plus final demand cannot exceed the domestic production supply and imports. Supply of each one of the products is derived from the primary supply function and the coefficients in A_r. The demand is a combination of the intermediate uses expressed in the MRIO tables, and the final demand as expressed in the demand functions. This condition guarantees equilibrium in the goods and services market. It is expressed mathematically as follows:

(15) $CAX + CY \leq X$

(16) $SQ < O$

Where:

(17) $$S = \left| \begin{array}{c:c:c} -I_g & A_r & O \\ \hdashline O & CA - I_{nm} & C \end{array} \right|$$

I_g = a (g x g) identity matrix

I_{nm} = a (mn x mn) identity matrix

X = a vector of total output of m sectors in n regions

C = a matrix of trade coefficients among regions

A = a (mn x mn) matrix of production coefficients in m sectors and n regions

Y = a (mn x 1) vector of final demand, including private consumption, government

The demand coefficients of the target function are acquired from estimates of the 'Roterdam demand system' for the Israeli economy. This system was chosen due to the simplicity of its use in the quadratic function. The mathematical structure of the quadratic function requires linear demand functions leading to a negative semi-definite slopes matrix. This negativity requirement can be arbitrary in other demand systems, but in the 'Roterdam model' the estimation of the demand equation under the negativity condition is in accordance with the negativity restriction for the Hicksian demand structure. The negativity constraint was achieved via a nonlinear constrained regression procedure.

The supply function coefficients were estimated under assumptions of producers' behavior. The model constraints are based on the multiregional input-output tables of the Israeli economic system (Freeman et al., 1987, 1985).

This model is used to estimate the effects of a policy favoring the southern Negev region in Israel. This region is mainly a desert with a low population density constituting 40% of the total area of Israel, but only 1.5% of the population. Over the years there have been continuous attempts to attract people to the region. A policy, currently in force, is a reduction in income tax to the residents of that region.

The interregional impact of a 15% tax reduction, aimed at the population of the southern Negev was studied, and a welfare-cost benefit analysis was carried out. The welfare for the southern Negev's consumers has increased, while the opposite occurs for the consumers of the other regions. The effects on production, consumption, and price level was also forecasted for each of the regions.

Such a comprehensive model, which computes the cross demand relationships intersectoral and interregional flows, is capable of measuring the economic impacts of different policies. It is a powerful tool for analyzing economic trade-offs among regions, stemming from policy choices.

4.3 Estimating future occupational make up

Further development for the MRIO model is the prediction of the different skills needed for a major development project. Each project generates demand for different skilled labour, directly and indirectly. This demand for labour is different in each of the regions according to its sectoral output.

Using information from the population census, one can build a matrix of number of workers in each sector by occupations for each region. Commuting patterns according to professional classifications have been used to adjust the census data. The second stage is to transfer the number of coefficients by dividing the number of workers by the total number of workers in each sector. Now one can predict the change in different occupations needed in the regions by the following operation:

$$\Delta SL = L'(\Delta O \times LO)$$

Where:

ΔSL = (mxr) matrix of change in number of people needed in different occupations and regions

m = number of occupations

L = Mx(nr) matrix of coefficients distributing the ratio of workers in the different occupations for n sectors and r regions

ΔO = ($n_r$xi) vector change in output

LO = ($n_r$xi) vector labour output ratio

This information is especially important in peripheral areas where the population density is low and the lack of skilled labour can create a bottleneck for a big investment project. This method has been utilized in Freeman et al., 1989.

5. Experience

This discussion is based on the experiences that we have gleaned in the usage of this model, for analytical purposes. The procedure is comprised of two stages:

a. Analysis of **economic feasibility** of the project under study, especially the magnitudes that would make it economically viable.

b. **Impact analysis** after the insertion of final uses, and the coefficients of intermediate purchases and the trade flows.

The working procedure comprises two steps:

1. A base MRIO model is constructed with the latest output and final-uses data available, and then its multipliers are derived.
2. Variations and extensions of the base model are applied to respond to differing analytical needs of projects.

In the course of the following discussion there are some insights into the multipliers, and outcomes of some of the projects.

So far, during the years, four base models were constructed, and then used in analyses. Mean multiplier magnitudes of them were compared. The comparisons were of both the **distributed** $[(I\text{-}CA)^{-1}\ C]$ and the **undistributed** $[(I\text{-}CA)^{-1}\ I]$ types, as well as the **'open'** and **'closed'** versions.

A multiplier of the **open** version is the value of output that must be produced in all branches and regions to supply one dollar worth of final demand in a specific region. The distribution of the multipliers into regional multipliers can be obtained by summing up the relevant entries of each column in the matrix $[(I\text{-}CA)^{-1}\ C]$ whose columns are 'region specific' and its rows are 'branch specific'. A **closed** model multiplier is calculated under the assumption that the **household** sector, as represented by the wage row and the private consumption column is a **labor producing branch**. This row and column, which are exogenous to the 'intermediate transaction' matrix, are endogenized into it. The basic assumption underlying this approach is that consumption in each region is a linear homogeneous function of the respective wage row. The multipliers which are derived from the closed model, represent the output which is required to produce one dollarworth of final demand, plus the additional personal consumption induced by the additional income generated. It follows that closed model multipliers should always be greater than open model multipliers.

Since 1973 up to 1987 the data base of I-O tables of Israel has changed four times. Thus four different MRIO models were constructed accordingly. They were used in the impact analysis of peripheral planning regions of the country and their interaction with the central major regions.

During this period the magnitude of the average of multiplier impacts has increased by approximately 15%. This points to an increase in the degree of sophistication in many of the branches, reflected by a greater proportion of intermediate uses bought from other branches. One might conclude that such an increase is as expected, due to changes in both, the technological make-up and the consumption patterns. The latter is due to the rise in the income level.

Ever since the first model, the objective of the team has been to evaluate the outcome of development spending in peripheral regions.

For each project a cost-benefit analysis and returns to scale evaluations were performed.

In a typical model, we have estimated the net and gross impacts of development spending in two extreme regions. The outcome was: (a) The direct final demand of the two regions depends heavily on output in the 'rest', or central region; (b) Any program to stimulate the peripheral region, will have to ponder the 'leakages' to the 'Rest' region, and attempt to overcome it by introducing branches that can bring with them some of their supportive branches as well.

Impacts of several major projects were studied by augmenting the I-O matrix with new columns that represented the purchases of the project. At first, only a single branch was implanted as an additional column of coefficients; later on several branches were added simultaneously, and finally 21 new branches were added, representing the major plants of the Israeli Chemical Corporation. The interaction between the new columns is represented in 21 rows, but no interactions perceived between the new branches and those of the base matrix; thus, these values are zeroes. This method was applied in some of our major studies. The elaboration of so many branches facilitated analysis of each branch separately, and its isolated impact with each change of one or several branches.

In our latest works, we have applied two extensions to this model: (1)

Integrating a Quadratic Programming algorithm with the MRIO model to allow for a general equilibrium solution; (2) Estimates of the occupational make-up.

All the foregoing variations and extensions are regularly used in our work.

6. Summary and Conclusions

This paper describes the development of a practical tool, a MRIO model, for the evaluation of regional projects. Different variations and extensions were introduced into the model; amongst them are: (1) the inclusion of the Rest of the World Region, (2) Markov Chain adjustment of trade flows to adjust for impact of middlemen, (3) substitution of new technology columns, representing new projects by extending matrix size, (4) quadratic programming, and (5) prediction of the regional make-up skills.

The foregoing changes advanced the MRIO model from the theoretical stage to a practical useful stage. Our team has been asked to evaluate with this model the comprehensive impact of some of the major development projects (e.g., the Mediterranean Dead-Sea Canal, Railway to Eilat, the Lavi Fighter-airplane, the economic relationships of Judea Samaria and Gaza with Israel, and some other critical issues of government policy).

In all the projects that were studied, the results have shown that the final demand in a development region generates about one-half to two-thirds of the multiplier impacts in the 'Rest of Israel' region. This means that products demanded by the region due to a development or a consumption expenditure create more output and employment in other parts of the country. The ability to estimate such impacts by branch could give planners a tool for evaluating the necessary development investments simultaneously with extra investments necessary in the rest of the country to supply the needs of the region.

The next stage would be the application of the model to evaluate the effects of regional projects on the income distribution in the region and other regions.

References

Alperovich, G., D. Freeman and I. Weksler (1987), An Evaluation System for Regional-Regional and Regional-National Growth Impact, Applied Economics 19:1367-1382.

Barten, A.P. (1966), Theorie en empirie van een volledig stelsel van vraagvergelijkingen, Doctoral dissertation, University of Rotterdam.

Fleischer, A., O. Laufman, D. Sivan, D. Freeman and H. Talpaz (1988), Total Impacts of Operating the Transmission Station of 'Voice of America' in the Arava: A Multi-Regional Input-Output Analysis, Final Report, mimeo, Rehovot Settlement Study Center (Hebrew).

Fleischer, A., H. Talpaz and D. Freeman (1989), An Optimal Quadratic Equilibrium Model Based on Israel's Multi-Regional Input-Output Tables, Journal of Socio-Economic Planning 22,5:195-199.

Freeman, D., G. Alperovich and I. Weksler (1985), Inter-Regional Input-Output Model - The Israeli Case, Applied Economics 17:381-393.

Freeman, D., G. Alperovich and I. Weksler (1987), Regional Macro Planning Using a Multi Regional Input-Output (MRIO) Model, Readings in Economics 1982-1984, Bank of Israel, pp. 123-137 (Hebrew).

Freeman, D., G. Alperovich and I. Weksler (1984), An Economic Analysis of Development Plans for the Rural Sector of the Eshkol Region, in: A. Rozenman (ed.), Human Settlement and Regional Development, pp. 273-290, Jerusalem, Keter Press (Hebrew).

Freeman, D. (1984), The Economic Feasibility and the Multiplier Impacts of the Mediterranean Dead Sea-Canal, Final Report, mimeo, The MDSC Co, Jerusalem (Hebrew).

Freeman, D. and H. Talpaz (1987), Economic Aspects and Benefit-Cost Analysis of the Lavi Airplane and its Multiplier Impacts, mimeo, The Israeli Air Industries (Hebrew).

Freeman, D., D. Sivan, O. Laufman and H. Talpaz (1988), Shipment ot the Red Sea by a Railway to Eilat versus the Suez Canal, working paper no. 20, Settlement Study Center, Rehovot.

Freeman, D., H. Talpaz and D. Sivan (1987), Project Impact Evaluation Using a Multi-Regional Input-Output Model, The Proceedings of the XXVII Congress of The Regional Science Association, Athens, Greece, 25-28 August.

Freeman, D., A. Fleischer and O. Laufman (1989), The Intra and Inter-Regional Impacts of a Regional Industrial Center in Northern Israel, (Hebrew), presented at the Galil Study Conference, University of Haifa.

Greytak, D. (1972), The Firm in Regional Input-Output Analysis, Regional Studies 6(3):327-329.

Hewings, G.J.D. (1972), Input-Output Models: Aggregation for Regional Impact Analysis, Growth and Change 3(1):15-19.

Polenske, K.R. (1970), A Multi-Regional Input-Output Model for the United States, EDA, USD of Commerce, Report 21.

Theil, H. (1965), The Information Approach to Demand Analysis, Econometrica 33:67-87.

Weksler, I., D. Freeman and G. Alperovich (1986), Estimation of Interregional Trade Flows: A Markov Chain Approach, Environment and Planning A 28:123-132.

Chapter 18

EX POST MEASUREMENT OF EFFECTS OF ENVIRONMENTAL POLICY BY MEANS OF LINEAR STRUCTURAL EQUATION MODELS

H. Cesar

H. Folmer

1. Introduction

Measurement of its effects is a vital element in the design and evaluation of environmental policy. It provides the feedback information necessary for the modification of current as well as for the design of future environmental policy. The importance of measurement of effects follows from the iterative nature of the process of policy-making, which consists of the following interrelated steps (see also Tinbergen, (1956, 1972)):

(i) Identification and classification of environmental policy goals[1];

(ii) Generation of environmental policy objectives from the policy goals;

(iii) Identification of a collective preference indicator from the individual preference indicators of the agents involved in a given environmental problem;[23]

(iv) Deduction of the targets of the policy objectives from the collective preference indicators;

(v) Identification of qualitative and quantitative instruments of environmental policy;

---

[1] It should be observed that there may not be consensus among politicians and policy makers with regard to the priority of policy goals and policy targets. In spite of that, impacts of policy interventions can be measured in terms of impact variables derived from the (disputed) goals and targets.

[2] The concept of a collective preference indicator has been criticized by, among others, students of the neoclassical approach to policymaking. They claim that government behaviour based on general interest and "collective preferences" is inconsistent with utility and profit maximizing behaviour of certain groups in the economy. This controversy, however, is of minor importance in the context of ex post measurement of effects of policy.

(vi) Ex ante assessment of effects and costs of the identified instruments;

(vii) Determination of the policy interventions;

(viii) Ex post measurement of effects and costs;

(ix) Adjustment of the preceding steps on the basis of the ex post measurement results.

This process is summarized in the following figure.

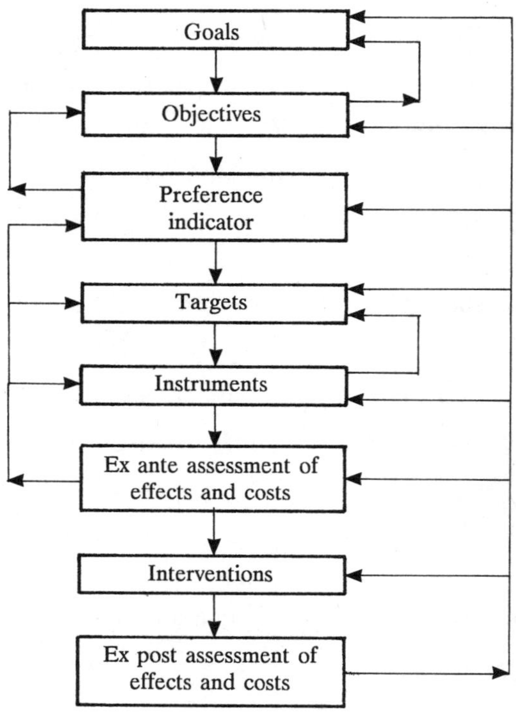

Figure 1. The process of environmental policymaking.

In spite of its importance, very few studies of effects of environmental policy have been undertaken. Most of the literature on environmental policy deals with the different instruments (their working, applicability and implementation) and the ex ante assessment of their effects and costs. Most prominent in this regard is the ex ante evaluation of regulatory instruments and economic incentives. In an overview article Bohm and Russell (1985) conclude that the latter are usually preferable (see also Siebert, (1987)). Another strand in the literature deals with the assessment of impacts of environmental policy on economic variables. In this regard partial and general equilibrium approaches could be distinguished. An example of the former is the study by Highton and Webb (1984) on the impact of emission standards on electricity prices. Another example is the measurement of the effects on agricultural production of ozone protection policies (Adams, Hamilton & McCarl (1986)). A recent example of the general equilibrium approach is the study of Merrifield (1988) on the impact of abatement strategies on transnational pollution, the terms of trade and factor rewards.

Another strand in the literature of assessment of effects of environmental policy could be grouped under the heading of valuation of environmental quality. In this context (possible) impacts of environmental policy are measured via the willingness to pay for a cleaner or healthier environment. Some recent examples are Johansson (1989), Navrud (1989), Dickie and Gerking (1989) and Shechter et al (1989).

As mentioned above, very few methodological and empirical studies of the ex post assessment of effects and costs of environmental policy have been undertaken. Some exceptions are the following. Box and Tiao (1975) estimated the effects of regulatory intervention on air quality by means of univariate time series analysis. Ashworth et al. (1987) assessed the impact of the EC directive on discharges of mercury on the British Chlor-Alkali industry by comparing actual data with expected values calculated on the basis of a regression equation.

The present paper deals with methodological issues of ex post measurement of effects of environmental policy. In the next section some relevant aspects of the nature of environmental policy will be described. Section 3 deals with requirements which an adequate measurement method should meet and various possible methods are evaluated on the basis of these requirements. In section 4 the LISREL approach is briefly described and it is shown that it is particularly appropriate for the purpose in question. The paper ends with a concluding section.

2. The Nature of Environmental Policy

Environmental policy consists of a specific set of institutional arrangements aimed at influencing the activities of households and firms. The institutional arrangements correspond to a given target of environmental quality in such a way that, given the restrictions, the probability of the target being reached is maximal. In the context of measurements of its effects the following aspects of environmental policy are relevant.

(i) Improvement of environmental quality or prevention of degradation is usually strived at by means of a combination of instruments or the application of similar instruments in different ways and intensities in different sectors. For instance, with regard to the abatement of acidification, different standards have been (or will be) developed for different sources (different NO_x emission standards for (subsectors of) industry and transport and traffic, standards for ammonia emissions in agriculture, etc.). Moreover, in the sector of traffic standards and incentives have been combined to reduce NO_x emissions (standards for the composition of fuel and incentives to stimulate the purchases of cars equipped with catalysts).

One reason for combining instruments and the application of similar instruments in different sectors follows from the nature of the generation of emissions. As can be concluded from the acidification example, a given kind of pollution is usually caused by a multiple of sources. Differences among the pollution generating sources usually require different instruments because of differences among the technologies involved. Moreover, the use of the different instruments may result from the fact that policy may simultaneously aim at different aspects of the generating source as in the case of the above example of pollution reduction in the sector of trafic. Finally, the application of similar instruments may differ in intensity because of e.g. differences in the marginal emission reduction cost functions. This follows from the well-known fact that efficient abatement implies that all emission generating sources control to the level at which their marginal emission reduction costs are equal.

(ii) Environmental policy is likely to be multi-effective. That is, it may not only have an impact on its target of environmental quality, but also on other variables which for the problem at hand do not belong to the set of policy targets. Referring again to the acidification example, policy measures in this context are likely to affect non-policy targets such as production volumes, investments, employment, etc, both in the sectors to which the policy measures refer and in other sectors.

(iii) The policy objectives are usually not only affected by environmental policy instruments but also by variables which do not belong to the set of institutional arrangements implemented by the government. This set consists of autonomous economic and technical developments.

The framework outlined so far is summarized in Figure 2 (see also Folmer and Nijkamp, 1986).

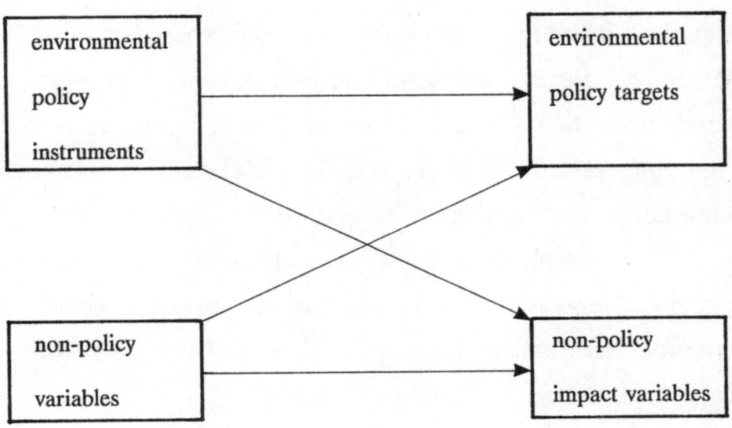

Figure 2. An environmental policy-response model.[4]

[4] Feedbacks such as from environmental policy targets on environmental policy instruments have been left out. In section 4, however, the policy instruments will be specified as endogenous variables.

In the context of Figure 2 the following remarks apply. First, it makes sense to distinguish between measurement of effects of environmental policy and environmental impact assessment. The latter focuses on changes in environmental quality as side effects of other policies than environmental policy, for instance regional policy. In this context environmental quality is a non-policy impact variable and e.g. investment subsidy a policy instrument. Secondly, adequate measurement requires that the effects of environmental policy on both environmental policy targets and non-policy impact variables should be taken into account.[5]

(iv) Environmental policy is likely to be differentiated over space and time. Let us first consider the spatial context. As the efficient level of environmental quality and, hence, environmental policy is determined by the intersection of the marginal benefit of control function and the marginal cost of control function[6], environmental policy is likely to differ in a system of regions with regional autonomy with regard to environmental quality if either or both of these functions differ. Also in a system in which equal environmental quality over the regions prevails environmental policy is likely to differ regionally because of differences in marginal emissions reduction costs functions. As pointed out above, efficient abatement requires that all emission generating sources control to the level at which their marginal emission reduction costs are equal. In the case of differences in environmental endowments this will lead to spatially differentiated environmental policy. So, environmental policy is likely to be differentiated according to regional conditions. Moreover, as the regional conditions change over time environmental policy is to be differentiated over space and

---

[5] It should be observed that ex ante studies, in particular the partial equilibrium approaches, usually focus on the relation between environmental policy variables on the one hand and the environmental policy targets or the non-policy impact variables on the other.

[6] The marginal costs of control function is uniquely determined by the shape and location of the marginal emission reduction costs function and the emission transfer function (which relates emissions to environmental quality).

time.

Three additional remarks apply in this context. First, there is substantial uncertainty in estimating marginal benefit and cost of control functions. These uncertainties relate to such matters as changes in processes, unanticipated technical developments and the fact that the impacts of pollutants on humans, plants and animals are still not well understood. Moreover, policy making does usually not take place in a sphere of total harmony. The trade-off between environmental quality and other goals is still a disputed matter, partly because of the uncertainties involved. The actual outcome of this process of policy making in disharmony under uncertainty is rather to increase than to reduce spatio-temporal differentiation. Secondly, both the environmental and the economic systems are known to be open systems. This applies to the local, regional and national levels. Hence, intra- and interregional effects as well as international effects of environmental policy should be taken into a account. Finally, responses of both environmental policy targets and additional impact variables may be considerably and differently lagged. When the time lag is not adequately taken into account the estimators of the policy effects may be substantially biased. For instance, the reduction of the emissions of cfc's would affect the ozone layer only after approximately a decade. Neglect of this delay in a time series model would give a negative effect of the policy instrument on the environmental policy target.

On the basis of the foregoing description the following requirements which an adequate ex post measurement method of environmental policy should meet, can be derived.

- Environmental policy usually consists of a combination of instruments or of a combination of applications of similar instruments in different sectors. Moreover, it may affect various indicators of both the environmental and the economic system. The former is likely to lead to multicollinearity in an ex post measurement approach of the regression type; the latter requires all the affected indicators of both the environmental and the economic system to be taken into account.

- Environmental policy targets may be affected by environmental policy instruments and by non-policy variables. Adequate measurement requires both kinds of variables to be included in the analysis.
- The spatio-temporal differentiation of environmental policy should be taken into account.
- Intra- and interregional as well as national and international effects should be considered.

In the next section some operational ex post measurement methods will be evaluated on the basis of these criteria.

3. Overview and Methodological Aspects of Some Operational Ex Post Measurement Approaches

The type of measurement approach to be used is dependent on inter alia the structure of the data available. Information is required at least with regard to one environmental policy variable and the policy instrument. The policy instrument may be a continuous variable or a (discrete) step variable representing a permanent change from a given level to a new level (for instance the implementation of a permanent standard) or a pulse variable representing a temporal change from a given level followed by the return to the original level (cf Bennett 1979, Fischer and Folmer, 1982). On the basis of Figure 2 the following types of data sets can be distinguished:

(i) A data set consisting of observations with regard to the environmental policy target and the policy intrument. If the policy target and the policy instrument have been measured at one location over time one has a bivariate time series; if both variables have been measured over time at different locations one has a spatio-temporal data set.

(ii) A second kind of data set consists of observations on both policy instruments and non-policy variables as well as the environmental policy target. Observation on the variables may be available in the form of multivariate time series or multivariate spatio-temporal data.

(iii) A third and last kind of data comprises observations on both policy and non-policy variables and on both policy targets and non-policy impact variables. It should be noted that there is usually a high degree of correspondence between the (economic) non-policy variables and the non-policy additional impact variables. For instance, the former may be a new kind of investment based on cleaner technology and the latter output or employment. Observations on the variables may be available again in the form of a multivariate time series or a spatio-temporal data set.

In the sequel of this section statistical models for these data sets will be described and evaluated. Before going into detail we will briefly discuss a method where the policy variable is a discrete step variable and where there are observations on the policy target for both the policy-off and the policy-on period. Because of the limited information required, this approach has become rather popular in other fields e.g. regional policy analysis (see amongst others Hart (1971), Frost (1975), Begg et al (1976), Folmer (1986)).

The method is basically an extrapolation of the policy target for the policy-on period on the basis of a univariate time series of this variable for the policy-off situation. It rests on the assumption that the development as described by the model in both the policy-on and the policy-off period is the same. This assumption may be violated, especially when a development from a short policy-off period is extrapolated over a long policy-on period. The difference between the actual development and the policy-off trend is defined as the policy effect.

In the case of a univariate time series of the goal variable, the extrapolation may be made by means of e.g. seasonal autoregressive integrated moving average models (cf Box and Jenkins, 1976). If there is evidence of adequate extrapolation this method is a simple and easy device to estimate policy effects. The difference between the actual development and the policy-off trend, however, is usually influenced by both policy and non-policy variables. In as far as the latter are not taken into account the method gives biased estimates of the policy effect. (cf. see among others Theil (1957), Dhrymes (1978), Folmer (1986).

The inadequacy encountered here can be handled by means of the application of two-stage time series analysis. This method goes as follows. Required are a univariate time series of the policy target for the pre-intervention period and a multivariate time series of the policy target and the instruments for the policy-on period. The first step is to model the pre-intervention series, e.g. as an ARIMA model (Box and Jenkins (1976)). At the second stage the effects of the non-policy variables are removed from the intervention series by means of a transformation of the (multivariate) intervention series on the basis of a model estimated for the pre-intervention series.

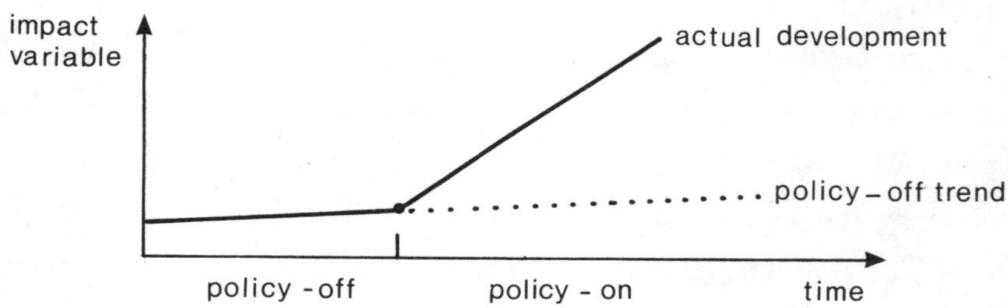

Figure 3. Schematic representation of the trend extrapolation ex post measurement method.

This step is based on the assumption that the relationship between the policy target and the non-policy variables in the intervention period is the same as in the pre-intervention period. Under the assumption of the independence of the policy instruments of the non-policy variables and an additive model structure, the effects of policy on the policy target can be estimated by standard techniques from the transformed multivariate time series. (For further details see Box and Tiao (1975) and Folmer (1986)).

In the case of the second kind of data sets multiple regression is in many cases an appropriate approach, as in Ashworth et al. (1987). If multiple time series for various regions are available spatio-temporal models ranging from models with all coefficients constant via models with constant slope coefficients and variable intercepts to models with variable intercept and slope coefficients could be applied. Moreover, dummy variable and error components models may be considered. Detailed information can be found in among others Judge et al (1985). Some alternative methods of handling spatial spill-over effects and, hence, spatially correlated residuals can be found in Folmer and Florax (1990).

The present methods make it possible to obtain estimates of the effects of the policy instruments on the policy target with the usual statistical properties. In the preceding section, however, it was argued that adequate measurement would require estimates of effects on both the policy target and the non-policy impact variables. The measurement methods discussed so far do not produce the latter. In that respect these methods are inadequate. In order to obtain estimates of both kinds of effects data sets of the third kind are required, which leads to the realm of simultaneous equation models. The distinction between time series and spatio-temporal models applies again. In order to estimate the direct effects of an instrument of policy on several variables (i.e. both policy targets and non-policy impact variables) all these variables should be incorporated into the models as endogenous variables. Each affected variable should be specified as a function of the instruments of policy and the non-policy variables. It should be observed that the additional impact variables may have impacts on the policy targets which implies indirect policy effects. For instance, a fuel tax may have a direct impact via a reduction of fuel consumption and an indirect effect via technical change. Adequate measurement requires that the indirect effects are also taken into account. (For further details see Folmer (1986)).

The linear structural equation model with latent variables belongs to the class of simultaneous equation models. It will be discussed in the next section.

4. Linear Structural Equation Models with Latent Variables

The linear structural equation model with latent variables (abbreviated as

LISREL models) has been developed by Jöreskog (1973a, 1973b), Goldberger (1972), Goldberger and Duncan (1973) and Jöreskog and Sörbom (1981, 1984, 1988). In this brief introduction we follow Jöreskog and Sörbom (1984, 1988).

A LISREL model is made up of two related submodels:

- A latent variables measurement model which represents the relationships between the latent variables and their observable indicators.[7]
- A structural model, representing the relationship between the latent variables.

The basic structure can formally be presented as follows. Let $y = (y_1, y_2,..., y_p)'$ and $x = (x_1, x_2,..., x_q)'$ be vectors of observable endogenous and exogenous variables, respectively. Furthermore, let $= (\eta_1,\eta_2,...,\eta_m)'$ be a vector of latent endogenous variables and $\xi = (\xi_1, \xi_2,..., \xi_n)'$ a vector of latent exogenous variables. Finally, $\epsilon = (\epsilon_1, \epsilon_2,..., \epsilon_p)'$ and $\delta = (\delta_1, \delta_2,..., \delta_q)'$ are defined as vectors of measurement errors of y and x, respectively. The relationships between the observed and latent variables are presented in the latent variables measurement models:

$$y = \Lambda_y \eta + \varepsilon \qquad (1)$$

and

$$x = \Lambda_x \xi + \delta \qquad (2)$$

where Λ_y and Λ_x are (pxm) and (qxn) matrices of factor loadings.
The structural model reads as

$$B\eta = \Gamma\xi + \zeta \qquad (3)$$

where

B is a (mxm) matrix with β_{ij} representing the effect of the j th endogenous variable on the i th endogenous variable ($\beta_{ii} = 1$).

---

[7] The variables of empirical sciences can be divided into observable and latent (or theoretical) variables. The former, such as age and income, possess direct empirical meanings derived from experience. Latent variables, however, refer to those phenomena that are supposed to exist but cannot be observed directly. The reason for this is either that those phenomena do not correspond directly to anything that is likely to be measured, or that observations of the phenomena concerned are contaminated with measurement errors. Examples of latent variables are "welfare", "socioeconomic status", "permanent income". Latent variables are given empirical meanings by means of correspondence statements which connect a latent variable with a set of observable indicators. For further details see among others Carnap (1956), Stegmüller (1970), Blalock (1971).

Γ is a (mxn) matrix with γ_{ij} representing the effect of the j th exogenous variable on the i th endogenous variable.
ς is a random vector of residuals.
The covariance matrices of ϵ and δ are Θ_ϵ (pxp) and Θ_δ (qxq) and the covariance matrices of ξ and γ are Φ (nxn) and Ψ (mxm), respectively.

The following remarks apply.

- It is usually assumed that the observed and latent variables are centralized, although this is not necessary.
- Usually, multiple observable variables for one latent variable are needed in order to avoid identification problem (see among others Goldberger (1972)).
- It is assumed that η and ξ are uncorrelated with ϵ and δ and that ϵ, δ and ς are mutually uncorrelated.
- Various less general models (e.g. factor analysis model, simultaneous equation model with observables only, the classical linear model) are contained in the general framework (1) - (3).
- Models (1) - (3) can be estimated by the LISREL VI or LISREL VII computer packages (Jöreskog and Sörbom (1984, 1988)). These programs also provide information about identification problems which may be used in addition to the usual conditions for identification. The LISREL VI and VII programs contain five kinds of estimators which are consistent in the case of fully identified models. The estimators are: instrumental variables; two-stage least squares; unweighted least squares, generalized least squares and maximum likelihood. Finally, the LISREL programs also produce several test statistics and modification indices which can be used when one is in search for an appropriate model.

Next we consider application of the LISREL model in the context of measurement of effects of environmental policy in a spatio-temporal framework. This implies that we have data on R regions over T periods. For the rth region the following vectors of variables are defined:

$\eta^{a}_{r,t}$ as a vector of jointly dependent endogenous affected variables (i.e. policy targets, and non-policy impact variables);

$\eta^{p}_{r,t}$ as a vector of endogenous latent policy variables;[8]

$\xi^{a}_{r,s}$ as a vector of lagged latent affected variables at time s, $s < t$;

$\xi^{p}_{r,s}$ as a vector of lagged latent policy variables at time s, $s < t$;

$\xi^{o}_{r,s}$ as a vector of exogenous latent non-policy variables (other than $\xi^{a}_{r's}$ and $\xi^{p}_{r's}$) at time s, $s < t$;

x^{c}_{r} as a vector of time invariant background variables.

It should be observed that concerning the observable variables a division similar to the one of the latent variables is made. That is, the vector $y^{a}_{r,s}$ corresponds to $\eta^{a}_{r,t}$, the vector $x^{p}_{r,s}$ to $\xi^{p}_{r,s}$ and so on.

With the notions formulated above the structural impact model for the periods max {m,u,v}, max {m,u,v} + 1,..., t reads as follows:

$$[B^{a}:B^{p}]\begin{bmatrix}\eta^{a}_{r,t}\\ \eta^{p}_{r,t}\end{bmatrix} = \sum_{s=t-m}^{t-1} \Gamma^{a}_{s}\,\xi^{a}_{r,s} + \sum_{s=t-u}^{t-1} \Gamma^{p}_{s}\,\xi^{p}_{r,s} + \sum_{s=t-v}^{t} \Gamma^{o}_{s}\,\xi^{o}_{r,s} + \Gamma^{c}\,\xi^{c}_{r} + \zeta_{r,t}\,, \quad r=1,2,\dots,R \qquad (4)$$

where

B^{a} represents the effects of $\eta^{a}_{r,t}$ on each other and on $\eta^{p}_{r,t}$;

B^{p} gives the effects of $\eta^{p}_{r,t}$ on each other and on $\eta^{a}_{r,t}$;

$\Gamma^{a}_{s}, \Gamma^{p}_{s}, \Gamma^{o}_{s}$ represent the effects, respectively, of $\xi^{a}_{r,s}$, $\xi^{p}_{r,s}$, $\xi^{o}_{r,s}$ on $h^{a}_{r,t}$ and $\xi^{p}_{r,t}$;

Γ^{c}_{t} represents the effects of ξ^{c}_{r} on $\eta^{a}_{r,t}$ and $\eta^{p}_{r,t}$.

The latent variables measurement models read as follows

$$\begin{aligned} y_{r,t} &= \Lambda_{y}\,\eta_{r,t} + \varepsilon_{r,t} \\ x_{r,t} &= \Lambda_{x}\,\xi_{r,t} + \delta_{r,t} \qquad\qquad r=1,2,\dots,R \qquad (5) \\ x^{c}_{r} &= \Lambda_{\xi^{c}}\,\xi^{c}_{r} + \delta^{c}_{r} \end{aligned}$$

For the sake of simplicity of notation model (4) is rewritten as

---

[8] Specification of the instruments as endogenous variables makes it possible to analyze the sensitivity of the instruments to the impact variables. If this is not wanted the instruments could be handled as exogenous variables.

$$B\,\eta_{r,t} = \Gamma^{a}\,\xi_{r}^{a} + \Gamma^{p}\,\xi_{r}^{p} + \Gamma^{o}\,\xi_{r}^{o} + \Gamma^{c}\,\xi_{r}^{c} + \zeta_{r,t} \qquad r=1,2,\dots,R \qquad (6)$$

where

$$B = [B^{a} : B^{p}]$$
$$\Gamma^{a} = [\,\Gamma_{t-1}^{a} : \Gamma_{t-2}^{a} : \dots : \Gamma_{t-m}^{a}\,]$$
$$\Gamma^{p} = [\Gamma_{t-1}^{p} : \Gamma_{t-2}^{p} : \dots : \Gamma_{t-u}^{p}\,]$$
$$\Gamma^{o} = [\,\Gamma_{t-1}^{o} : \Gamma_{t-1}^{o} : \dots : \Gamma_{t-v}^{o}\,]$$
$$\eta_{r,t} = [\,(\eta_{r,t}^{a})' , (\eta_{r,t}^{p})'\,]'$$
$$\xi_{r}^{a} = [\,(\xi_{r,t-1}^{a})' , (\xi_{r,t-2}^{a})', \dots, (\xi_{r,t-m}^{a})']'$$
$$\xi_{r}^{p} = [\,(\xi_{r,t-1}^{p})' , (\xi_{r,t-2}^{p})', \dots, (\xi_{r,t-u}^{p})']'$$
$$\xi_{r}^{o} = [\,(\xi_{r,t}^{o})' , (\xi_{r,t-1}^{o})', \dots, (\xi_{r,t-v}^{o})']'$$

Similar modifications are assumed to apply to the observable variables. Estimation of the spatio-temporal model outlined above has been extensively described in Folmer (1986).

We now turn to the problem why the LISREL approach is appropriate with regard to measurement of effects of environmental policy. As mentioned in section 2, environmental policy is usually made up of several instruments and/or standards. These instruments are usually highly correlated which is likely to lead to the problem of multicollinearity, which causes the estimated variances of the estimators of the coefficients of the collinear explanatory variables to increase. This may lead one to drop variables incorrectly from an equation. The structure of LISREL models may mitigate the consequences of multicollinearity. This can be seen as follows. Collinear explanatory variables are dependent variables in one of the latent variables measurement models and therefore are not removed from one of these models because of their collinear nature. Furthermore, in the structural model the latent variables appear instead of their corresponding observable variables. So, collinear observable variables are neither removed from the structural model inspite of the fact that they are collinear.

The policy targets and the non-policy impact variables may (also) be measured by means of various indicators. For instance, "air quality" may be indexed by means of NO_x, SO_2 and CO_2. The LISREL framework makes it possible to handle these variables simultaneously as indicators of the endogeneous latent variable air quality. In addition and related to these features is the possibility of the LISREL framework to handle spatial spillover effects adequately (see Folmer, 1986).

A limitation of the Lisrel approach is that it is restricted to linear-additive relationships. However, when non-linear or non-additive relationships occur they can often be transformed into linear-additive ones. Furthermore, if tansformations are not possible or not wanted the range of the variables is often small enough to allow for linear approximation.

To illustrate the LISREL approach, a very simple model is given here. Current and lagged output (OUT) in region r affect contemporaneous air quality (AIR). Air quality is operationalized by emissions of SO_2, NO_x and NH_3. In order to improve air quality two policy variables are used: money spent on catalysts (CAT) and money spent on combustion filters (FIL). These variables are the observable indicators of the latent variable instruments of policy (INS). This model is graphically represented in Figure 4.

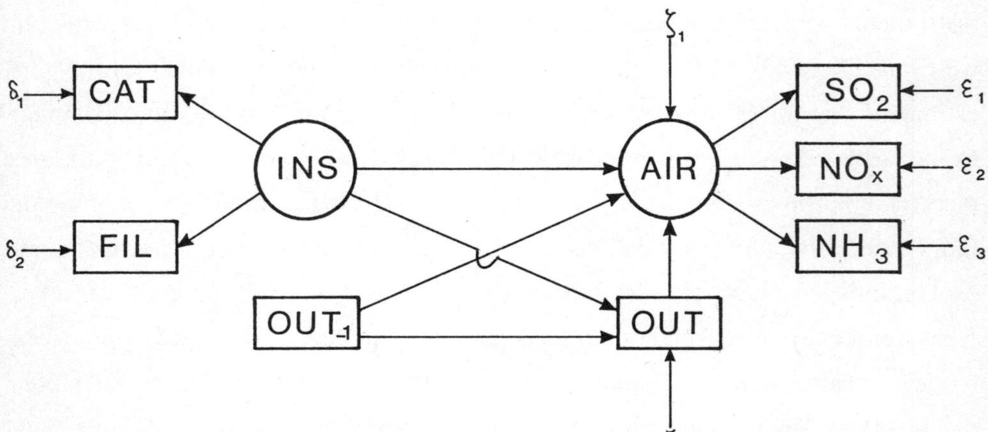

Figure 4. A simple example of a LISREL model.

In the notation introduced above, CAT and FIL are x-variables and INS is a ξ variable. Further, SO_2, NO_x and NH_3 are y-variables and AIR is a η variable. Output is assumed to be directly observable. The relationships among the latent variables, their observable indicators and the measurement errors make up the measurement models. The relationships among the latent variables AIR, INS, the directly observable variables OUT_{-1} and OUT and the ς-variables make up the structural model.

From the parameter matrices of the Lisrel model various kinds of effects can be derived:

- The direct effects of the η_t variables on each other by paths of length k, k > 1 are given by the matrix B^k.
- The total effects of the η_t variables on each other are

$$\sum_{k=1}^{\infty} B^k = (I-B)^{-1} - I. \quad (7)$$

  where I is the unitary matrix.

  (7) is only valid if the infinite series converges which is the case for all recursive models and for non-recursive models if and only if all eigenvalues of (I-B) are within the unit circle.
- The direct causal effects of the η_t variables on the y_t variables are given by Λ_y and the indirect effects by Λ_y $[(I-B)^{-1} - I]$.
- The direct effects of the ξ^p variables lagged for e.g. u periods on the η variables are given by $\Gamma^p{}_{t-u}$ and the total effects by B-1jG$^p{}_{t-u}$. Similar relationships hold for the other ξ variables.
- The total effects of ξ on y are given by Λ_y $(I-B)_{-1}$ Γ (Graff and Schmidt (1982)). This implies that the effects of e.g. $\xi^p{}_{t-u}$ on y_t are given by

  Λ_y $(I-B)^{-1}$ $\Gamma^p{}_{t-u}$.
- We will now turn to the effects of environmental policy over several periods. For the derivation to be presented below it is assumed that the effects on observable y variables are wanted. Effects on the η^a variables can be obtained in a similar way.

  Suppose the instruments were applied in period t^*, $t^* < t$. The effects may be realized in the periods t^*, $t^* + 1$, ..., t. As mentioned above, the total effects of the η variables (including the instruments of environmental policy) on y_{t*} are given by Λ_y $(I-B)^{-1}$ whereas the direct effects on y_{t^*+1}, y_{t^*+2}, ..., y_t without the intervention of latent policy targets or latent non-policy impact varibles are given by Λ_y $(I-B)^{-1}$ $\Gamma^{p}_{t^*+1}$, Λ_y $(I-B)^{-1}$ $\Gamma^{p}_{t^*+2}$, ..., Λ_y $(I-B)^{-1}$ Γ^{p}_{t} , respectively. The indirect effects of policy by way of latent policy targets or latent non-policy impact variables are given by $\Lambda_y(I-B)^{-1}\Gamma^{a}_{t^*+1}Q$, $\Lambda_y(I-B)^{-1}\Gamma^{a}_{t^*+2}Q$, ..., $\Lambda_y(I-B)^{-1}\Gamma^{a}_{t}Q$, where Q is the block of the matrix $(I-B)^{-1}$ - I which represents the total effect of the latent policy instruments on the latent impact variables.

- It may also be of interest to know what the effect is of an individual instrument on a latent or observable policy target or non-policy impact variable (η_1, say). Because the effects of the indicators on their corresponding latent variables are not contained in the parameter matrices of the Lisrel model, these effects are to be obtained indirectly. The simplest procedure to obtain these effects is by adding an additional equation to the structural model in which the individual instrument concerned is separated from the combination. As an example, consider the latent policy instrument η_2 made up by the observables y_3, y_4, y_5. If the effect of, say y_5, is wanted, the following structural model is specified:

  $\eta_1 = \beta_{12}\eta_2 + \beta_{13}\eta_3 + \xi_1$

  where η_3 corresponds to y_5 and y_5 is left out as indicator of η_2.

5. Conclusions

Ex post measurement of effects of policy is an essential step in the process of environmental policymaking. Very few studies, however, have focused on this step up to now.

In the first part of the paper various requirements which an adequate measurement method should meet are analyzed. Next various measurement methods are reviewed and evaluated on the basis of these requirements. It has been argued that in order to meet all requirements simultaneous equation models are needed. Because of the variation of environmental policy over space and time, spatio temporal data should preferably be analyzed.

It has also been argued in this paper that the LISREL model is particularly appropriate to measure effects of environmental policy because in the framework of this model latent and observable variables can be handled simultaneously. This implies that the frequently encountered problem of multicollinearity, which is likely to occur in the context of measurement of effects of policy, can be circumvented or mitigated. Multicollinearity is likely to occur in measurement studies because of the combination of policy instruments or the application of similar instruments in different sectors in different degrees. Finally, LISREL models are appropriate to handle spatial spill-over effects.

References

Adams, R.M., S.A. Hamilton and B.A. McCarl (1986), The Benefits of Pollution Control: The Case of Ozone and U.S. Agriculture, American Journal of Agricultural Economics 68,4:886-93.

Ashworth, J., I. Papps and D. Storry (1987), Assessing the Impact upon the British Chlor-Alkali Industry of the EEC Directive on Discharges of Mercury into Waterways, Land Economics 63,1:72-8.

Begg, H.M., C.M. Lythe and D.R. McDonald (1976). The Impact of Regional Policy on Investment in Manufacturing Industry. Scotland 1960-71. Urban Studies 13,2:171-179.

Bennett, R.J. (1979) Spatial Time Series. Pion, London.

Blalock, H.M. Jr. (1971). Theory Building and Causal Influences, in: H.M. Blalock, Jr. and A.B. Blalock (eds) Methodology in Social Research, London, McGraw-Hill.

Bohm, P. and C.S. Russell (1985), Comparative Analysis of Alternative Policy Instruments, in: A.V. Kneese and J.L. Sweeney (eds.), Handbook of Natural Resource and Energy Economics, Amsterdam, North-Holland.

Box, G.E.P. and G.C. Tiao (1975), Intervention Analysis with Applications to Economic and Environmental Problems, Journal of the American Statistical Association 70,3:70-9.

Box, G.E.P. and G.M. Jenkins (1976), Time Series Analysis: Forecasting and Control, San Fransisco, Holden Day.

Carnap, K. (1956), The Methodological Character of Theoretical Concepts, in: H. Feigl and M. Scriven (eds.), Minnesota Studies in the Philosophy of Science I, Minneapolis, University of Minnesota Press.

Dhrymes, Ph. J. (1978), Introductory Econometrics, Berlin, Springer.

Dickie, M. and S. Gerking (1989), Benefits of Reduced Morbidity from Air Pollution Control: A. Survey, in: H. Folmer and E.C. van Ierland (eds.), Valuation Methods and Policy Making in Environmental Economics, Amsterdam, North-Holland.

Fischer, M.M. and H. Folmer (1982), Measurement of effects of regional policy by means of time series analysis. Papers of the Regional Science Association 49:133-151.

Folmer, H. (1986), Regional Economic Policy. Measurement of its Effects, Boston, Martinus Nijhoff.

Folmer, H. and R. Florax (1990), The Detection of Spatially Correlated Disturbances in Linear Models. Paper available at the Department of General Economics, Agricultural University Wageningen, The Netherlands.

Folmer. H. and P. Nijkamp (1986), Methodological Aspects of Impact Analysis of Regional Economic Policy, Papers of the Regional Science Association 57:165-181.

Frost, M.E. (1975), The Impact of Regional Policy: A Case Study of Manufacturing Employment in the Northern Region, Progress in Planning 4:169-237.

Goldberger, A.S. (1972), Structural Equation Models in the Social Sciences, Econometrica, 40,6:979-1002.

Goldberger, A.S. and O.P. Duncan (Eds) (1973), Structural Equation Models in the Social Sciences, New York, Seminar Press.

Graff, M.A. and P. Schmidt (1982). On decomposition of effects in causal models, in: K.G. Jöreskog and H.A.O. Wold (eds.), Proceedings of the Conference Systems under Indirect Observation. Causality-Structure-Prediction, Amsterdam, North Holland.

Hart, R.A. (1971), The Distribution of New Industrial Buildings in the 1960s, Scottish Journal of Political Economy, 18,2:181-197.

Highton, N.H. and M.G. Webb (1984), The Effects on Electricity Prices in England and Wales of National Sulphur Dioxide Emissions for Power Sta-tions, Journal of Environmental Economics and Management 11,1:70-84.

Johansson, P.O. (1989), Valuing Public Goods in a Risky World: An Experiment, in: H. Folmer and E. van Ierland. Op. cit.

Jöreskog, K.G. (1973a). A General Method for Estimating a Linear Structural Equation System, in: A.S. Goldberger and O.D. Duncan (eds.), Structural Equation Models in the Social Sciences, New York, Seminar Press.

Jöreskog K.G. (1973b), Analysis of Covariance Structures, in: M. Krishnaiah (ed.), Multivariate Analysis III, New York, Academic Press.

Jöreskog K.G. and D. Sörbom (1978), LISREL IV. A General Computer Program of Estimating of Linear Structural Equation Systems by Maximum Likelihood Methods. Department of Statistics, University of Uppsala, Sweden.

Jöreskog, K.G. and D. Sörbom (1984), Lisrel VI. LISREL VI. Analysis of Linear Structural Relationships by the Method of Maximum Likelihood. Users Guide. Department of Statistics, University of Uppsala, Sweden.

Jöreskog, K.G. and D. Sörbom (1988), LISREL VII. A Guide to the Program and Applications. Department of Statistics, University of Uppsala, Sweden.

Judge, G.G. and W.E. Griffiths, et al. (1985), The Theory and Practice of Econometrics, 2 nd Edition, New York, Wiley.

Kolstad, C.D. (1986), Empirical Properties of Economic Incentives and Command-and-Control Regulations for Air Pollution Control, Land Economics 62,3:250-268.

Merrifield, J.D. (1988), The Impact of Selected Abatement Strategies on Transnational Pollution, the Terms of Trade, and Factor Rewards: A General Equilibrium Approach, Journal of Environmental Economics and Management 15,3:259-84.

Navrud, S. (1989), Estimating Social Benefits of Environmental Improvements from Reduced Acid Rain Deposition: A Contingent Valuation Survey, in: H. Folmer and E. van Ierland, Op. cit.

Shechter, M., M. Kim and L. Golan (1989), Valuing a Public Good: Direct and Indirect Valuation Approaches to the Measurement of the Benefits from Air Pollution Abatement, in: H. Folmer and E. van Ierland, Op. cit.

Siebert, H. (1987), Economics of the Environment, Berlin, Springer.

Stegmüller, W. (1970), Probleme und Resultate der Wissenschaftstheorie und Analytischen Philosophie 2, Berlin, Springer.

Theil, J. (1957), Specification Errors and the Estimation of Economic Relationships, Review of the International Statistical Institute 25,1:41-51.

Tinbergen, J. (1956), Economic Policy: Principles and Design, Amsterdam, North Holland.

Tinbergen, J. (1977), On the Theory of Economic Policy, Amsterdam, North Holland.

Chapter 19

REGIONAL MEDICOMETRY: HEALTH EXPENDITURES, REGIONAL DISPARITIES, PROBLEMS AND POLICIES

A.S. Bailly
W.J. Coffey

1. From Medical Geography to Regional Medicometry

The fields of medical geography and of health economics have had long and fruitful histories. The analysis of the spatial distribution of specific illnesses and of the variations in levels of physical well-being is an ancient practice, having been initiated at the time of Hippocrates. In the 18th century, in order to better understand the antecedents of certain illness, physicians began investigating the relationships between the type and frequency of illnesses, on the one hand, and aspects of the human and physical environment on the other hand. In the 20th century, epidemiologists and geographers have further refined the understanding of the diffusion of various illnesses and of their associated causes and consequences.

In the 20th century as well, an important conceptual transition has been made in the general way in which medical issues are treated; there has been a shift in emphasis from the notion of illness to that of health. Rather than emphasizing the spatial variations in illness or its causes, researchers in various fields have begun to become more interested in the areas of preventive medicine and public health. The preamble to the constitution of the World Health Organization, written in 1947, exemplifies this approach, defining health as 'a state of complete well-being: physical, mental and social'.

In a similar manner, the investigation of the impacts of illness upon the efficiency of production systems, and of the costs to individuals and to society of providing adequate levels of health care has evolved continuously over the past two centuries; health economics has emerged as a major field within the discipline of economics. In its modern form, it is largely characterized by the measurement of health care expenditures and the modelling of the financial impacts of various health care policies.

Both of these perspectives, in spite of their inherent utility, may be regarded as partial approaches, ones that address limited aspects of the complex issues related to spatial variations in levels of health and in the provision and cost of health care. Given the high degree of interdependency between questions involving levels of health and those involving the costs of providing adequate health care, it is necessary that these two issues be considered together. Further, neither the medical geography nor the health economics approach is based upon an original theoretical framework.

In this paper, we sketch the outlines of the newly emerging field of **regional medicometry**, an interdisciplinary undertaking which employs the combined viewpoints and the methods of geography, epidemiology, economics, sociology, and regional science to attempt to develop a comprehensive analysis of the multifaceted aspects of modern health care systems. It should be stressed that, since the field is still in its infancy and is developing rapidly, the portrait that we present is perhaps a transitory one. We begin by attempting to identify the theoretical underpinnings of the regional medicometric approach. We then stress the importance of adopting a broad analytic perspective, one that explicitly recognizes the principal actors in the system and their specific objectives, as well as the complex interrelations between the system and its environment. This is followed by the application of such a broad perspective, using the case of Switzerland. Finally, the importance of health care activities in the context of regional development is emphasized.

2. Regional Medicometry: Towards a Theoretical Framework

The strengths and weaknesses of the approaches briefly reviewed above provided the foundation for the reflections made by J. Paelinck and A. Bailly, who in 1980 proposed to break out of the constraints of disciplinary boundaries in order to advance a new branch of regional science - regional medicometry. The newly emerging field was initially defined by these researchers as 'the application of mathematical and statistical methods to the testing of medical regularities in space'. Rather than attempting to create a global theory, early research in regional medicometry sought to identify a number of logical methodological stages which could contribute to the process of medical decision making. These stage include:

- the **specification** of a model (or of a functional relationship, or a set of functional relationships), indicating the manner in which medical variables interact;
- the **estimation** of the parameters of the model, taking into account the manner in which these parameters are in themselves a product both of the modelling framework and of the manner in which observations have been combined;
- the **testing** of the elements of the model and of the complete model, in order to estimate its robustness;
- finally, the identification of the model's **utility for forecasting**, given the specific uses for which it is intended.

This definition was rapidly expanded by integrating the notions of human beings, and of the territories in which they live, into medicometric analysis. This evolution was based upon five principles (P1-P5 below) presented by J. Paelinck at the first Conference on Regional Medicometry (Bailly and Periat, 1984).

P1 The models of regional medicometry should, a priori, be formulated in a spatially interdependent manner. This interdependence is fundamental, being based on the spatial mobility of human beings; the diffusion of epidemics is one classic example. More generally, however, it is acceptable that there be a spatial autocorrelation between health indicators and those of morbidity. This means that, in the specification of our models, interdependence should be a central principle for formulating relationships for which the parameters are to be estimated and the applicability is to be tested.

P2 A certain number of spatial relationships will be asymmetric. This principle is illustrated by the differing demographic structures of regions and by their varying endowments of pathogenic factors. Further, certain regions are very well equipped to provide health care, and others much less so.

P3 Medical-related phenomena located in one area will often be explained by causal factors located in other areas. The examples of this principle are numerous: the possibility of catching an illness that does not manifest itself in the local area due to travel, the location of hospitals that specialize in non-indigenous diseases, etc.

P4 Spatial diffusion is a fundamental principle of regional medicometric analysis. A process or a phenomenon may have its strongest impacts in a non-contiguous, and perhaps far-removed, area. Thus it becomes necessary to verify the nature of spatial interdependencies and the spatial distribution of impacts. The movement and interaction of the human population plays a major role in the process of spatial diffusion.

P5 Two-dimensional space is the fundamental locus of medical activities and phenomena. To this must be added the third dimension of time. Investments, the effects of policies, the diffusion of innovations, and all forms of human behaviour and interaction, for example, all occur in both space and time.

Our experience in regional science, in quantitative and medical geography, and in health economics has shown us that, in order to respect these principles, it is necessary to employ not only our disciplinary perspectives, but above all to incorporate the viewpoints of all persons participating in the medical health system - physicians, trained health personnel of all types, hospital directors, the administrators of health insurance programmes and of pharmaceutical firms, and the direct consumers of health services. Such a global approach is based upon the desire to discover the underlying interdependencies that characterize the health care system, and the ability to rigorously measure and to model these interdependencies. Useful medicometric research which is able to contribute to the orientation of the medical and health care policies of our society must be based upon such a multidimensional approach, for it must not neglect any of the actors in the health system. This consideration leads us to propose a broader and more appropriate definition of regional medicometry: **the application of mathematical and statistical methods to the testing, the evaluation, and the prediction of medical regularities in space, in such a manner as to incorporate the viewpoints of all the actors** in the health system, and with an overall concern for issues of efficiency and spatial equity.

3. The Health Care System: A Global Perspective

Figure 1 is a simple representation which summarizes our conception of the field of regional medicometry, taking into account four types of conflicting criteria:

- economic:
  . medical costs vs. social costs;
  . health care supply vs. patient demand;
  . group payments vs. individual payments;
- social:
  . collective vs. individual preferences;
  . equality-solidarity vs. individualism-subjectivism
  . qualitative goals vs. quantitative goals.
- spatial
  . national vs. regional/local objectives;
  . national vs. regional/local needs and performance.
- temporal
  . short-term vs. long-term objectives;
  . continuation of past/present practices vs. innovation.

Even if all the actors in the health care system are aware of the utility of establishing objectives dealing with education, efficiency, and prevention, as soon as one begins to speak of the allocation of scarce resources serious differences of opinion appear concerning these four types of criteria. Which should be given the greatest priority: the economic, the social, the spatial or the temporal? Between the extremes represented by the notion of an unlimited growth capacity and that of the limits to growth, serious problems of choice appear. This is true even if, following the objective of economic efficiency, one attempts to maximize the health care impact of a

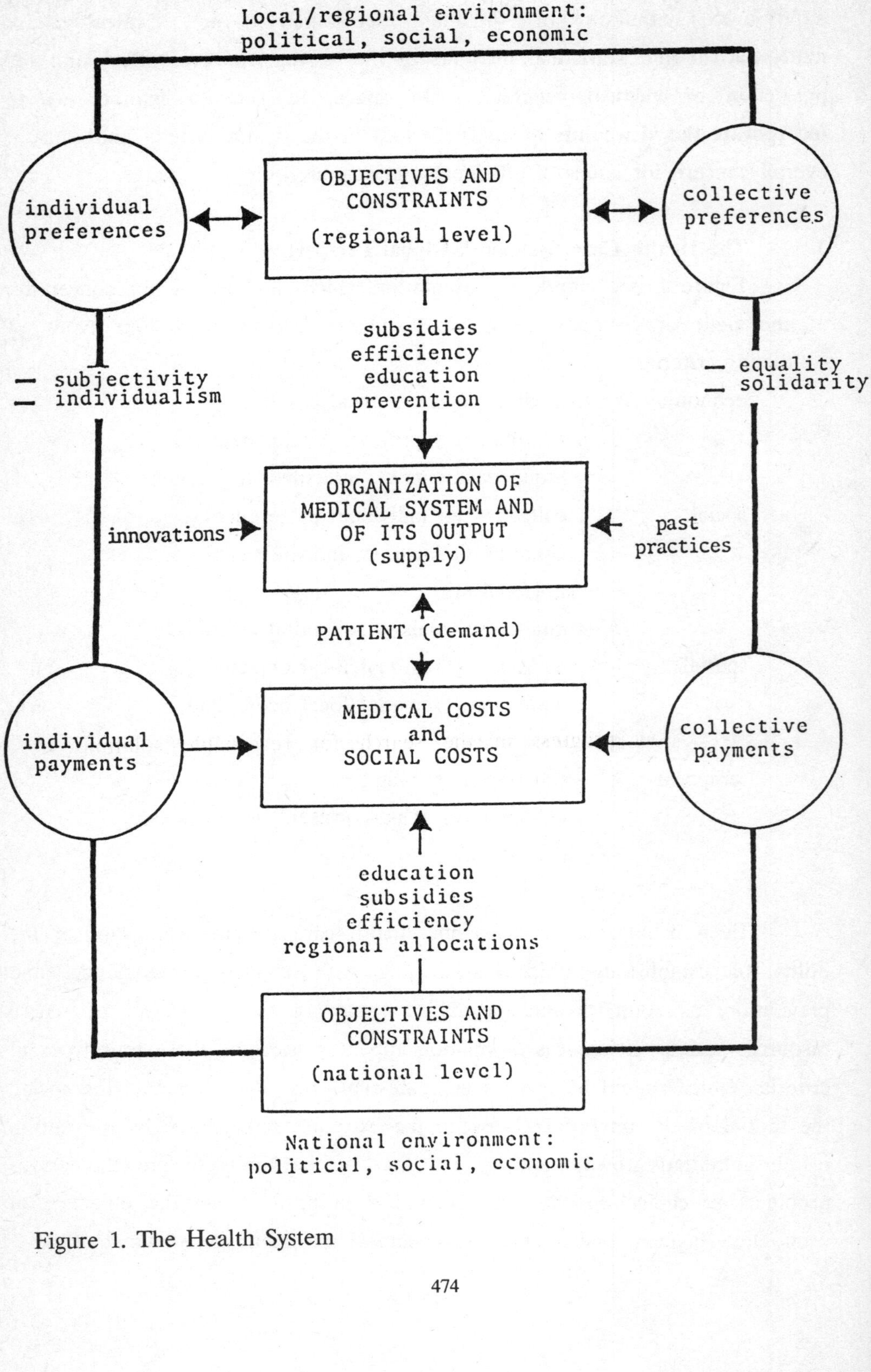

Figure 1. The Health System

given amount of spending. Should one favour that which is considered as indispensable from the viewpoint of society or of the individual? And at what geographical scale: that of the nation or that of the region? Should one attempt to create systems that will endure, or one that will efficiently solve a problem in the short term? In the words of Rostand (1954), 'too often the choice that reality offers us is of a type which makes us wish to avoid choosing.'

Regional medicometry refuses to accept this unwillingness to act. The problem of allocating expenditures is not only one of knowing which objectives to favour, and thus of what type of action to take; it is also a question of knowing where to act and for what period. This is true not only for policies related to hospitals and health insurance, but also for all of this interrelated set of elements which comprise the health care system.

Regional medicometry is based upon four perspectives that are global in nature and that recognize the important role played by space and by society:

1. **The explicit recognition of the system's environment**. Due to the insertion of the health care system within the broader context of economic and social systems, it is necessary to conceptualize it in terms of alternative allocations of scarce resources and of highly complex interrelations. Economic, social and ethical criteria are operative at this level, and may indeed override considerations that are strictly based upon health care criteria.
2. **A broad interpretation of efficiency**. The notion of efficiency requires one to examine questions dealing with medical investments and expenditures from the perspectives of alternative strategies and of social responsibilities, as well as from the perspective of pure economic efficiency. In addition, this notion requires us to analyze the demand for health services in the framework of flexibility of choice, and to regard the supply of health care as a function of both the present and future states of the system of health care.
3. **The geographical scale of intervention**. All health-related policies must form a coherent system in which objectives and efficiency criteria must

be defined in a global manner, taking into consideration the various spatial scales at which intervention can occur: a top-down approach that gives priority to broad national objectives and constraints; or a bottom-up approach that gives priority to local objectives and constraints. The allocation of expenditures can only approach an optimal form when the spatial context is specified.

4. **The temporal scale of intervention**. Similarly, the objectives and efficiency criteria of all health-related policies must be defined with reference to the temporal scales at which intervention can occur: a short-term approach that emphasizes the resolution of immediate problems; or a long-term approach that gives priority to the development and continuing support of infrastructure and comprehensive programmes, and to addressing major structural problems such as the aging of the population.

These four perspectives make it readily apparent that all narrow approaches to the problem - narrow in the spatial, temporal or professional sense - have a high probability of failure due to their inability to reflect the complexity of the system and the complex motivations of the actors in the system. This type of narrow approach is one of the principal factors underlying the slow progress in the search for reasonable solutions to questions of health care and of health insurance. In Switzerland, for example, the system is so complex that the agreement of all partners must be obtained before any definitive decision can be taken. Such a consensus is impossible when the actors adhere to their narrowly defined perspectives. If the association of health insurance organizations is opposed to a certain policy, it can request a national referendum on the question by obtaining the required number of signatures. If rates for certain medical services have limits imposed upon them, the Federation of Swiss Physicians will seek a referendum. If certain measures of health care planning are suggested, the Association of Private Clinics or the Swiss Association of Hospitals may be opposed. The democratic system thus becomes paralysed, especially since the decisions of the national legislative body are not always accepted by the citizens. 'When the base can no longer establish contact with the summit, and when the summit no longer knows its base, the lack of concern of each element for its counterpart will lead to all sorts of excesses' (Bailly and Periat, 1987), all the more so since each professional group, fearing the loss of acquired advantages, behaves in a highly conservative fashion, effectively blocking any possible dynamism in the system. The example cited here is by no means exceptional.

4. An Alternative Perspective: The Example of Switzerland

Regional medicometry offers an alternative to this highly fragmented set of approaches, based upon a set of competing narrow viewpoints, by conceptualizing the health care system as an integrated whole. This alternative perspective is illustrated in Figure 2, using the specific example of the Swiss system. This model explicitly identifies the methods of financing the system (payments by individuals, 84%; contributions by the federal government, 11%; and by cantons and communes, 5%) and the principal actors in the system (hospitals and clinics, physicians, pharmaceutical industry, para-medical services, health insurance organizations, other medical-related services). Such an approach, one which also integrates the spatial dimension, enables one to examine the evolution of the health care system.

What types of changes have occurred in the system over the past 20 years? In the case of Switzerland, the payments by individuals have been increased in order to balance the deficits incurred by health insurance organizations. Such an approach represents the control of individuals' payments by an outmoded administrative structure; these payments, devoted not only to the daily medical problems of individuals, but also to the maintenance of the entire health care system, are expensive. As will be seen below, however, they also represent an important source of jobs and of various types of economic activities.

The general perception of the Swiss health care system is of a very expensive undertaking in which the individual bears the majority of medical-related costs. This approach ignores the economic importance of the sector

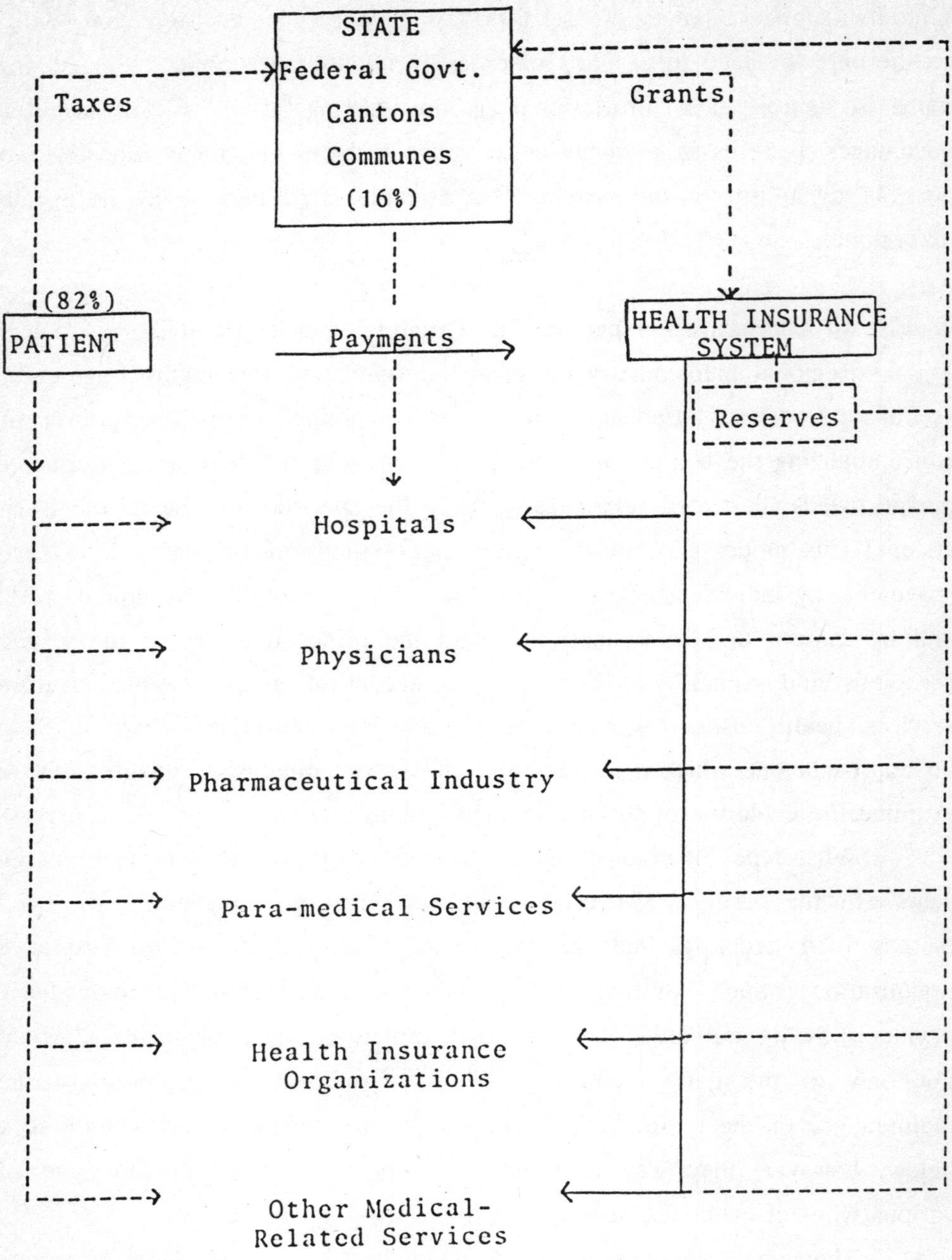

Figure 2. The Swiss Health System

which directly or indirectly supports one person out of six[1], and which constitutes one of the principal bases of the quality of life in Switzerland. What would happen to the thousands of jobs directly linked to health care in the event of the restriction of health care expenditures, especially in peripheral regions? When one takes account of the fact that approximately 80% of the costs of health care are spent on salaries, it is clear that health care expenditures have an important impact upon government revenues at various levels, and upon local business. Only 20% of health care expenditures are destined to the purchase of medicines, health care products and materials; but these industries pay taxes too. Thus to ignore this important economic sector, or to fail to take into consideration its real contribution to the national economy, constitutes an economic heresy. The provision of health services is expensive; but even if only the financial benefits of these services are considered, it is clear that they produce enormous benefits. The direct taxes coming from the 'medical world' finance a large proportion of the expenditures made by the state (Figure 3). Further, a similar argument can be made concerning all the benefits directly inherent in medical progress.

Regional medicometry approaches these types of questions from a global perspective, and this is one of the field's major points of originality. That which is often considered as an expense is in fact an investment in the economic system and in the quality of life of a country.

The classic conflict between the long-term objectives of health professionals and the short-term objectives of political authorities is presently one of the most serious problems for health systems. By seeking to reduce expenses there is the danger of cutting back on essential high level research. To cite a specific example, attempts to reduce expenditures by buying less expensive generic medicines constrain the ability of pharmaceutical firms to

---

[1] The fact is sometimes overlooked that in 1987 this economic sector, as it should properly be called, directly employed mor than 200,000 persons, 140,000 of which worked in hospitals and 20,000 of which were physicians. Using a conservative assumption of a multiplier with a value of 1.6, direct and indirect employment related to the health care system amounts to 320,000 persons; with the dependents of these persons included, 1,000,000 persons out of a population of 6,000,000 are supported by health care activities. This figure illustrates the economic significance of the health care sector in an affluent country that provides a high level of care to its population.

Expenditures:

Federal Government:	894 million francs
Cantons:	4,566 million francs
Communes:	2,883 million francs
TOTAL:	8,343 million francs

Revenues: (direct taxes; conservative estimates)

Taxes from salaries of hospital personnel:	1,030 million francs
Taxes from physicians' incomes:	751 million francs
Taxes from dentists' incomes:	423 million francs
Taxes from the sales of pharmaceutical products:	425 million francs
Taxes from the salaries of other personnel:	260 million francs
TOTAL:	2,889 million francs

To this total must be added the taxes paid by pharmaceutical firms, by personnel working in other health-related sectors, and by all persons indirectly supported by the health sector in Switzerland. Our estimate is that the addition of these revenues produces a balance between total expenditures and total revenues.

Figure 3. Direct Government Expenditures and Revenues Related to Health Care, Switzerland 1983

conduct further research. Long-term welfare is thus sacrificed to the exigencies of the moment; investment is forgotten in the concern for balancing today's budget.

It must be remembered that the concepts of well-being and of social justice imply that all persons have an equal access to health and to health care. Thus society is confronted by a choice: it can accept to pay the price of progress and of improved performance in the area of health care; or it can return to a less advanced system of health care, one that does not make full use of modern medical technologies. The concern for improved levels of well-being and of quality of life that is so apparent in our modern society is not compatible with this second alternative; rather, this concern is manifested in the establishment of health care policies which put the principal emphasis upon the well-being of the individual, and then upon the role of health care professionals in providing this well-being.

The objective of health care policies should not be the reduction of costs, but rather the promotion of **rationalization and efficiency** in the system. To seek to reduce the costs of health care is to ignore the economic and social impacts of health care related activities upon the entire nation, and especially upon the peripheral regions where the jobs in this sector are relatively numerous. On the other hand, to rationalize the health care system is to put in place a more efficient delivery system, one appropriate to a society based upon technology and upon the provision of high order services. It is through seeking this efficiency that all the actors in the system will assume their appropriate role. Efficiency can only be achieved by a complementarity and cooperation between the actors, not by the competition between them. It is through the use of this broad perspective that regional medicometry is able to propose frameworks and models that permit narrow corporatist approaches to the health care system to be surpassed.

5. Health Care and Regional Development

The set of activities that constitute elements of the health care system have considerable potential for stimulating economic development, not only in metropolitan areas but in peripheral regions as well. The spatial distribution of these activities is significantly different from that which characterizes the other consumer services; the distribution of the population is not necessarily the principal locational factor where health services are concerned. This difference may be attributed, in certain cases, to specializations based upon physical amenities (e.g., sanatoria located in mountain regions, clinics near thermal waters); in other cases, the difference may be attributed to functional health care specializations developed over time in particular regions.

Health care services can constitute an important element of the economic base of a region. Not only are regional hospital complexes often the principal employer in peripheral regions, but their presence also stimulates the creation of a set of related activities, e.g., accounting, computer services, and research functions (Coffey et al., 1981). In addition to often being the most rapidly growing sector of the economy, health care functions often also constitute export-oriented activities which can produce important injections of funds into the local economy; through the multiplier mechanism and the circular flow of income, these health care activities serve a propulsive function, stimulating local economic growth. Thus, rather than being relegated to the set of 'nonproductive' residentiary activities, following the traditional Fisher-Clark typology, a significant proportion of health care related functions must be regarded as basic, in that they are not only exportable but are also highly responsive to external demand.

If regional development is defined in terms of job creation, growth stimulation, and improvement in the quality of life, health care functions are capable of making significant contributions in this area, generating the production of earned market income and the creation of structural change. Government expenditures on health care activities will thus not only enhance the level of well-being of local residents, but will also have a significant impact upon the level of economic development in the local region. It must be noted, however, that because of their perception as expenses rather than as investments, the notion of health care expenditures as instruments of regional development policy is still relatively novel. An incomplete understanding of the economics of the health care system, of the interrelations between the various actors in the system, and of the complex interactions between the health care system and the broader social-economic environment in which it is imbedded is the principal reason that the traditional, narrow perspective remains dominant. The expansion of this perspective, and the recognition of the role of health care activities as elements of economic development is one of the benefits of adopting the regional medicometric approach.

6. Conclusions

Considerations of social justice and of well-being imply equality of health care. It therefore becomes necessary to accept the price of progress and to change the manner in which we make our health care expenditures. As we have shown, such expenditures can be considered as a basic investment in the future of society. Health care decisions need to be made in such a way as to improve both the quality of our life and the efficiency of our economic system. The spatial and temporal dimensions are very important elements of this decision-making process. By making health care decisions at a regional level where hierarchies of health, social and economic needs can be defined, both improved health care and enhanced economic development will result. Similarly, it is necessary for health care professionals, policy-makers, and social science researchers to engage in debate concerning the value of a long-term perspective on health care investment, and concerning methods of maximizing the impact of this highly productive element of all advanced economies.

Regional medicometry is, simultaneously, a rigorous technical field that utilizes sophisticated modelling procedures, and an applied field concerned with the formulation, and with the criticism, of public policy. The principal distinguishing characteristic of regional medicometry is its broad, integrative perspective. As has been demonstrated, the field incorporates the analysis of three aspects of the health care system: from its origins in the mathematical modelling of the variation of medical regularities in space, regional medicometry has expanded to include issues related to the financing and to the efficiency of health care delivery systems; once again the spatial and temporal dimensions are preeminent. These two aspects - variations in level of well-being and questions of financing/efficiency - are complementary, involving issues of health service supply and demand. Regional medicometry has also come to include a third aspect: the analysis of the health care system as a major economic sector, and as such explicitly treats the implications of this sector for economic development, especially in peripheral regions.

After four Regional Medicometry conferences held in Switzerland since 1983, it is evident that considerable progress is being made in understanding the causes of rising health care costs, and of regional disparities in well-being and in the provision of health care services. Similarly, the social and economic costs of illness, the economic impact of the health care system, and the changes in system that are currently being made due to new medical technology are now better understood. A large and complex research agenda remains to be addressed, however. This agenda involves: first, the refinement of the epistemological bases of the field, so as to better understand the role of health and health care in their broad social and economic context; second, the further development of modelling capabilities that are able to treat the multifaceted aspects of the health care system, as well as the broad environment in which it is embedded; and, third, the establishment of data bases that will permit a careful analysis of the set of issues raised here.

Regional medicometry is a new and important field. The growing literature based upon this perspective indicates that it is well on its way to becoming an established subdiscipline of regional science.

References

Bailly, A., F. Bridel and M. Periat (1987), La Santé: Perspectives Médicales et Economiques, Geneva and Paris, ERESA-Economica.

Bailly, A., F. Bridel and M. Periat (1989), L'assurance Maladie: Solutions Actuelles et Futures, Geneva and Paris, ERESA-Economica.

Bailly, A. and M. Periat (1984), Médicométrie Régionale, Paris, Economica-Anthropos.

Bailly, A. and M. Periat (1985), Médicométrie Régionale, Revue d'Economie Régionale et Urbaine, special issue.

Coffey, W.J., W.S. Macdonald, K.S. Wood, D. Liew, H. Galarneau, G. Karaska, W. Cochrane and H. Giese (1981), The Feasibility of a Nova Scotia Economic Activity Complex Based in the Medical Health Sector. Report prepared for the Canadian Department of Regional Economic Expansion and the Nova Scotia Department of Development.

Gilliand, P. (ed.) (1986), Les Coûts et l'Assurance, Lausanne, Réalités Sociales.

Gilliand, P. (ed.) (1986), Les Professionels, Lausanne, Réalités Sociales.

Mechanic, D. (1980), Readings in Medical Sociology, New York, Free Press.

Paelinck, J.H.P. (1982), in: K. Smith and B.V. French (eds.), Proceedings of a Cost-Benefit Symposium, Rotterdam, Netherlands Economic Institute.

Pharma Information (1986), La Santé Publique en Suisse: Presentations, Coûts, Prix, Basle, Pharma Information.

Riley, M. (1982), Accessibility to Hospitals: A Practical Application, Environment and Planning A 14:1107-1111.

Rostand, J. (1954), Pensées d'un Biologiste, Paris, Stock.

Schacter, G., D. Hellman and M. Horowitz (1981), Location Hierarchy in Health Planning. Paper presented at the European Congress of the Regional Science Association, Barcelona.

Sommer, J. and H. Gutzwiller (1986), Economie et Efficacité du Système Suisse de Santé, Berne, Hans Huber.

Chapter 20

ON THE SIGNIFICANCE OF ECONOMIC AND NON-ECONOMIC COMPONENTS OF REGIONAL DEVELOPMENT[1]

S. Saltzman
I. Viashima

1. Introduction

It is probably true that the majority of regional scientists, economists and planners believe that the most important component of development, whether it be at the national or regional level, is economic rather than 'social' (i.e., non-economic). Even many of those who believe that the social component can be highly significant would probably agree that economics is the 'engine' that drives development and that the higher the rate of economic growth in a nation or region, the better. Other things being equal, we expect there would be more resources available for social investments in nations with higher economic growth than those with lower growth rates.

The problem arises, of course, when we face the question of how best to allocate scarce resources across the many budget demands facing national and regional planners (and politicians). In most governments, these important decisions are usually made on the basis of political and institutional considerations rather than in an attempt to achieve some degree of optimality in the allocation process. Although this approach works in a practical sense, it still leaves unanswered the question of how resources should be allocated in order to best achieve some common national or regional goal.

One of the difficulties of doing research on this problem is obtaining agreement on what are the common goals to be achieved by national or regional development. This difficulty exists both in the 'real world' and in

[1] We are indebted to George Kassinis for his assistance in carrying out various aspects of the statistical analysis in section 2 and in the presentation of those results.

academia and is usually resolved by falling back on (maximum) economic growth as the common criteria that is understood by and (usually) satisfactory to all interested constituencies. After all, the greater the growth of the economic 'pie', the larger the piece that everyone can have.

Another difficulty arises in defining which development activities are economic and which are non-economic. To some extent, these definitions are arbitrary. For example, expenditures on education or health might be considered social by some analysts whereas others might define them as investments in human capital and, therefore, primarily economic.

In spite of these difficulties, it is possible to obtain preliminary results that cast some light on measuring the impacts and the relative importance of economic and non-economic variables on regional development. This can be accomplished by explicitly defining appropriate metrics for measuring development and its various components and by suitably categorizing the variables of interest in the analyses. At a minimum, this modelling approach would at least focus on these specific issues some of the discussion about the relative significance of economic and non-economic components of regional development.

The results of two exploratory studies which attempt to cast some light on the issue of the relative significance of the economic and non-economic components of development are presented in the remainder of this paper. Section 2 presents the results of testing a relatively simple model with data from most of the nations of the world. Section 3 deals with a more sophisticated model that is tested with national data from the continent of Africa. The last section (4) briefly summarizes the more important conclusions and limitations of these studies and outlines potentially useful directions for further research.

2. The Life Expectancy Model

The motivation for this model exploring the relationship between development and its economic and non-economic components grew out of earlier research dealing with the concept of 'structural violence' as found in the peace science literature (for example, see Alcock and Kohler, 1979;

Hoivik, 1977; and Kohler and Alcock, 1976, among others). It is worthwhile to look at certain aspects of these earlier efforts because they provide an interesting framework within which to view the life expectancy data used in this study.

If we examine the relationship between life expectancy and per capita gross national product by nation (for say 1965), we find, not unexpectedly, that the data break down into two main groups: low income countries with relatively low life expectancy and high income countries with relatively high life expectancy.[2] If one were to fit a curve to these two sets of data, we would find that a regression line for the low income, developing nations of the world would have a much steeper slope than would a line for the high income, developed nations which, as a group, have the highest life expectancies but a relatively flat curve (see Alcock and Kohler, 1976). This overall relationship of life expectancy increasing more slowly as per capita gross national product increases is reinforced when examining such data for specific nations over time, e.g, for 1948, one year in the mid-1960s, and one year in the 1970s (Russett, 1978).

As interesting as these two sets of data are, they represent a very simplified view of the relationship between life expectancy and economic growth. Life expectancy is clearly a function of many other variables in addition to economic growth and efforts to estimate the quantitative relationship between only these two variables will result in biased estimates of the parameters. Nevertheless, this is not an unreasonable starting point in efforts to analyze the relationships between development and its economic and non-economic components.

Thus, in the simplified linear regression model used in this study, the dependent variable, development (in a nation or region), is defined and measured in terms of life expectancy in years (LE) at the time of birth for the resident population. Among the independent variables, all of the economic components explaining development (i.e., life expectancy) are assumed to be represented by one proxy variable, per capita gross national product (GNP/C).

---

[2] Kohler and Alcott (1976) call this relationship 'the economic law of life'.

In a similar fashion, all of the non-economic components also contributing to development are assumed to be represented by a single proxy variable, in this case, the literacy rate (LR). The form of the simple model to be estimated can, therefore be represented as:

$$LE = B_o + B_1(GNP/C) + B_2(LR)$$

This model was tested against data for virtually all of the nations of the world for two periods (i.e., 1965 with 130 observations and 1982 with 135 observations) in order to determine the extent to which the observed relationships may have changed over time. These results are summarized in Table 1, which presents three types of outcomes of the regression analyses in which the national data were categorized by continent.

Part (a) of Table 1 presents the values of the estimated coefficients for the independent variables, literacy rate (LR) and per capita gross national product (GNP/C). The magnitudes of these estimates are dependent upon the measurement scales of their respective variables and are, therefore, not suitable for comparison with each other. Parts (b) and (c), on the other hand, present standardized values of these coefficients which can be used for comparative purposes. Standardized coefficients (sometimes called 'Beta-weights') are shown in part (b) while the respective elasticities are presented in part (c).

For four of the five continents listed in Table 1 (i.e., except for Europe in 1982), the values of the standardized coefficients for the literacy variable (i.e., the proxy for non-economic components) exceeds those for per capita gnp, in most cases by a substantively significant amount. In fact, again except for Europe, on the average, the standardized coefficients for literacy are approximately 3.3 times larger than those for per capita gnp in 1965 and approximately 2.2 times larger in 1982.

With respect to elasticities, again ignoring Europe, the average for literacy is significantly larger than the average for per capita gnp in 1965 and in 1982. These averages reflect the results obtained for the world as a whole

Table 1. Regression Results on Nations Grouped by Continent for Selected Years.

	1965		1982	
	LR	GNP/C	LR	GNP/C
	(a) Coefficients (ß)			
WORLD	.304	.002	.303	.002
AFRICA	.215	.013	.173	.002
N. AMERICA	.307	.001	.264	.001
S. AMERICA	.205	.010	.542	.003
ASIA & OCEANA	.257	.004	.294	.002
EUROPE	.105	.000	-.106	.001
	(b) Beta-Weights			
WORLD	.81	.15	.70	.32
AFRICA	.40	.32	.54	.30
N. AMERICA	.86	.10	.65	.36
S. AMERICA	.47	.43	.79	.30
ASIA & OCEANA	.67	.34	.70	.42
EUROPE	.44	.01	-.21	.95
	(c) Elasticities			
WORLD	.29	.03	.35	.05
AFRICA	.09	.06	.17	.02
N. AMERICA	.35	.01	.33	.03
S. AMERICA	.25	.10	.73	.05
ASIA & OCEANA	.24	.04	.33	.05
EUROPE	.14	.00	-.14	.07

[Source: Sawada (1987)]

for both the standardized coefficients and the elasticities.

Similar results are found for the nations of the world when categorized by income level. If we disaggregate the low income countries by continents the results are also similar but the differences between the standardized coefficients of the economic and non-economic components are less pronounced (Sawada, 1987).

In this section, we have used data for virtually all of the nations of the world to test a relatively simple regression model in order to determine whether the economic components of development appear to have a greater or lesser effect on life expectancy than do the non-economic components. The available evidence from this very tentative experiment indicates that, in general, the non-economic components have a greater effect in most regions of the world. However, there are severe limitations in using such simple models to test relatively sophisticated hypothesis. These and other issues are discussed in more detail in the final section (4) of this paper. In section 3, a more complex model is used to examine a similar hypothesis but in only one region of the world.

3. The Quality of Life Model

The model discussed in this section is more complex than that described in the previous section in both the formulation of the proxy variable for development and in the formulation of the explanatory variables for the economic and the non-economic components assumed to contribute to development. Furthermore, this model has been tested with national data for the period 1975 to 1979 only for nations of Sub-Saharan Africa.

3.1 The measure of development

There is an extensive literature with compelling arguments for using variables other than gross national product as the most desirable or useful measure of development (for example, see Adelman and Morris, 1973; McGranaham et al., 1972; Paukert, 1973; Chenery and Syrquin, 1975; Seers, 1977; Hicks, 1982; and Viashima, 1987; among many others). A proposal of Morris (1979) to use what he called the 'physical quality of life index' is adopted as the dependent variable in this study.

Morris argued that this composite index, consisting of life expectancy, infant mortality and adult literacy, complements gross national product as a measure of development and emphasizes the distributional characteristics of a society rather than its aggregate economic growth. Interestingly, studies show that nations ranked high in an aggregate economic metric (e.g., per capita gross national product) may not rank high, for example, in the three measures which comprise Morris' index. Also, since infant mortality and life expectancy draw on different aspects of a nation's performance and the allocation of its resources, these two variables are also not necessarily highly correlated.

The choice of a method to combine these three variables (life expectancy, infant mortality and adult literacy) into a single index to measure development can be somewhat arbitrary. For the purposes of this study, we would expect that the results obtained in regressing this index against the selected independent variables would, at least, be invariant with respect to the method used.

In his study, Morris (1979) normalized each of the three variables on an index of 0 to 100 (from worst to best) and then averaged the three normalized variables in order to obtain a single overall index for each nation in his study. Another approach is to normalize each observation by subtracting its mean and dividing by its standard deviation and then taking a (weighted?) average of the three normalized variables. A third approach, and the one taken in this study, is to use the principle component of the three variables (i.e., the weighted combination of the three variables that accounts for the maximum of their total variance, which in this case is 80%).

The correlation coefficients among these three methods of computing the physical quality of life index varies from 0.70 to 0.93 while their correlation coefficients with per capita gross national product varies from 0.10 to 0.42. Thus, the use of this index would appear to be more or less invariant to the method used to combine these three variables into one index and, also, to be relatively independent of a nation's primary economic measure, per capita gross national product.

3.2 The explanatory factors

For discussion purposes, the independent variables in the model are organized into two categories: structural and policy variables. Structural variables are generally considered to describe characteristics of a nation which are difficult to change in a significant way, even in the relatively long term. On the other hand, policy variables describe characteristics of a nation which can be changed in the shorter term by policy decisions of a national government. As will be obvious, the final results obtained from the analysis are independent of the way the explanatory variables have been categorized.

Space limitations do not allow a full description of the theoretical underpinnings of the model. However, its conceptual framework is summarized in the 19 operational variables actually used in the model and for which data were available for 34 nations of Sub-Saharan Africa (see Table 2). Of these 19 variables, nine are classified as structural variables (number 1 through 9) and ten are classified as policy variables (number 10 through 19). Policy variable numbers 10 through 14 are health services related, number 15 through 17 deal with other types of social services, number 18 is related to government intervention in the agricultural sector and number 19 is military. The identification of the structural variables (number 1 through 9) is clear from the variable names used in Table 2.

There are basically three reasons why it would be desirable to reduce the number of independent variables actually used in the regression model to less than nineteen. First, it is likely that the 19 variables really represent a smaller number of basic underlying conceptual phenomena which affect the dependent variable, development. Second, we can expect a number of these 19 variables to be highly correlated with each other and the results of a regression analysis would provide at least some parameter estimates with large standard errors. Third, with only 34 observations on which to test the model, there could be estimation problems as a result of the relatively small sample size.

A factor analysis was therefore applied to the 19 independent variables in the model and six major factor loadings were derived from these

No.	Variable Name	Factor Loading	Type*
	Factor I - Economic		
1	Per capita GNP	0.73	S
6	Settlement index	-0.79	S
7	% population urban	0.65	S
9	Population in agriculture	-0.77	S
14	Population per pharmacist	-0.50	P
17	Secondary school enrollment	0.72	P
	Factor II - Health and Basic Education		
10	Population per doctor	0.56	P
11	Ratio of doctors to total land area	0.81	P
12	Population per hospital bed	0.84	P
13	Population per midwife	0.64	P
16	Primary school enrollment	-0.73	P
18	Agricultural input supplier scale	0.45	P
	Factor III - Demographic		
2	Population growth rate	-0.50	S
3	Age dependency	-0.97	S
8	Working age population (%)	0.97	S
	Factor IV - Urban Bias		
4	Population density	0.76	S
13	Population per midwife	0.61	P
14	Population per pharmacist	0.58	P
15	Population with water	0.51	P
	Factor V - Nutrition		
2	Population growth rate	0.76	S
5	Caloric supply	0.69	S
	Factor VI - Military		
19	Armed forces per 1,000 population	0.89	P

*Type of variable: P = Policy, S = Structure (Source: Viashima, 1987) data (Viashima, 1987).[3] Depending on the application of factor analysis, the conceptual interpretation of the factor loadings may be somewhat controversial (even though the statistical methodology for extracting them is sound). The six factors with their respective variables with the highest loadings are summarized in Table 2. Space limitations preclude a detailed interpretation of the composition of each factor. However, the title assigned to each of the six factors is intended to summarize our interpretation of the basic nature of the underlying phenomena represented by the variables included in each factor. As we shall see, a strict, unequivocal interpretation of these factors is not necessary in order to consider these results as preliminary evidence regarding the underlying question of the relative significance of economic and non-economic components of regional development.

The results of the multiple regression analysis of the dependent variable, physical quality of life, against the six independent variables derived from the factor analysis are presented in Table 3. The computed F-statistic of 10.66 for the whole equation has a p-value of less than 0.0001. The dependent variable is explained primarily by four of the six factors: economic (X1); health and basic education (X2); urban bias (X4); and military (X6). The estimated coefficients of these four factors each have their expected sign and, in addition, the military factor has a p-value of 0.07 while the other three have p-values of 0.0002 or less. On the other hand, the estimated coefficients for nutrition (X5) and for demographic (X3) are not statistically significant and, in addition, the negative sign of X3 is not as expected. Approximately two-thirds of the variance in the physical quality of life variable is explained by this model.[4]

---

[3] From a statistical viewpoint, the factor analysis appears to be satisfactory. For the most part, the more important factor loadings are clearly identifiable and only three variables (population growth rate, population per midwife and population per pharmacist) have high factor loadings with more than one factor. The first three factors together explain approximately two-thirds of the variance. Furthermore, the variance of each variable explained by its six factors (i.e., the communality measure, h^2) tend also to be relatively high. In summary, the factor analysis appears to summarize the data adequately.

[4] These results are in line with those of previous studies which used similar but not necessarily the same variables and did not use factor analysis. The much smaller sample size of this study (approximately one quarter that of earlier studies) suggests that this model is more robust than earlier ones.

Table 3. Multiple Regression Analysis

	Factor	Estimated Coefficient	t-statistic	P-value	Standardized Beta Estimate
X1	Economic	0.46	4.40	0.0002	0.46
X2	Health and Basic Education	0.47	4.50	0.0001	0.47
X3	Demographic	-0.03	-0.28	0.78	-0.03
X4	Urban Bias	0.47	4.44	0.0001	0.465
X5	Nutrition	0.12	1.18	0.25	0.12
X6	Military	-0.20	-1.89	0.07	-0.20

a = -0.00407 R^2 = 0.70 ADJ. R^2 = 0.64 F = 10.66 (p-value <0.0001)

(Source: Viashima, 1987)

The standardized coefficients of the economic factor (X1), health and basic education (X2) and urban bias (X4) are approximately the same value (i.e., in the range 0.46 to 0.47). Even if one does not agree with the interpretation given to the extracted factors, it is clear that the non-economic components explaining our measure of development (i.e., the physical quality of life) are as statistically significant as the economic component. Thus, we have some additional evidence supporting the very tentative results from the simple model presented in section 2 that non-economic components can be as significant as the economic components in explaining the dependent variable, development.

4. Closing Comments

Obvious limitations in both the simple and the more sophisticated models proscribe their being used in detail for policy formulation. For example, the use of single equation models in which relevant explanatory variables have been neglected will provide biased estimates of the coefficients as is probably the case with the simple model in section 2 and may even be true in the more sophisticated model of section 3. Also, because both models were tested with cross-sectional data, it is not clear that the empirical results are not measures of co-variation rather than causality.

In addition, the interpretation of the factor analysis for the second model, while apparently rational within the conceptual framework within which it is developed, may be somewhat arbitrary. Furthermore, with both models, the units of analysis are groups of nations and the empirical results would not necessarily apply to any one nation. Finally, it is not obvious that the direction of causality in both models is necessarily in the directions indicated. More realistic assumptions would recognize the interrelationships that exist among the dependent variables and, also, that some components of the dependent variable in the second model, such as literacy, probably influence some right hand side variables as well.

These and other shortcomings in the use of single equation statistical models in this type of application suggest that we gather more evidence related to the underlying hypothesis (on the significance of non-economic components of development) before deciding to accept or reject it. The most compelling and useful additional evidence would probably be that obtained from the application of multi-equation, simultaneous, time series models of individual countries or subnational regions in which important non-economic variables are considered to be endogenous.

In spite of these limitations, the preliminary and tentative but consistent evidence in these two experiments that non-economic components of development are statistically significant indicates that additional empirical research on this issue would be worthwhile. Although the data used in both of these models are national data, it is not unreasonable to expect that similar inferences about the importance of the non-economic components of development can be made with some degree of confidence for subnational regions as well as for nations and multi-nation regions.

Bibliography

Adelman, I. and C. Morris (1973), Economic Growth and Social Equity in Developing Countries, Stanford, CA, Stanford University Press.

Alcock, N. and G. Kohler (1979), Structural Violence at the World Level: Diachronic Findings, Journal of Peace Research 16:215-262.

Chenery, H. and M. Syrquin (1975), Patterns of Development, 1950-1970, Oxford, Oxford University Press.

Hicks, N. (1982), Sector Priorities in Meeting Basic Human Needs: Some Statistical Evidence, World Development 10:489-499.

Hoivink, T. (1977), The Demography of Structural Violence, Journal of Peace Research 14:59-73.

Kohler, G. and N. Alcock (1976), An Empirical Table of Structural Violence, Journal of Peace Research 32,4:913-928.

McGranahan, D.V., C. Richard-Proust, N.V. Sovani and M. Subramanian (1972), Contents and Measurement of Socio-Economic Development, New York, N.Y, Praeger Publishers.

Morris, D. (1979), Measuring the Condition of the World's Poor: The Physical Quality of Life Index, New York, N.Y., Pergamon Press.

Paukert, F. (1973), Income Distribution at Different Levels of Development: A Survey of Evidence, International Labor Review CVIII,2-3:97-125.

Russett, B. (1978), The Marginal Utility of Income Transfers to the Third World, International Organization 32,4:913-928.

Sawada, S. (1987), A Multiple Regression Analysis of Structural Violence in Developing Countries, A Master thesis, Ithaca, N.Y., Cornell University, pp. 42-55.

Seers, D. (1977), Life Expectancy as an Integrating Concept in Social and Demographic Analysis and Planning, The Review of Income and Wealth 23:195-203.

Viashima, I. (1987), A Cross-National Study of the Physical Quality of LIfe in Sub-Saharan Africa: Determinants, Policy, and Planning Implications, Ph.D. dissertation, Ithaca, N.Y., Cornell University, January.

List of Contributors

L. Anselin
University of California
Department of Geography
Santa Barbara
California 93106
USA

A. Bailly
Université de Genève
Département de Géographie
7, Route de Drize
1227 Carouge
Switzerland

D.E. Boyce
Urban Transportation Center (MK 357)
Suite 700 South
1033 West Van Buren St
Chicago, Illinois 60607-9940
USA

R.P. Camagni
Università Commerciale Luigi Bocconi
via R. Sarfatti 25
20136 Milano
Italy

R. Cappellin
Università Commerciale Luigi Bocconi
via R. Sarfatti 25
20136 Milano
Italy

H. Cesar
Wageningen Agricultural University
Department of General Economics
P.O. Box 8130
6700 EW Wageningen
The Netherlands

K. Choi
University of Illinois at Urbana-Champaign
324 Coble Hall
801 South Wright Street
Champaign, IL 61820
USA

W.J. Coffey
Département d'Etudes Urbaines
Université du Québec
Montréal
Canada

L. Diappi
Politecnico di Milano
Facoltà di Architettura
via Bonardi 3
20133 Milano
Italy

O. Fisch
Lehman College
The City University of New York
Bedford Park Boulevard West
Bronx, N.Y. 10468-1589
USA

M.M. Fischer
Vienna University of Economics
and Business Administration
Department of Economic and Social Geography
Augasse 2-6
A-1090 Vienna
Austria

A. Fleischer
Agricultural Research Organization
The Volcani Center
Department of Statistics
P.O. Box 6
Bet-Dagan 50250
Israel

H. Folmer
Wageningen Agricultural University
Department of General Economics
P.O. Box 8130
6700 EW Wageningen
The Netherlands

D. Freeman
Agricultural Research Organization
The Volcani Center
Department of Statistics
P.O. Box 6
Bet-Dagan 50250
Israel

M. Giaoutzi
National Technical University
Department of Geography & Regional Planning
Zographou Campus
Athens 175 10 GR
Greece

T.J. Kim
University of Illinois at Urbana-Champaign
324 Coble Hall
801 South Wright Street
Champaign, IL 61820
USA

O. Laufman
Agricultural Research Organization
The Volcani Center
Department of Statistics
P.O. Box 6
Bet-Dagan 50250
Israel

S.H. Levine
Regional Science Research Center
18 Wheeler Road
Lexington, Massachusetts 02173
USA

P. Nijkamp
Free University
Department of Economics
De Boelelaan 1105
1081 HV Amsterdam
The Netherlands

D.A. Plane
The University of Arizona
Faculty of Social and Behavioral Sciences
Department of Geography and Regional Development
Tucson, Arizona 85721
USA

P.A. Rogerson
The University of Arizona
Faculty of Social and Behavioral Sciences
Department of Geography and Regional Development
Tucson, Arizona 85721
USA

E. Romanoff
Regional Science Research Center
18 Wheeler Road
Lexington, Massachusetts 02173
USA

S. Saltzman
Graduate Field of Regional Science
Sibley Hall
Cornell University
Ithaca, N.Y. 14850
USA

D. Shefer
Centre for Urban and Regional Studies
Technion - Israel Institute of Technology
Haifa
Israel

M. Sonis
Universität Stuttgart
Institut für Theoretische Physik
Pfaffenwaldring 57/III
D-7000 Stuttgart 80
FRG

M. Spiegel
Bar-Ilan University
Department of Economics and
Business Administration
52900 Ramat-Gan
Israel

U. Spiegel
Bar-Ilan University
Department of Economics and
Business Administration
52900 Ramat-Gan
Israel

A. Stratigea
National Technical University
Department of Geography & Regional Planning
Zographou Campus
Athens 175 10 GR
Greece

L. Suarez-Villa
Program in Social Ecology
University of California
Irvine, California 92717
USA

H. Talpaz
Agricultural Research Organization
The Volcani Center
Department of Statistics
P.O. Box 6
Bet-Dagan 50250
Israel

I. Viashima
Graduate Field of Regional Science
Sibley Hall
Cornell University
Ithaca, N.Y. 14850
USA

W. Weidlich
Universität Stuttgart
Institut für Theoretische Physik
Pfaffenwaldring 57/III
D-7000 Stuttgart 80
FRG

E. Werczberger
The Public Policy Program
Tel Aviv University
Tel Aviv 69978
Israel

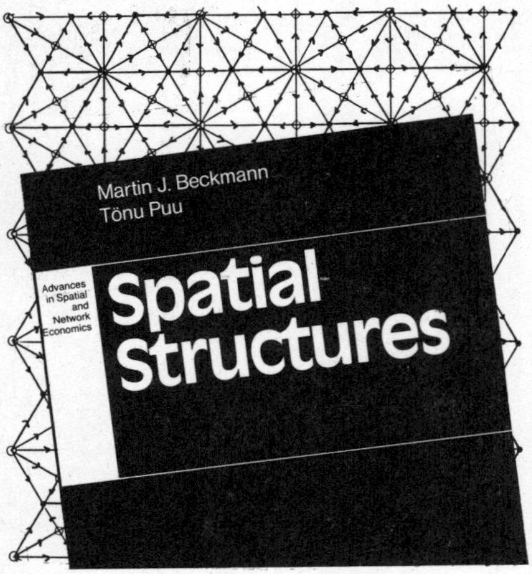

M. J. Beckmann, Technical University Munich; **T. Puu,** Umeå University, Umeå

Spatial Structures

1990. IX, 139 pp. 40 figs. (Advances in Spatial and Network Economics) Hardcover DM 65,– ISBN 3-540-51957-2

The theme of this book is the emergence and survival of spatial economic structures. How are economic structures created in an otherwise homogeneous environment? The answer is found through an analysis of economic forces that operate in the two dimensional continuum of space. And here economies of scale are found to be paramount.
First, the genesis of economic centers is examined by means of a bifurcation model. This leads to a reexamination of the von Thünen model of city and hinterland. Next, the shape and size of Löschian market areas is studied. At the core of the book is a detailed analysis of Hotelling's migration model. Solutions are given for the steady state. The general theme of order from disorder is taken up in the final chapters of this book.

K. Peschel, University of Kiel (Ed.)

Infrastructure and the Space-Economy

Essays in Honor of Rolf Funck

1990. XIV, 448 pp. 58 figs. 34 tabs. Hardcover DM 148,– ISBN 3-540-52388-X

This volume contains a broad spectrum of research in regional economics. The individual articles cover a wide range of subjects in this field. The main subjects are the research contributing to the advancement of spatial economics, analyses dealing with the impact of material infrastructure on regional development, and contemporary problems of regional policy. **Infrastructure and the Space-Economy** is dedicated to Professor Rolf Funck on the occasion of his 60th birthday.

The Collected Economics Articles of Harold Hotelling

Edited by and with an Introduction by A. C. Darnell

1990. IX, 178 pp. 10 figs. Hardcover DM 98,– ISBN 3-540-97011-8

Contents: The Life and Economic Thought of Harold Hotelling. – Bibliography of Harold Hotelling. – Reprints of the Published Economics Papers by Harold Hotelling.

Springer-Verlag
Berlin
Heidelberg
New York
London
Paris
Tokyo
Hong Kong
Barcelona

The Europe Asia Pacific Connection

T. Leuenberger, Institut für Technologiemanagement, St. Gallen (Ed.)

From Technology Transfer to Technology Management in China

1990. VIII, 283 pp. 6 figs. 20 tabs. (Europe-Asia-Pacific Studies in Economy and Technology) Hardcover DM 98,– ISBN 3-540-52478-9

This book deals with the Chinese economy and ways in which modern technology can improve its efficiency. Access to technology, especially the new technologies, will be crucial if China's modernisation efforts are to succeed. But access is only the first barrier. What is then needed is its thorough assimilation throughout the economy. China's science and technology resources are considerable but are very ineffectively utilised. A major constraint to an effective use of imported technology is the still largely planned nature of Chinese industry, the role of the defence-oriented sector and the lack of demand for technology by Chinese enterprises. The book analyses the difficulties in respect to the assimilation of modern technology and proposes ways to overcome them.

M. Kulessa, German Foundation for International Development, Berlin (Ed.)

The Newly Industrializing Economies of Asia

Prospects of Co-operation

1990. XII, 360 pp. 17 figs. (Europe-Asia-Pacific Studies in Economy and Technology) Hardcover DM 120,– ISBN 3-540-52578-5

This volume deals with the Newly Industrializing Economies (NIEs) of Asia, namely Hong Kong, Korea, Singapore and Taiwan. Based on an international conference held in Berlin in September 1989, the book contains contributions by active economists, politicians, and government officials from Asia and Europe.
In Part One, country-related issues and questions of the Asia-Pacific region are presented and discussed. In Part Two, the relations between the NIEs and other regions, especially Europe, are critically examined.
Against the background of recent developments in both continents, perspectives on future cooperation are analysed and discussed.

H. Hax, W. Kraus, T. Matsuda, T. Nakamura (Eds.)

Pacific Cooperation from the Japanese and the German Viewpoint

With contributions by numerous experts

1990. XIV, 175 pp. 16 figs. 46 tabs. Hardcover DM 59,– ISBN 3-540-51694-8

The Pacific Rim includes highly industrialized countries like the USA and Japan and a number of developing countries ranging from South Korea to Indonesia. This economic region has over the last decade shown the highest dynamic growth of any region in the world. The consequences for the world economy are far reaching, and affect not only Japan as a part of the region but also the countries of the European Community.
The papers in this volume are the result of a Japanese-German seminar. They deal with some of the most important aspects of the new situation, such as the reaction of commodity markets and corresponding trade flows, the impact of regional cooperation upon free international trade, the role and development of financial markets, and management problems arising in connection with direct investments. Each of these points is examined by a Japanese and a German expert. Their partly contrasting views result in a comprehensive survey of the whole set of problems.

Springer-Verlag
Berlin
Heidelberg
New York
London
Paris
Tokyo
Hong Kong
Barcelona